de Gruyter Studies in Mathematics 30

de Gruyter Studies in Mathematics

1 Riemannian Geometry, 2nd rev. ed., *Wilhelm P. A. Klingenberg*
2 Semimartingales, *Michel Métivier*
3 Holomorphic Functions of Several Variables, *Ludger Kaup and Burchard Kaup*
4 Spaces of Measures, *Corneliu Constantinescu*
5 Knots, 2nd rev. and ext. ed., *Gerhard Burde and Heiner Zieschang*
6 Ergodic Theorems, *Ulrich Krengel*
7 Mathematical Theory of Statistics, *Helmut Strasser*
8 Transformation Groups, *Tammo tom Dieck*
9 Gibbs Measures and Phase Transitions, *Hans-Otto Georgii*
10 Analyticity in Infinite Dimensional Spaces, *Michel Hervé*
11 Elementary Geometry in Hyperbolic Space, *Werner Fenchel*
12 Transcendental Numbers, *Andrei B. Shidlovskii*
13 Ordinary Differential Equations, *Herbert Amann*
14 Dirichlet Forms and Analysis on Wiener Space, *Nicolas Bouleau and Francis Hirsch*
15 Nevanlinna Theory and Complex Differential Equations, *Ilpo Laine*
16 Rational Iteration, *Norbert Steinmetz*
17 Korovkin-type Approximation Theory and its Applications, *Francesco Altomare and Michele Campiti*
18 Quantum Invariants of Knots and 3-Manifolds, *Vladimir G. Turaev*
19 Dirichlet Forms and Symmetric Markov Processes, *Masatoshi Fukushima, Yoichi Oshima and Masayoshi Takeda*
20 Harmonic Analysis of Probability Measures on Hypergroups, *Walter R. Bloom and Herbert Heyer*
21 Potential Theory on Infinite-Dimensional Abelian Groups, *Alexander Bendikov*
22 Methods of Noncommutative Analysis, *Vladimir E. Nazaikinskii, Victor E. Shatalov and Boris Yu. Sternin*
23 Probability Theory, *Heinz Bauer*
24 Variational Methods for Potential Operator Equations, *Jan Chabrowski*
25 The Structure of Compact Groups, *Karl H. Hofmann and Sidney A. Morris*
26 Measure and Integration Theory, *Heinz Bauer*
27 Stochastic Finance, *Hans Föllmer and Alexander Schied*
28 Painlevé Differential Equations in the Complex Plane, *Valerii I. Gromak, Ilpo Laine and Shun Shimomura*
29 Discontinuous Groups of Isometries in the Hyperbolic Plane, *Werner Fenchel and Jakob Nielsen*

Liviu I. Nicolaescu

The Reidemeister Torsion of 3-Manifolds

Walter de Gruyter
Berlin · New York 2003

Author
Liviu Nicolaescu
Department of Mathematics
University of Notre Dame
Notre Dame, IN 46556-4618
USA

Mathematics Subject Classification 2000: 57-02; 43A40; 55N25, 57M05, 57M12, 57M25, 57M27, 57Q10, 57R15, 57R57, 58J28

Keywords: 3-manifolds; Reidemeister torsion; Turaev's refined torsion; spin-c structures; linking forms; harmonic analysis on discrete Abelian groups; knots and links; surgery formulae; plumbings; Seiber–Witten invariants; surface singularities

♾ Printed on acid-free paper which falls within the guidelines of the ANSI to ensure permanence and durability.

Library of Congress – Cataloging-in-Publication Data

Nicolaescu, Liviu I.
The Reidemeister torsion of 3-manifolds / Liviu I. Nicolaescu.
p. cm. – (De Gruyter studies in mathematics ; 30)
Includes bibliographical references and index.
ISBN 3-11-017383-2 (cloth : alk. paper)
1. Three-manifolds (Topology). 2. Reidemeister torsion. I. Title: Reidemeister torsion of three-manifolds. II. Title. III. Series.
QA613.2.N53 2003
514′.3–dc21 2002041469

ISBN 3-11-017383-2

Bibliographic information published by Die Deutsche Bibliothek

Die Deutsche Bibliothek lists this publication in the Deutsche Nationalbibliografie; detailed bibliographic data is available in the Internet at <http://dnb.ddb.de>.

Printed in Germany.
Cover design: Rudolf Hübler, Berlin.
Typeset using the authors' TEX files: I. Zimmermann, Freiburg.
Printing and binding: Hubert & Co. GmbH & Co. KG, Göttingen.

To the memory of my father, Ion I. Nicolaescu,
who shared with me his passion for books.

Introduction

The torsion of a cellular (simplicial) complex was introduced in the 30s by W. Franz [29] and K. Reidemeister [90] in their study of lens spaces. The lens spaces $L(p, q)$ (p fixed) have the same fundamental groups and thus the same homology groups. However, they are not all homeomorphic. They are not even homotopically equivalent. This can be observed by detecting some below the radar interactions between the fundamental group and the simplicial structure. The torsion captures some of these interactions. In particular, it is able to distinguish lens spaces which are homotopically equivalent but not homeomorphic, and moreover completely classify these spaces up to a homeomorphism. This suggests that this invariant is reaching deep inside the topological structure.

What is then this torsion? What does it compute? These are the kind of questions we try to address in these notes, through many examples and various equivalent descriptions of this invariant.

From an algebraic point of view, the torsion is a generalization of the notion of determinant. The most natural and general context to define the torsion would involve the Whitehead group and algebraic K-theory as in the very elegant and influential Milnor survey [72], but we did not adopt this more general point of view. Instead we look at what Milnor dubbed R-torsion.

This invariant can be viewed as a higher Euler characteristic type invariant. Much like the Euler characteristic, the torsion satisfies an inclusion–exclusion (a.k.a. Mayer–Vietoris) principle which can be roughly stated that for any topological spaces A and B we have

$$\mathrm{Tors}(A \cup B) = \mathrm{Tors}(A) + \mathrm{Tors}(B) - \mathrm{Tors}(A \cap B)$$

which suggests that the torsion could be interpreted as counting something.

The classical Poincaré–Hopf theorem states that the Euler–Poincaré characteristic of a smooth manifold counts the zeros of a generic vector field. If the Euler-Poincaré characteristic is zero then most vector fields have no zeroes but may have periodic orbits. The torsion counts these closed orbits, at least for some families of vector fields. As D. Fried put it in [34], "the Euler characteristic counts points while the torsion counts circles".

One of the oldest results in algebraic topology equates the Euler-Poincaré characteristic of simplicial complex, defined as the alternating sums of the numbers of simplices, with a manifestly combinatorial invariant, the alternating sum of the Betti numbers. Similarly, the R-torsion can be given a description in terms of chain complexes or, a plainly invariant description, in homological terms. Just like the Euler characteristic, the R-torsion of a smooth manifold can be given a Hodge theoretic description, albeit much more complicated.

More recently, this invariant turned up in 3-dimensional Seiberg–Witten theory, in the work of Meng–Taubes ([68]). This result gave us the original impetus to understand the meaning of torsion.

This is a semi-informal, computationally oriented little book which grew out of our efforts to understand the intricacies of the Meng–Taubes–Turaev results, [68, 115]. For this reason a lot of emphasis is placed on the Reidemeister torsion of 3-manifolds. These notes tried to address the author's own struggle with the overwhelming amount of data involved and the conspicuously scanty supply of computational examples in the traditional literature on the subject. We considered that at an initial stage a good intuitive argument or example explaining why a certain result could be true is more helpful than a complete technical proof. The classical Milnor survey [72] and the recent introductory book [117] by V. Turaev are excellent sources to fill in many of our deliberate foundational omissions.

When thinking of topological issues it is very important not to get distracted by the ugly looking but elementary formalism behind the torsion. For this reason we devoted the entire first chapter to the algebraic foundations of the concept of torsion. We give several equivalent definitions of the torsion of an acyclic complex and in particular, we spend a good amount of time constructing a setup which coherently deals with the torturous sign problem. We achieved this using a variation of some of the ideas in Deligne's survey [18].

The general algebraic constructions are presented in the first half of this chapter, while in the second half we discuss Turaev's construction of several arithmetically defined subrings of the field of fractions of the rational group algebra of an Abelian group. These subrings provide the optimal algebraic framework to discuss the torsion of a manifold. We conclude this chapter by presenting a dual picture of these Turaev subrings via Fourier transform. These results seem to be new and simplify substantially many gluing formulæ for the torsion, to the point that they become quasi-tautological.

The Reidemeister torsion of an arbitrary simplicial (or CW) complex is defined in the second chapter. This is simply the torsion of a simplicial complex with Abelian local coefficients, or equivalently the torsion of the simplicial complex of the maximal Abelian cover. We present the basic properties of this invariant: the Mayer–Vietoris principle, duality, arithmetic properties and an Euler–Poincaré type result. We compute the torsion of many mostly low dimensional manifolds and in particular we explain how to compute the torsion of any 3-manifold with $b_1 > 0$ using the Mayer–Vietoris principle, the Fourier transform, and the knowledge of the Alexander polynomials of links in S^3. Since the literature on Dehn surgery can be quite inconsistent on the various sign conventions, we have devoted quite a substantial appendix to this subject where we kept an watchful eye on these often troublesome sings.

The approach based on Alexander polynomials has one major drawback, namely it requires a huge volume of computations. We spend the whole section §2.6 explaining how to simplify these computation for a special yet very large class of 3-manifolds, namely the graph manifolds. The links of isolated singularities of complex surfaces are included in this class and the recent work [75, 76] proves that the Reidemeister torsion captures rather subtle geometrical information about such manifolds. We conclude this chapter with some of the traditional applications of the torsion in topology.

Chapter 3 focuses on Turaev's ingenious idea of Euler structure and how it can be used to refine the concept of torsion by removing the ambiguities in choosing the bases needed for computing the torsion. Turaev later observed that for a 3-manifold a choice of an Euler structure is equivalent to a choice of spin^c-structure. After we review a few fundamental properties of this refined torsion for 3-manifolds we then go on to present a result of Turaev which in essence says that the refined torsion of a 3-manifold with positive b_1 is uniquely determined by the Alexander polynomials of links in S^3 and the Mayer–Vietoris principle.

This uniqueness result does not include rational homology spheres, and thus offers no indication on how to approach this class of manifolds. We spend the last part of this chapter analyzing this class of 3-manifolds.

In §3.8 we describe a very powerful method for computing the torsion of such 3-manifolds, based on the complex Fourier transform results in Chapter 1, and an extremely versatile holomorphic regularization technique. These lead to explicit formulæ for the Fourier transform of the torsion of a rational homology sphere in terms of surgery data. These formulæ still have the two expected ambiguities: a sign ambiguity and a spin^c ambiguity. In §3.9 we describe a very simple algorithm for removing the spin^c ambiguity. This requires a quite long topological detour in the world of quadratic functions on finite Abelian groups, and surgery descriptions of spin and spin^c structures, but the payoff is worth the trouble. The sign ambiguity is finally removed in §3.10 in the case of plumbed rational homology spheres, relying on an idea in [75], based on the Fourier transform, and a relationship between the torsion and the linking form discovered by Turaev.

Chapter 4 discusses more analytic descriptions of the Reidemeister torsion: in terms of gauge theory, in terms of Morse theory, and in terms of Hodge theory. We discuss Meng–Taubes theorem and the improvements due to Turaev. We also outline our recent proof [83] of the extension of the Meng–Taubes–Turaev theorem to rational homology spheres. As an immediate consequence of this result, we give a new description of the Brumfiel–Morgan [7] correspondence for rational homology 3-spheres which associates to each spin^c structure a refinement of the linking form.

On the Morse theoretic side we describe Hutchings–Lee–Pajitnov results which give a Morse theoretic interpretation of the Reidemeister torsion. We barely scratch the Hodge theoretic approach to torsion. We only provide some motivation for the ζ-function description of the analytic torsion and the Cheeger–Müller theorem which identifies this spectral quantity with the Reidemeister torsion.

* * *

Acknowledgements.[1] I spent almost two years thinking about these issues and I was helped along the way by many people. I am greatly indebted to Frank Connolly who

[1]This work was partially supported by NSF grant DMS-0071820.

introduced me to this subject. His patience and expertise in answering my often half baked questions have constantly kept me on the right track. I also want to thank my colleagues Bill Dwyer, Stephan Stolz, Larry Taylor and Bruce Williams for many illuminating discussions which made this scientific journey so much more enjoyable.

This survey started up as a set of study notes to myself with no large audience in mind. They may have not seen the "light of day" were it not for Andrew Ranicki's interest and support. I want to thank him for the insightful comments concerning these notes which have substantially helped me to improve the quality of the presentation.

I want to thank Vladimir Turaev whose work had a tremendous influence on the way I think about this subject. I appreciate very much his interest in these notes, and his helpful suggestions.

The sections §3.8, §3.9 and §3.10 are a byproduct of the joint work with András Némethi, [75, 76]. I want use this opportunity to thank him for our very exciting and stimulating collaboration.

I performed the final revision of the manuscript during my sabbatical leave at the Ohio State University. I want to express my gratitude to Dan Burghelea for his very warm and generous hospitality, and his contagious enthusiasm about mathematics, and culture in general.

And last but not the least I want to thank my family for always being there for me.

Notre Dame, Indiana 2002 *Liviu I. Nicolaescu*

Notations and conventions

• For simplicity, unless otherwise stated, we will denote by $H_*(X)$ the homology with integral coefficients of the topological space X.

• For $\mathbb{K} = \mathbb{R}, \mathbb{C}$, we denote by $\underline{\mathbb{K}}^n_X$ the trivial rank n, $\mathbb{K}$-vector bundle over the space X.

• $\mathbf{i} := \sqrt{-1}$.

• $\underline{u}(n)$ = the Lie algebra of $\mathrm{U}(n)$, $\underline{su}(n)$= the Lie algebra of $\mathrm{SU}(n)$ etc.

• $\mathbb{Z}_+ = \mathbb{Z}_{\geq 0} := \{n \in \mathbb{Z};\ n \geq 0\}$.

• For all integers $m < n$ we set $\overline{m,n} := \mathbb{Z} \cap [m,n]$.

• For any Abelian group G we will denote by $\mathrm{Tors}(G)$ its torsion subgroup. We will use the notation $\mathbb{Z}_n := \mathbb{Z}/n\mathbb{Z}$.

• If R is a commutative ring with 1, then $R^\times$ denotes the group of invertible elements of R.

• Also, we will strictly adhere to the following orientation conventions.

• If M is an oriented manifold with boundary then the induced orientation of ∂M is determined by the *outer-normal-first* convention

$$\mathbf{or}\,(M) = \textbf{outer normal} \wedge \mathbf{or}\,(\partial M).$$

• If $F \hookrightarrow E \twoheadrightarrow B$ is a smooth fiber bundle, where F and B are oriented, then the induced orientation of E is determined by the *fiber-first* convention

$$\mathbf{or}\,(E) = \mathbf{or}\,(F) \wedge \mathbf{or}\,(B).$$

Contents

Chapter 1
Algebraic preliminaries

§1.1 The torsion of acyclic complexes of vector spaces

The notion of torsion is a multifaceted generalization of the concept of determinant of an isomorphism of vector spaces. We begin with a baby example to give the reader a taste of the ingredients which enter into the fabric of torsion, and of the type of issues it addresses.

Example 1.1. Suppose that U_0 and U_1 are finite dimensional real vector spaces and $S, T : U_1 \to U_0$ are two linear isomorphisms. If we take into account only the vector space structures then we could consider S and T to be equivalent, i.e. there exist $A_i \in \mathrm{Aut}(U_i)$, $i = 0, 1$, such that

$$T = A_0 S A_1^{-1}. \tag{1.1}$$

Suppose now that $\Lambda_i \subset U_i$, $i = 0, 1$ are lattices and S, T are compatible with them, i.e. $S(\Lambda_1) \subset \Lambda_0$, $T(\Lambda_1) \subset \Lambda_0$. We could then ask whether there exist $A_i \in \mathrm{Aut}(\Lambda_i) \subset \mathrm{Aut}(U_i)$, $i = 0, 1$, such that (1.1) holds. We can easily construct an invariant to show that S and T need not always be equivalent in this more restricted sense. Consider for example the finite Abelian groups

$$I(S) = \Lambda_0 / S(\Lambda_1), \quad I(T) = \Lambda_0 / T(\Lambda_1).$$

If S is equivalent to T then $|I(S)| = |I(T)|$ and we see that the quantity $S \mapsto |I(S)|$ is an invariant of this restricted equivalence relation. It is very easy to compute it. Pick $\mathbb{Z}$-bases of Λ_i, $i = 0, 1$. We can then identify S and T with integral matrices and, up to a sign, $|I(S)|$ and $|I(T)|$ are the determinants of these matrices.

The torsion tackles a slightly more general question than this. This generality entails several aspects, all motivated by topological issues. First, instead of lattices in real vector spaces we will work with free modules over a commutative ring R. Instead of the field of real numbers we will work with a field $\mathbb{K}$ related to R via a nontrivial ring morphism $\chi : R \to \mathbb{K}$. If F is a free R module then $F \otimes_\chi \mathbb{K}$ is a $\mathbb{K}$-vector space. The role of the groups $\mathrm{Aut}(\Lambda_i)$ we will played by certain subgroups of $\mathrm{Aut}_R(F)$, which act in an obvious way on $F \otimes_\chi \mathbb{K}$. Finally, instead of morphisms of R-modules we will consider chain complexes of R-modules. □

We will begin our presentation by discussing the notion of torsion (or determinant) of a chain complex of finite dimensional vector spaces. In the sequel, $\mathbb{K}$ will denote a field of characteristic zero. A *basis* of a $\mathbb{K}$-vector space will be *a totally ordered* generating set of linearly independent vectors.

Suppose $f : U_1 \to U_0$ is an *isomorphism* of n-dimensional $\mathbb{K}$-vector spaces. Once we fix bases $\boldsymbol{u}_i = (u_{i,1}, \dots, u_{i,n})$ of U_i, $i = 0, 1$, we can represent f as an $n \times n$-matrix

$$A = A(\boldsymbol{u}_0, \boldsymbol{u}_1) \in \mathrm{GL}_n(\mathbb{K}),$$

and as such it has a determinant $\det A \in \mathbb{K}^*$.

Suppose additionally that $\mathfrak{A}$ is a group[1] acting linearly on U_i, $i = 0, 1$. We denote by $\mathrm{Fr}(U_i)$, the set of bases of U_i. There is an obvious right action of $\mathfrak{A}$ on $\mathrm{Fr}(U_i)$, and we will denote by $[\boldsymbol{u}_i]_{\mathfrak{A}}$ the $\mathfrak{A}$-orbit of $\boldsymbol{u}_i \in \mathrm{Fr}(U_i)$, $i = 0, 1$. If we change $\boldsymbol{u}_i$ by elements in $\mathfrak{A}$,

$$\boldsymbol{u}_i \mapsto \boldsymbol{u}_i \cdot g_i,$$

then the matrix description of f changes according to the rule

$$A = A(\boldsymbol{u}_1, \boldsymbol{u}_0) \mapsto A(\boldsymbol{u}_1 g_1, \boldsymbol{u}_0 g_0) := (\boldsymbol{u}_1 g_1/\boldsymbol{u}_1)^{-1} A(\boldsymbol{u}_0 g_0/\boldsymbol{u}_0), \tag{1.2}$$

where for any $\boldsymbol{u}, \boldsymbol{v} \in \mathrm{Fr}(U)$ we denoted by $\boldsymbol{v}/\boldsymbol{u}$ the matrix describing the base change $\boldsymbol{u} \mapsto \boldsymbol{v}$,

$$\boldsymbol{v} = \boldsymbol{u} \cdot (\boldsymbol{v}/\boldsymbol{u}), \quad (\boldsymbol{u}/\boldsymbol{v}) = (\boldsymbol{v}/\boldsymbol{u})^{-1}.$$

Also, we set

$$[\boldsymbol{v}/\boldsymbol{u}] := \det(\boldsymbol{v}/\boldsymbol{u})$$

$$\det \mathfrak{A} = \big\{[\boldsymbol{u}g/\boldsymbol{u}];\ g \in \mathfrak{A},\ \boldsymbol{u} \in \mathrm{Fr}(U_0) \cup \mathrm{Fr}(U_1)\big\}.$$

Observe that $\det \mathfrak{A}$ is a subgroup of the multiplicative group $\mathbb{K}^*$. In particular, the determinant of A changes by an element in the subgroup $\det(\mathfrak{A}) \subset \mathbb{K}^*$.

Definition 1.2. The correspondence

$$(\boldsymbol{u}_0, \boldsymbol{u}_1) \mapsto 1/\det A(\boldsymbol{u}_1, \boldsymbol{u}_0) \in \mathbb{K}^* \twoheadrightarrow \mathbb{K}^*/\det(\mathfrak{A})$$

defines an element in $\mathbb{K}^*/\det(\mathfrak{A})$ which depends only on the $\mathfrak{A}$-orbits of $\boldsymbol{u}_i$, $i = 0, 1$. We denote it by

$$\mathcal{T}(f, [\boldsymbol{u}_0]_{\mathfrak{A}}, [\boldsymbol{u}_1]_{\mathfrak{A}})$$

and we call it the *torsion of the map* f with respect to the $\mathfrak{A}$-equivalence classes of bases $\boldsymbol{u}, \boldsymbol{v}$. □

To ease the presentation, in the remainder of this section we will drop the group $\mathfrak{A}$ from our notations since it introduces no new complications (other than notational).

Observe that an isomorphism $f : U_1 \to U_0$ can be viewed as a very short acyclic chain complex

$$0 \to U_1 \xrightarrow{f} U_0 \to 0.$$

[1] Intuitively, $\mathfrak{A}$ is the group of *a*mbiguities. All the vectors in the same orbit of $\mathfrak{A}$ are equal partners and in a given concrete setting there is no canonical way of selecting one vector in a fixed orbit.

The notion of torsion described above extends to acyclic chain complexes of arbitrary sizes. Suppose that

$$\underline{C} := 0 \to C_n \xrightarrow{\partial_n} C_{n-1} \xrightarrow{\partial_{n-1}} \cdots \xrightarrow{\partial_1} C_0 \to 0$$

is an acyclic complex of finite dimensional $\mathbb{K}$-vector spaces. Fix bases $\boldsymbol{c}_i$ of C_i. Because $\underline{C}$ is acyclic there exists an algebraic contraction, i.e. a degree one map

$$\eta : C_i \to C_{i+1}$$

such that

$$\partial\eta + \eta\partial = \mathbf{1}_{\underline{C}}.$$

(See Appendix §A.1.) Set $\hat{\eta} = \eta\partial\eta$.

Exercise 1.1. (a) Prove that $\hat{\eta}$ is an algebraic contraction satisfying $\hat{\eta}^2 = 0$.

(b) Show that if $\eta^2 = 0$ then $\eta = \hat{\eta}$. □

Consider the operator

$$\partial + \hat{\eta} : \underline{C} \to \underline{C}.$$

It satisfies

$$(\partial + \hat{\eta})^2 = \partial\hat{\eta} + \hat{\eta}\partial = \mathbf{1},$$

so that it is an isomorphism. Moreover, with respect to the direct sum decomposition

$$\underline{C} = \underline{C}_{\text{even}} \oplus \underline{C}_{\text{odd}}$$

it has the block form

$$\partial + \hat{\eta} = \begin{bmatrix} 0 & T_{01} \\ T_{10} & 0 \end{bmatrix}, \quad T_{01} : \underline{C}_{\text{odd}} \to \underline{C}_{\text{even}}, \quad T_{10} : \underline{C}_{\text{even}} \to \underline{C}_{\text{odd}}.$$

We deduce that T_{10} is an isomorphism of vector spaces and $T_{10}^{-1} = T_{01}$. We can define

$$\begin{aligned}\mathcal{T}(\underline{C}, \underline{c}) := \mathcal{T}(\partial + \hat{\eta}, [\underline{c}_{\text{even}}], [\underline{c}_{\text{odd}}]) &= \det\big(\partial + \eta : \underline{C}_{\text{odd}} \to \underline{C}_{\text{even}}\big)^{-1} \\ &= \det\big(\partial + \hat{\eta} : \underline{C}_{\text{even}} \to \underline{C}_{\text{odd}}\big).\end{aligned}$$

We need to be more specific about $\underline{c}_{\text{odd}}$ and $\underline{c}_{\text{even}}$. If we denote by $2m+1$ (resp. 2ν) the largest odd (resp. even) number not greater than the length of $\underline{C}$ then

$$\underline{c}_{\text{odd}} = \boldsymbol{c}_{2m+1} \cup \cdots \cup \boldsymbol{c}_3 \cup \boldsymbol{c}_1, \quad \underline{c}_{\text{even}} = \boldsymbol{c}_{2\nu} \cup \cdots \cup \boldsymbol{c}_2 \cup \boldsymbol{c}_0. \tag{1.3}$$

Proposition 1.3 ([19]). $\det\big(\partial + \eta : \underline{C}_{\text{even}} \to \underline{C}_{\text{odd}}\big)$ *is independent of the choice of* η.

Proof. Suppose η_0, η_1 are two algebraic contractions. Set $\dot\eta := \eta_1 - \eta_0$, $\eta_t = \eta_0 + t\dot\eta$. Observe that $\dot\eta\partial = -\partial\dot\eta$ and η_t is an algebraic contraction of $\underline{C}$. Moreover

$$\hat\eta_{t+s} = (\eta_t + s\dot\eta)\partial(\eta_t + s\dot\eta) = \hat\eta_t + s(\dot\eta\partial\eta_t + \eta_t\partial\dot\eta) + s^2\dot\eta\partial\dot\eta.$$

We set $T_t := \partial + \hat\eta_t$ and $\hat\eta'_t = \frac{d}{ds}|_{s=0}\hat\eta_{t+s} = \dot\eta\partial\eta_t + \eta_t\partial\dot\eta$. Derivating[2] the identities

$$\hat\eta_t^2 = 0, \quad \partial\hat\eta_t + \hat\eta_t\partial = \mathbf{1},$$

we deduce that

$$\hat\eta_t\hat\eta'_t = -\hat\eta'_t\hat\eta_t, \quad \partial\hat\eta'_t = -\hat\eta'_t\partial.$$

This shows that $T_t\hat\eta'_t = -\hat\eta'_t T_t$. Using the identity $T_t^2 = \mathbf{1}$ we obtain

$$T_{t+s} = \partial + \hat\eta_{t+s} = T_t + s\hat\eta_t = T_t\big(\mathbf{1} + sT_t\hat\eta'_t\big).$$

To prove that $\det(T_t : \underline{C}_{\text{even}} \to \underline{C}_{\text{odd}})$ is independent of t it suffices to show that

$$\operatorname{tr}(T_t\hat\eta'_t : \underline{C}_{\text{even}} \to \underline{C}_{\text{even}}) = 0.$$

Observe that

$$T_t\hat\eta'_t = (\hat\eta_t\dot\eta\partial\eta_t + \hat\eta_t\eta_t\partial\dot\eta) + \partial\dot\eta =: A + B.$$

Since $A(C_k) \subset C_{k+2}$, we deduce $\operatorname{tr}(A) = 0$. Next, consider the filtration $\underline{C}_{\text{even}} \supset \ker\partial \supset \operatorname{Im}\partial \supset 0$. Observe that $B\underline{C}_{\text{even}} \subset \operatorname{Im}\partial$ and B acts trivially on $\ker\partial$. This shows that $\operatorname{tr} B = 0$ and completes the proof of the proposition. □

Definition 1.4. The quantity $\mathcal{T}(\underline{C}, \underline{c})$ is called the *torsion of the acyclic complex* $\underline{C}$ with respect to the bases $\underline{c}$. □

Observe that if $\underline{c}'$ is another basis of $\underline{C}$ then using (1.2) we deduce

$$\mathcal{T}(\underline{C}, \underline{c}') = \mathcal{T}(\underline{C}, \underline{c})\prod_{i=0}^{n}[\boldsymbol{c}'_i/\boldsymbol{c}_i]^{(-1)^i}. \tag{1.4}$$

Convention. *When the complex* $\underline{C}$ *is not acyclic we define its torsion to be* 0.

We can alternatively define the torsion as follows. Choose finite, totally ordered collections $\boldsymbol{b}_i \subset C_i$ of such that the restriction of ∂ to $\boldsymbol{b}_i$ is one-to-one for all i, $\boldsymbol{b}_0 = \emptyset$, and

$$\partial\boldsymbol{b}_{i+1} \cup \boldsymbol{b}_i \text{ is a basis of } C_i. \tag{†}$$

(This condition uses the acyclicity of $\underline{C}$.) Now set $\underline{c} := \oplus_i\underline{c}_i$, and define the torsion of $\underline{C}$ with respect to the bases $\underline{c}_i$ by

$$\mathcal{T}(\underline{C}, [\underline{c}]_{\mathfrak{A}}) := \prod_{i=0}^{n}[(\partial\boldsymbol{b}_{i+1})\boldsymbol{b}_i/\boldsymbol{c}_i]^{(-1)^{i+1}} \in \mathbb{K}^*/\det(\mathfrak{A}). \tag{1.5}$$

[2]The derivatives are understood in the formal sense, as linearizations.

The relationship between these two definitions is very simple. Let us first introduce a notation. If X is a basis of a vector space U, and u is a vector in U, then the decomposition of u along this basis will be denoted by

$$u := \sum_{x \in X} \langle u|x\rangle x.$$

Given collections $\boldsymbol{b}_i$ as above we define a contraction $\eta : \underline{C} \to \underline{C}$,

$$C_i \ni u = \sum_{b \in \boldsymbol{b}_{i+1}} \langle u|\partial b\rangle \partial b + \sum_{b' \in \boldsymbol{b}_i} \langle u|b'\rangle b' \mapsto \sum_{b \in \boldsymbol{b}_{i+1}} \langle u|\partial b\rangle b \in C_{i+1}.$$

We define $\boldsymbol{c}_i' = \partial \boldsymbol{b}_{i+1} \cup \boldsymbol{b}_i$. Then

$$\mathcal{T}(\underline{C}, \underline{\boldsymbol{c}}') = \det\big(\partial + \eta \colon (\underline{C}_{\mathrm{even}}, \underline{\boldsymbol{c}}'_{\mathrm{even}}) \to (\underline{C}_{\mathrm{odd}}, \underline{\boldsymbol{c}}'_{\mathrm{odd}})\big) = 1.$$

The equality (1.5) now follows by invoking the transition formula (1.4).

We present below another simple and effective way of performing concrete computations.

Proposition 1.5 ([38, 110]).[3] *Suppose $\underline{C}$ is an acyclic complex of finite dimensional $\mathbb{K}$-vector spaces. Denote by ℓ the length of $\underline{C}$, fix a basis $\underline{\boldsymbol{c}}$ of $\underline{C}$ and denote by D_i the matrix of the linear operator*

$$\partial \colon C_{i+1} \to C_i$$

with respect to the chosen bases. Set

$$n_i := \dim_{\mathbb{K}} C_i, \quad s_i := \dim_{\mathbb{K}} \ker(\partial \colon C_i \to C_{i-1}).$$

*Assume there exists a τ-*chain, *i.e. a collection*

$$\big\{(S_i, \tilde{D}_i);\ S_i \subset \overline{1, n_i},\ s_i = |S_i|,\ i = 0, 1, \dots, \ell - 1,\ \tilde{D}_i \colon \mathbb{K}^{n_{i+1} - s_{i+1}} \to \mathbb{K}^{s_i}\big\}$$

such that the matrix $\tilde{D}_i$ obtained from D_i by deleting the columns belonging to S_{i+1} and the rows belonging to $\overline{1, n_i} \setminus S_i$ is quadratic and nonsingular (see Figure 1.1). Then

$$\mathcal{T}(\underline{C}, \underline{\boldsymbol{c}}) = \prod_{i=0}^{\ell-1} \det(\tilde{D}_i)^{(-1)^{i+1+\nu_i}},$$

where

$$\nu_i := \big|\{(x, y) \in \mathbb{Z} \times \mathbb{Z};\ 1 \le x < y, x \in \overline{1, n_i} \setminus S_i,\ y \in S_i\}\big|.$$

---

[3]This result has a long history, going back to A. Cayley [12].

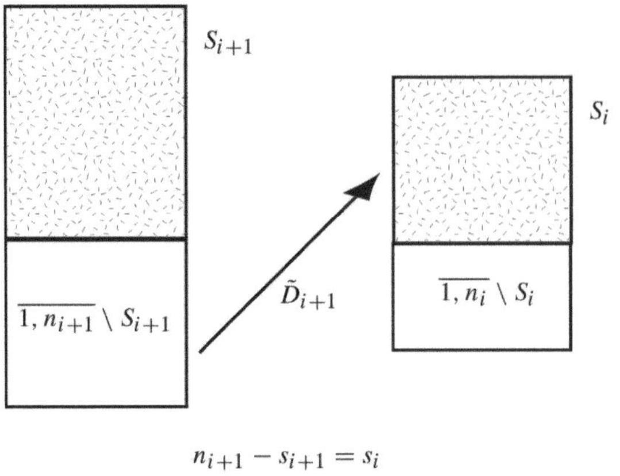

Figure 1.1. Visualizing a τ-chain.

Proof. Let

$$\boldsymbol{c}_i := \{c_{i,1}, \ldots, c_{i,n_i}\}.$$

Define

$$\boldsymbol{b}_i := \{c_{i,j};\ j \notin S_i\}$$

where the above vectors are arranged in the increasing order given by j. The bases $\boldsymbol{b}_i$ satisfy the condition (†) and moreover,

$$[\partial \boldsymbol{b}_{i+1} \boldsymbol{b}_i / \boldsymbol{c}_i] = (-1)^{\nu_i} \det(\tilde{D}_i). \qquad \square$$

Example 1.6 (Algebraic mapping torus, [33, 34]). A useful operation one can perform on chain complexes is the *algebraic mapping torus construction*, [33]. More precisely, suppose $(\underline{C}, \partial)$ is a chain complex of $\mathbb{K}$-vector spaces, $\underline{\boldsymbol{c}}$ is a basis of $\underline{C}$ and

$$f : \underline{C} \to \underline{C}$$

is a chain morphism, i.e. a degree zero map commuting with ∂. The algebraic mapping torus of $\underline{C}$ with respect to f is the chain complex

$$(T(f), \partial_f), \quad T(f)_k := C_k \oplus C_{k-1},$$

$$\partial_f : \begin{matrix} C_k \\ \oplus \\ C_{k-1} \end{matrix} \to \begin{matrix} C_{k-1} \\ \oplus \\ C_{k-2} \end{matrix}, \quad \begin{bmatrix} u \\ \\ v \end{bmatrix} \mapsto \begin{bmatrix} \partial & (\mathbf{1} - f) \\ & \\ 0 & -\partial \end{bmatrix} \cdot \begin{bmatrix} u \\ \\ v \end{bmatrix}$$

The bases $\underline{\boldsymbol{c}}$ define bases $T(\underline{\boldsymbol{c}})$ in $T(f)$, unique up to ordering. *Assume* $\det(1 - f) \in \mathbb{K}^*$. Then the map

$$\eta : T(f)_k \to T(f)_{k+1}, \quad \eta = \begin{bmatrix} 0 & 0 \\ & \\ (\mathbf{1} - f)^{-1} & 0 \end{bmatrix}$$

is an algebraic contraction, and the operator

$$\partial + \eta\colon T(f)_{\text{odd}} = \underline{C}_{\text{odd}} \oplus \underline{C}_{\text{even}} \to T(f)_{\text{even}} = \underline{C}_{\text{even}} \oplus \underline{C}_{\text{odd}}$$

is given by

$$\begin{aligned}\partial + \eta &= \begin{bmatrix} \partial & (\mathbf{1} - f) \\ (\mathbf{1} - f)^{-1} & -\partial \end{bmatrix} \\ &= \begin{bmatrix} \partial & (\mathbf{1} - f) \\ (\mathbf{1} - f)^{-1} & 0 \end{bmatrix} \cdot \begin{bmatrix} \mathbf{1} & -\partial(\mathbf{1} - f) \\ 0 & \mathbf{1} \end{bmatrix}\end{aligned}$$

We conclude

$$\mathcal{T}\big(T(f), T(\underline{\mathbf{c}})\big) = \pm \det{}_s\big((\mathbf{1} - f)\colon \underline{C} \to \underline{C}\big)^{-1} = \pm \hat{\zeta}_f(1) \tag{1.6}$$

where the s-determinant $\det_s$ and the s-zeta function $\hat{\zeta}_f(t)$ are discussed in §A.2. □

§1.2 The determinant line of a chain complex

We want to offer yet another interpretation for the torsion, in terms of determinant lines, [18, 23, 38, 53]. This has the conceptual advantage that it deals in a coherent way with the thorny issue of signs. Assume again that $\mathbb{K}$ is a field of characteristic zero.

Definition 1.7. A *weighted* $\mathbb{K}$*-line* is a pair (L, w), where w is an integer called the *weight*, and L is a one-dimensional $\mathbb{K}$-vector space L together with a linear action of $\mathbb{K}^*$ on L of the form

$$\mathbb{K}^* \times L \ni (t, u) \mapsto t * u := t^w \cdot u \in L$$

An isomorphism of weighted lines (L_i, w_i), $i = 0, 1$, is a an isomorphism $L_0 \to L_1$ which commutes with the $\mathbb{K}^*$-action. □

Example 1.8. Suppose V is $\mathbb{K}$-vector space of dimension d. Then the one-dimensional space $\Lambda^d V$ is naturally a weighted line of weight d. The pair $(\Lambda^d V, d)$ is called the *determinant line* of V and is denoted by $\mathrm{Det}(V)$. The trivial line equipped with the weight w-action of $\mathbb{K}^*$ will be denoted by $(\mathbb{K}, w)$. By definition $\mathrm{Det}(0) = (\mathbb{K}, 0)$. □

We can define the tensor product of two weighted spaces (L_i, w_i), $i = 0, 1$

$$(L_0, w_0) \hat{\otimes} (L_1, w_1) = (L_0 \otimes L_1, w_0 + w_1).$$

The dual of a weighted line (L, w) is the weighted line

$$(L, w)^{-1} := (L^*, -w).$$

We can organize the collection of weighted lines as a category where

$$\operatorname{Hom}((L_0, w_0), (L_1, w_1)) = \begin{cases} 0 & \text{if } w_0 \neq w_1 \\ \operatorname{Hom}(L_0, L_1) & \text{if } w_0 = w_1. \end{cases}$$

Remark 1.9 (Koszul's sign conventions). We would like to discuss an ubiquitous but quite subtle problem concerning signs. Suppose $(L_i, w_i), i = 0, 1$, are weighted lines. The tensor products

$$U = L_0 \hat{\otimes} L_1 \quad \text{and} \quad V = L_1 \hat{\otimes} L_0$$

are *not equal* as sets but are isomorphic as vector spaces. We will identify them, but not using the obvious isomorphism. We well use instead the *Koszul transposition*

$$\Upsilon_{L_0, L_1} : U \to V, \quad \ell_0 \otimes \ell_1 \mapsto (-1)^{w_0 w_1} \ell_1 \otimes \ell_0.$$

Similarly, given a weighted line (L, w), we will identify the tensor product $(L, w)^{-1} \hat{\otimes} (L, w)$ with $\operatorname{Det}(0)$ using in place of the obvious pairing, the *Koszul contraction* $\operatorname{Tr}_L : (L, w)^{-1} \hat{\otimes} (L, w) \to \operatorname{Det}(0)$ defined by

$$L \otimes L^* \ni (u, u^*) \mapsto u^* \lrcorner\, u := (-1)^{w(w-1)/2} \langle u^*, u \rangle \in \mathbb{K},$$

where $\langle \bullet, \bullet \rangle : L^* \times L \to \mathbb{K}$ denotes the canonical pairing.

Note that $(L^{-1})^{-1} \neq L$ but we will identify them using the tautological map

$$\boldsymbol{\imath}_L : L \longleftrightarrow (L^{-1})^{-1}.$$

The identifications Υ_{L_0, L_1}, Tr_L and $\boldsymbol{\imath}_L$ are compatible in the sense that the diagram below is commutative.

$$\begin{array}{ccccc} (L^{-1})^{-1} \hat{\otimes} L^{-1} & \xrightarrow{\boldsymbol{\imath}_L \otimes \mathbf{1}} & L \hat{\otimes} L^{-1} & \xrightarrow{\Upsilon_{L, L^{-1}}} & L^{-1} \hat{\otimes} L \\ & \searrow^{\operatorname{Tr}_{L^{-1}}} & & \swarrow_{\operatorname{Tr}_L} & \\ & & \operatorname{Det}(0). & & \end{array}$$

Finally note that $L_0 \hat{\otimes} (L_1 \hat{\otimes} L_2) \neq (L_0 \hat{\otimes} L_1) \hat{\otimes} L_2$ but we will identify them via the tautological isomorphism

$$L_0 \hat{\otimes} (L_1 \hat{\otimes} L_2) \to (L_0 \hat{\otimes} L_1) \hat{\otimes} L_2, \quad \ell_0 \otimes (\ell_1 \otimes \ell_2) \mapsto (\ell_0 \otimes \ell_1) \otimes \ell_2.$$

The tautological identification $(L_0 \hat{\otimes} L_1)^{-1} \stackrel{\iota}{\longleftrightarrow} L_0^{-1} \hat{\otimes} L_1^{-1}$ is compatible with the above rules in the sense that the diagram below is commutative.

$$\begin{array}{ccc} & (L_0^{-1}\hat{\otimes} L_1^{-1})\hat{\otimes}(L_0\hat{\otimes} L_1) & \\ \nearrow^{\iota\hat{\otimes}\mathbf{1}} & & \searrow^{\Upsilon_{L_1^{-1},L_0}} \\ (L_0\hat{\otimes} L_1)^{-1}\hat{\otimes}(L_0\hat{\otimes} L_1) & & (L_0^{-1}\hat{\otimes} L_0)\hat{\otimes}(L_1^{-1}\hat{\otimes} L_1) \\ \searrow_{\mathrm{Tr}_{L_0\hat{\otimes} L_1}} & & \swarrow_{\mathrm{Tr}_{L_0}\hat{\otimes}\mathrm{Tr}_{L_1}} \\ & \mathrm{Det}(0)\,. & \end{array}$$

To simplify the presentation we will use the following less accurate descriptions of the above rules.

$$L_0 \hat{\otimes} L_1 = (-1)^{w_0 w_1} L_1 \hat{\otimes} L_1, \quad L^{-1} \hat{\otimes} L = (-1)^{w(w-1)/2}\,\mathrm{Det}(0), \quad (L^{-1})^{-1} = L$$
$$(L_0 \hat{\otimes} L_1) \hat{\otimes} L_2 = L_0 \hat{\otimes} (L_1 \hat{\otimes} L_2), \quad (L_0 \hat{\otimes} L_1)^{-1} = L_0^{-1} \hat{\otimes} L_1^{-1}.$$

Two weighted lines U, V are said to be equal up to permutation, and we write this $U =_p V$, if there exist weighted lines (L_i, w_i), $i = 1, \dots, n$ and a permutation

$$\varphi\colon \{1, \dots, n\} \to \{1, \dots, n\}$$

such that

$$U = \widehat{\bigotimes}_{i=1}^{n} L_i, \quad V = \widehat{\bigotimes}_{i=1}^{n} L_{\varphi(i)}.$$

We denote by $\Upsilon = \Upsilon_\varphi$ the composition of Koszul transpositions which maps U to V.

We can generalize the Koszul contraction to the following more general context. Suppose for example that (L_i, w_i), $i = 0, 1, 2, 3$, are weighted lines. Then define

$$\mathrm{Tr}\colon U := L_1 \hat{\otimes} L_0^* \hat{\otimes} L_2 \hat{\otimes} L_0 \hat{\otimes} L_3 \to V := L_1 \hat{\otimes} L_2 \hat{\otimes} L_3,$$
$$u_1 \hat{\otimes} u_0^* \hat{\otimes} u_2 \hat{\otimes} u_0 \hat{\otimes} u_3 \mapsto (-1)^{w_0 w_2} (u_0^* \lrcorner\, u_0) \cdot u_0 \hat{\otimes} u_2 \hat{\otimes} u_3.$$

This contraction continues to be compatible with the Koszul transpositions in the following sense. For any permutation φ of the five factors

$$L_1, \quad L_0^*, \quad L_2, \quad L_0, \quad L_3$$

we get a new line $\Upsilon_\varphi(U)$ equipped as above with a trace

$$\mathrm{Tr}\colon \Upsilon_\varphi(U) \to V$$

and the diagram below is commutative

$$\begin{array}{ccc} U & \xrightarrow{\ \Upsilon_\varphi\ } & \Upsilon_\varphi(U) \\ & \searrow_{\mathrm{Tr}} \quad \swarrow_{\mathrm{Tr}} & \\ & V. & \end{array}$$

□

Example 1.10. Suppose that L_i, $i = 0, 1$ are weighted lines with the same weight w and $u \in L_0^{-1} \hat{\otimes} L_1$ is a nontrivial element. Then u defines an element in $\mathrm{Hom}(L_0, L_1)$

$$u_0 \mapsto u \lrcorner u_1.$$

If $u = u_0^* \hat{\otimes} u_1$ then

$$u \lrcorner u_0 = (-1)^{w(w+1)/2} \langle u_0^*, u_0 \rangle u_1. \qquad \square$$

If $W_* = \bigoplus_{j \in \mathbb{Z}} W_j$ is a finite dimensional $\mathbb{Z}$-graded $\mathbb{K}$-vector space then its *determinant line* is the weighted line

$$\mathrm{Det}(W) = \widehat{\bigotimes}_{j \in \mathbb{Z}} \mathrm{Det}(W_{-j})^{(-1)^j}.$$

Its weight is the Euler characteristic

$$\chi(W) = \sum_{j \in \mathbb{Z}} (-1)^j \dim W_j.$$

For example, if $W = W_0 \oplus W_1 \oplus W_2$, then

$$\mathrm{Det}(W) = \mathrm{Det}(W_2) \hat{\otimes} \mathrm{Det}(W_1)^{-1} \hat{\otimes} \mathrm{Det}(W_0).$$

To perform *numerical* computations we need to work with richer objects, namely *based vector spaces* and *based weighted lines*. All the tensorial operations on vector spaces have a based counterpart. The dual of a based vector space $(W, \boldsymbol{w})$ is the based vector space $(W^*, \boldsymbol{w}^*)$, where $\boldsymbol{w}^*$ denotes the dual basis. The dual of a based weighted line (L, w, δ) is the based weighted line $(L, w, \delta)^{-1} := (L^*, -w, \delta^*)$ where δ^* denotes the basis of L^* dual to δ,

$$\langle \delta^*, \delta \rangle = 1.$$

We can define the *ordered* tensor product of based weighted lines

$$(L_0, w_0, \delta_0) \hat{\otimes} (L_1, w_1, \delta_1) = (L_0 \hat{\otimes} L_1, w_0 + w_1, \delta_0 \wedge \delta_1).$$

To any based vector space $(W, \boldsymbol{w})$ we can associate in a tautological fashion a based weighted line $(\mathrm{Det}(W), \dim W, \det \boldsymbol{w})$. If

$$(W, \underline{\boldsymbol{w}}) = \bigoplus_{n \in \mathbb{Z}} (W_n, \boldsymbol{w}_n)$$

is a based graded vector space, the associated *based determinant line* is

$$(\mathrm{Det}(W), \chi(W), \det \underline{\boldsymbol{w}}) = \bigotimes \big(\mathrm{Det}(W_{-n}), \dim W_{-n}, \det \boldsymbol{w}_{-n}\big)^{(-1)^n}.$$

Given two *based* weighted lines (L_i, w_i, δ_i), $i = 0, 1$, and a morphism $f : (L_0, w_0) \to (L_1, w_1)$ we define its *torsion* to be the scalar $\langle \delta_1 | f | \delta_0 \rangle \in \mathbb{K}$ uniquely determined by the equality

$$f(\delta_0) = \langle \delta_1 | f | \delta_0 \rangle \delta_1.$$

Example 1.11. Suppose $W = \bigoplus_{n\in\mathbb{Z}} W_n$ is a finite dimensional $\mathbb{Z}$-graded $\mathbb{K}$-vector space, $d_n := \dim_{\mathbb{K}} W_n$. Define

$$\mathrm{Det}_+(W) := \bigotimes_{n\in\mathbb{Z}} \mathrm{Det}(W_{-2n}),\ \mathrm{Det}_-(W) := \bigotimes_{n\in\mathbb{Z}} \mathrm{Det}(W_{-2n-1}),$$

$$\mathrm{Det}_s(W) = \mathrm{Det}_+(W) \hat{\otimes} \mathrm{Det}_-(W)^{-1} =_p \mathrm{Det}(W).$$

For example, if $W = W_0 \oplus W_1 \oplus W_2$ then

$$\begin{aligned}\mathrm{Det}_s(W) &= \mathrm{Det}(W_2) \hat{\otimes} \mathrm{Det}(W_0) \hat{\otimes} \mathrm{Det}(W_1)^{-1}\\ &= (-1)^{d_0d_1}\,\mathrm{Det}(W) = (-1)^{d_0d_1}\,\mathrm{Det}(W_2) \otimes \mathrm{Det}(W_1)^{-1} \hat{\otimes} \mathrm{Det}(W_0). \quad \square\end{aligned}$$

We have the following important result.

Proposition 1.12. *Suppose*

$$0 \to A \xrightarrow{f} C \xrightarrow{g} B \to 0$$

is a short exact sequence of finite dimensional $\mathbb{K}$-vector spaces. Then there exist natural isomorphisms

$$\mathrm{Tr}_{f,g}\colon\ \mathrm{Det}(A) \hat{\otimes} \mathrm{Det}(C)^{-1} \hat{\otimes} \mathrm{Det}(B) \to \mathrm{Det}(0),$$

and

$$\det_{f,g}\colon\ \mathrm{Det}(A) \hat{\otimes} \mathrm{Det}(B) \to \mathrm{Det}(C).$$

Proof. Fix an isomorphism $h\colon C \to A\oplus B$ such that the diagram below is commutative

$$\begin{array}{ccccccccc} 0 & \longrightarrow & A & \xrightarrow{f} & C & \xrightarrow{g} & B & \longrightarrow & 0\\ & & \uparrow{\scriptstyle \mathbf{1}_A} & & \uparrow{\scriptstyle h} & & \uparrow{\scriptstyle \mathbf{1}_B} & & \\ 0 & \longrightarrow & A & \hookrightarrow & A\oplus B & \twoheadrightarrow & B & \longrightarrow & 0. \end{array}$$

We obtain an isomorphism

$$\mathrm{Det}(A) \hat{\otimes} \big(\mathrm{Det}(A) \hat{\otimes} \mathrm{Det}(B)\big)^{-1} \hat{\otimes} \mathrm{Det}(B) \stackrel{[f\otimes g^{-1}]}{\longrightarrow} \mathrm{Det}(A) \hat{\otimes} \mathrm{Det}(C)^{-1} \hat{\otimes} \mathrm{Det}(B)$$

which is independent of the choice h. Define $\mathrm{Tr}_{f,g}$ as the composition

$$\mathrm{Tr}_{f,g} = \mathrm{Tr} \circ [f \otimes g^{-1}]^{-1},$$

i.e.

$$\begin{array}{ccc} \mathrm{Det}(A)\,\hat{\otimes}\,\big(\mathrm{Det}(A)\,\hat{\otimes}\,\mathrm{Det}(B)\big)^{-1}\,\hat{\otimes}\,\mathrm{Det}(B) & & \\ {\scriptstyle [f\otimes g^{-1}]}\big\downarrow & \searrow^{\mathrm{Tr}} & \\ \mathrm{Det}(A)\,\hat{\otimes}\,\mathrm{Det}(C)^{-1}\,\hat{\otimes}\,\mathrm{Det}(B) & \underset{\mathrm{Tr}_{f,g}}{\dashrightarrow} & \mathrm{Det}(0), \end{array}$$

where the map

$$\mathrm{Tr}\colon\ \mathrm{Det}(A)\,\hat{\otimes}\,\big(\mathrm{Det}(A)\,\hat{\otimes}\,\mathrm{Det}(B)\big)^{-1}\,\hat{\otimes}\,\mathrm{Det}(B) \to \mathrm{Det}(0)$$

is the Koszul contraction. By taking tensor products we obtain an isomorphism

$$\mathrm{Det}(A)\,\hat{\otimes}\,\mathrm{Det}(C)^{-1}\,\hat{\otimes}\,\mathrm{Det}(B)\,\hat{\otimes}\,\mathrm{Det}(C) \to \mathrm{Det}(C),$$

and if we take the Koszul contraction on the left hand side we obtain another isomorphism

$$\mathrm{Det}(A)\,\hat{\otimes}\,\mathrm{Det}(B) \xleftarrow{\mathrm{Tr}} \mathrm{Det}(A)\,\hat{\otimes}\,\mathrm{Det}(C)^{-1}\,\hat{\otimes}\,\mathrm{Det}(B)\,\hat{\otimes}\,\mathrm{Det}(C)$$

The definition of $\det_{f,g}$ is now obvious. □

Proposition 1.13. *Suppose $(\underline{C}, \partial)$ is a finite dimensional chain complex. Then there exists a natural isomorphism*

$$\det_{\partial}\colon\ \mathrm{Det}\big(H_*(\underline{C}, \partial)\big) \to \mathrm{Det}(\underline{C}).$$

Proof. We have short exact sequences

$$0 \to R_{i+1}(:= \mathrm{Im}\,\partial_{i+1}) \overset{\imath}{\hookrightarrow} K_i(:= \ker \partial_i) \overset{\pi}{\to} H_i(\underline{C}, \partial) \to 0,$$

and

$$0 \to K_i \overset{\imath}{\hookrightarrow} C_i \overset{\partial}{\to} R_i \to 0.$$

Using Proposition 1.12 we have isomorphisms

$$\det{}_{\imath,\pi}^{(-1)^i}\colon\ \mathrm{Det}(R_{i+1})^{(-1)^i}\,\hat{\otimes}\,\mathrm{Det}\,\big(H_i(\underline{C}, \partial)\big)^{(-1)^i} \to \mathrm{Det}(K_i)^{(-1)^i},$$

and

$$\det{}_{\imath,\partial}^{(-1)^i}\colon\ \mathrm{Det}(K_i)^{(-1)^i}\,\hat{\otimes}\,\mathrm{Det}(R_i)^{(-1)^i} \to \mathrm{Det}(C_i)^{(-1)^i}.$$

By taking tensor products we obtain isomorphisms

$$U_i := \mathrm{Det}(R_{i+1})^{(-1)^i}\,\hat{\otimes}\,\mathrm{Det}\,\big(H_i(\underline{C}, \partial)\big)^{(-1)^i}\,\hat{\otimes}\,\mathrm{Det}(R_i)^{(-1)^i} \to \mathrm{Det}(C_i)^{(-1)^i},$$

and

$$[\det \partial]\colon \widehat{\bigotimes}_{-n=-\infty}^{\infty} U_n \to \operatorname{Det}(\underline{C}).$$

Taking the Koszul contractions of the pairs (R_i^{-1}, R_i) in the left-hand-side we obtain an isomorphism

$$\operatorname{Tr}\colon L \to \operatorname{Det}\big(H_*(\underline{C}, \partial)\big).$$

Then $\det_\partial$ is the unique isomorphism which renders commutative the diagram below.

$$\begin{array}{ccc} & L & \\ {}^{\operatorname{Tr}}\swarrow & & \searrow^{[\det\partial]} \\ \operatorname{Det}\big(H_*(\underline{C}, \partial)\big) & \xrightarrow[\det_\partial]{} & \operatorname{Det}(\underline{C}). \end{array}$$

□

Definition 1.14. The inverse of the above isomorphism is known as the *Euler isomorphism* and will be denoted by $\boldsymbol{Eul}_{\underline{C}} = \boldsymbol{Eul}_{(\underline{C},\partial)}$. □

Example 1.15. Consider the elementary complex

$$0 \hookrightarrow C_1 = V \xrightarrow{\mathbf{1}_V} V = C_0 \to 0.$$

Then the Euler isomorphism

$$\operatorname{Det}(V)^{-1} \hat{\otimes} \operatorname{Det}(V) \to \operatorname{Det}(0)$$

coincides with the Koszul contraction. This simple fact lies at the core of the remarkable compatibility between the Euler isomorphism and the various Koszul identifications, and keeps in check what Deligne called in [18] *"le cauchemar de signes"*. □

Example 1.16. Suppose that $(\underline{C}, \underline{c}, \partial)$ is a finite dimensional acyclic complex. We can choose as in the previous section linearly ordered finite collections $\boldsymbol{b}_i \subset C_i$ such that the restriction of ∂ to the span of $\boldsymbol{b}_i$ is one-to-one, and the linearly ordered set $\partial\boldsymbol{b}_{i+1} \cup \boldsymbol{b}_i$ is a basis of C_i. We then get a basis

$$\delta := \cdots \wedge \big(\det \partial\boldsymbol{b}_{i+1} \wedge \det \boldsymbol{b}_i\big)^{(-1)^i} \wedge \big(\det \partial\boldsymbol{b}_i \wedge \boldsymbol{b}_{i-1}\big)^{(-1)^{i-1}} \wedge \cdots \in \operatorname{Det}(\underline{C}).$$

The Euler isomorphism maps $\operatorname{Det}(\underline{C})$ to $\operatorname{Det}(0)$, and the basis δ to the canonical basis $(-1)^\nu$ of $\operatorname{Det}(0)$, where

$$\nu = \sum_{i=1}^{n} \frac{|\boldsymbol{b}_i|(|\boldsymbol{b}_i| + (-1)^i)}{2}.$$

□

Exercise 1.2. Prove that the isomorphism $\det_{f,g}$ constructed in the above proposition has the following compatibility properties.

(a) Consider the elementary acyclic complex

$$0 \hookrightarrow 0 = C_2 \overset{\iota}{\hookrightarrow} \mathbb{K} = C_1 \overset{\pi}{\twoheadrightarrow} \mathbb{K} = C_0 \to 0.$$

Then $\det_{\iota,\pi}$ is the tautological isomorphism $\mathrm{Det}(\mathbb{K}) \to \mathrm{Det}(\mathbb{K})$.

(b) Consider the commutative diagram

$$\begin{array}{ccccccccc} 0 & \longrightarrow & A_0 & \overset{f_0}{\hookrightarrow} & C_0 & \overset{g_0}{\twoheadrightarrow} & B_0 & \longrightarrow & 0 \\ & & \downarrow{\scriptstyle \alpha} & & \downarrow{\scriptstyle \gamma} & & \downarrow{\scriptstyle \beta} & & \\ 0 & \longrightarrow & A_0 & \overset{f_1}{\hookrightarrow} & C_1 & \overset{g_1}{\twoheadrightarrow} & B_1 & \longrightarrow & 0 \end{array}$$

in which the vertical arrows are isomorphisms. Then the diagram below is commutative

$$\begin{array}{ccc} \mathrm{Det}(A_0) \hat{\otimes} \mathrm{Det}(B_0) & \overset{\det_{f_0,g_0}}{\longrightarrow} & \mathrm{Det}(C_0) \\ \downarrow{\scriptstyle \det\alpha \otimes \det\beta} & & \downarrow{\scriptstyle \det\gamma} \\ \mathrm{Det}(A_1) \hat{\otimes} \mathrm{Det}(B_1) & \overset{\det_{f_1,g_1}}{\longrightarrow} & \mathrm{Det}(C_1). \end{array}$$

□

Exercise 1.3. Show that for any short exact sequence of vector spaces

$$0 \to A \overset{\alpha}{\to} B \overset{\beta}{\to} C \to 0,$$

and for every $s, t \in \mathbb{K}^*$ we have

$$\det_{s\alpha,t\beta} = s^{\dim A} t^{-\dim C} \det_{\alpha,\beta}.$$

□

Exercise 1.4. Show that for every $t \in \mathbb{K}^*$ we have

$$\boldsymbol{Eul}_{(\underline{C},t\partial)} = t^{\zeta} \boldsymbol{Eul}_{(\underline{C},\partial)}, \quad \zeta = \sum_{n \in \mathbb{Z}} (-1)^n n \big(\dim C_n - \dim H_n(\underline{C})\big).$$

□

Suppose now that $(\underline{C}, \underline{\boldsymbol{c}}, \partial)$ is a based acyclic complex. $\mathrm{Det}(\underline{C})$ is a based weighted line with basis $\det \underline{\boldsymbol{c}}$. Since

$$\mathrm{Det}(H_*(\underline{C}, \partial)) = \mathrm{Det}(0).$$

we deduce that $\mathrm{Det}(H_*(\underline{C}, \partial))$ has a natural basis, $\mathbf{1}$.

Proposition 1.17.

$$\mathcal{T}(\underline{C}, \underline{c}) = (-1)^{\nu}\langle \mathbf{1} | \boldsymbol{Eul}_{\underline{C}} | \det \underline{\boldsymbol{c}} \rangle.$$

where ν *is defined as in Example* 1.16.

Exercise 1.5. Prove the above equality. □

When $\mathbb{K} = \mathbb{R}$ we can be even more explicit. More precisely, fix Euclidean metrics on each of C_i. Then, as explained in §A.1, we can explicitly write down a generalized contraction (Definition A.7), i.e. a degree one map

$$\eta \colon C_k \to C_{k+1}$$

such that $\eta^2 = 0$ and $P = \partial\eta + \eta\partial = (\partial + \eta)^2$ is a projector onto a perfect subcomplex with the same homology as $\underline{C}$. More precisely, we can choose η of the form

$$\eta = (ii^* + \Delta)^{-1}\partial^*$$

where $\Delta = (\partial + \partial^*)^2$, and i is the natural inclusion $i\colon \ker \Delta \to \underline{C}$. The formal Hodge theorem shows that $\ker \Delta \cong H_*(\underline{C})$. Consider the linear operator

$$\ker \Delta_{\mathrm{odd}} \oplus \underline{C}_{\mathrm{even}} \to \ker \Delta_{\mathrm{even}} \oplus \underline{C}_{\mathrm{odd}}, \quad \begin{bmatrix} k_{\mathrm{odd}} \\ c_{\mathrm{even}} \end{bmatrix} \mapsto \begin{bmatrix} 0 & i^*_{\mathrm{even}} \\ i_{\mathrm{odd}} & \partial + \eta \end{bmatrix} \cdot \begin{bmatrix} k_{\mathrm{odd}} \\ c_{\mathrm{even}} \end{bmatrix}.$$

We thus get an isomorphism

$$\mathrm{Det} \ker \Delta_{\mathrm{odd}} \hat{\otimes} \mathrm{Det}(\underline{C}_{\mathrm{even}}) \to \mathrm{Det} \ker \Delta_{\mathrm{even}} \hat{\otimes} \mathrm{Det}(\underline{C}_{\mathrm{odd}})$$

This yields the isomorphisms

$$\mathrm{Det} \ker \Delta_{\mathrm{odd}} \hat{\otimes} \mathrm{Det}_s\,(\underline{C}) \to \mathrm{Det} \ker \Delta_{\mathrm{even}},$$

and

$$I\colon \mathrm{Det}_s\,(\underline{C}) \to \mathrm{Det}(\ker \Delta_{\mathrm{odd}})^{-1} \hat{\otimes} \mathrm{Det} \ker \Delta_{\mathrm{even}} \to \mathrm{Det}_s\big(H_*(\underline{C})\big).$$

Up to a permutation, this is the Euler isomorphism. More precisely, we have a commutative diagram

$$\begin{array}{ccc} \mathrm{Det}_s\,(\underline{C}) & \xrightarrow{\ I\ } & \mathrm{Det}_s\big(H_*(\underline{C})\big) \\ \Big\downarrow \Upsilon & & \Big\downarrow \Upsilon \\ \mathrm{Det}(\underline{C}) & \xrightarrow{\ \boldsymbol{Eul}_{\underline{C}}\ } & \mathrm{Det}\big(H_*(\underline{C})\big). \end{array}$$

§1.3 Basic properties of the torsion

The torsion behaves nicely with respect to the basic operations on chain complexes. Consider first a short exact sequence of chain complexes

$$0 \to (\underline{A}, \partial_A) \xrightarrow{f} (\underline{C}, \partial_C) \xrightarrow{g} (\underline{B}, \partial_B) \to 0.$$

Using Proposition 1.12 we obtain canonical isomorphisms

$$\det_{f_n, g_n} \in \operatorname{Det}(A_n) \hat{\otimes} \operatorname{Det}(B_n) \to \operatorname{Det}(C_n),$$

and thus an isomorphism

$$\det_{f,g} \colon \widehat{\bigotimes}_{n=-\infty}^{\infty} \big(\operatorname{Det}(A_n) \hat{\otimes} \operatorname{Det}(B_n)\big)^{(-1)^n} \to \operatorname{Det}(\underline{C}).$$

Now observe that

$$\widehat{\bigotimes}_{-n=-\infty}^{\infty} \big(\operatorname{Det}(A_n) \hat{\otimes} \operatorname{Det}(B_n)\big)^{(-1)^n} =_p \operatorname{Det}(A) \hat{\otimes} \operatorname{Det}(\underline{B}).$$

We get an isomorphism

$$\det_{f,g} \colon \operatorname{Det}(A) \hat{\otimes} \operatorname{Det}(\underline{B}) \to \operatorname{Det}(\underline{C})$$

compatible with the Koszul permutation identifying the two weighted lines,

$$\operatorname{Det}(A) \hat{\otimes} \operatorname{Det}(\underline{B}), \quad \widehat{\bigotimes}_{-n=-\infty}^{\infty} \big(\operatorname{Det}(A_n) \hat{\otimes} \operatorname{Det}(B_n)\big)^{(-1)^n}.$$

On the other hand, we have a long exact sequence

$$\cdots \xrightarrow{\partial} H_q(\underline{A}) \xrightarrow{f_*} H_q(\underline{C}) \xrightarrow{g_*} H_q(\underline{B}) \xrightarrow{\partial} H_{q-1}(\underline{A}) \to \cdots.$$

We can regard this sequence as an acyclic complex which we denote by $H(A, B, C)$. The Euler isomorphism of this acyclic complex induces an isomorphism

$$\boldsymbol{Eul}_{H(A,B,C)} \colon \operatorname{Det}(H(A, B, C)) \to \operatorname{Det}(0).$$

Taking the tensor product of $\operatorname{Det}(H(A, B, C))$ with $\operatorname{Det}(H_*(\underline{C}))$ and then applying the Koszul contraction to the pair

$$\operatorname{Det}(H_*(\underline{C}))^{-1}, \quad \operatorname{Det}(H_*(\underline{C}))$$

we obtain an isomorphism

$$H(\det_{f,g}) \colon \operatorname{Det}(H_*(\underline{A})) \hat{\otimes} \operatorname{Det}(H_*(\underline{B})) \to \operatorname{Det}(H_*(\underline{C})).$$

Proposition 1.18. *The diagram below is commutative.*

$$\begin{array}{ccc} \operatorname{Det}(\underline{A}) \hat{\otimes} \operatorname{Det}(\underline{B}) & \xrightarrow{\det_{f,g}} & \operatorname{Det}(\underline{C}) \\ \Big\downarrow {\scriptstyle \boldsymbol{Eul}_{\underline{A}} \otimes \boldsymbol{Eul}_{\underline{B}}} & & \Big\downarrow {\scriptstyle \boldsymbol{Eul}_{\underline{C}}} \\ \operatorname{Det}\big(H_*(\underline{A})\big) \hat{\otimes} \operatorname{Det}\big(H_*(\underline{B})\big) & \xrightarrow{H(\det_{f,g})} & \operatorname{Det}\big(H_*(\underline{C})\big). \end{array} \tag{1.7}$$

To better understand the meaning of the above result suppose we fix bases $\underline{\boldsymbol{a}}$, $\underline{\boldsymbol{b}}$, $\underline{\boldsymbol{c}}$ of $\underline{A}$, $\underline{B}$ and respectively $\underline{C}$, and bases $[\underline{\boldsymbol{a}}]$, $[\underline{\boldsymbol{b}}]$, $[\underline{\boldsymbol{c}}]$ of $H_*(\underline{A})$, $H_*(\underline{B})$ and respectively $H_*(\underline{C})$. We assume that

$$\underline{\boldsymbol{c}} = f(\underline{\boldsymbol{a}}) \cup \underline{\boldsymbol{b}}', \quad g(\underline{\boldsymbol{b}}') = \underline{\boldsymbol{b}}.$$

We can now identify $\boldsymbol{Eul}_{\underline{A}}$, $\boldsymbol{Eul}_{\underline{B}}$, $\boldsymbol{Eul}_{\underline{C}}$ with scalars in $\mathbb{K}^*$. $H(\det_{f,g})$ can also be identified with a scalar, the torsion of the acyclic complex $H(A, B, C)$. Then (1.7) implies

$$\boldsymbol{Eul}_{\underline{C}} \cdot \mathcal{T}^{-1}_{H(A,B,C)} = \pm \boldsymbol{Eul}_{\underline{A}} \cdot \boldsymbol{Eul}_{\underline{B}}\,. \tag{1.8}$$

Exercise 1.6. Prove (1.7) and (1.8). □

The above result implies immediately that the torsion is multiplicative with respect to direct sums. More precisely, we have the following elementary, but extremely versatile result.

Theorem 1.19. *Consider a short exact sequence of based acyclic complexes of* $\mathbb{K}$-*vector spaces*

$$0 \to (\underline{A}, \underline{\boldsymbol{a}}) \xrightarrow{f} (\underline{C}, \underline{\boldsymbol{c}}) \xrightarrow{g} (\underline{B}, \underline{\boldsymbol{b}}) \to 0,$$

such that

$$\underline{\boldsymbol{c}} = f(\underline{\boldsymbol{a}}) \cup \underline{\boldsymbol{b}}', \quad g(\underline{\boldsymbol{b}}') = \underline{\boldsymbol{b}}.$$

Then $H(\det_{f,g}) = 1$,

$$\big\langle \det \underline{\boldsymbol{c}} | \det_{f,g} | \det \underline{\boldsymbol{a}} \hat{\otimes} \det \underline{\boldsymbol{b}} \big\rangle = \pm 1,$$

and

$$\big\langle \mathbf{1} |\, \boldsymbol{Eul}_{\underline{A}} \,|\det \underline{\boldsymbol{a}} \big\rangle \cdot \big\langle \mathbf{1} |\, \boldsymbol{Eul}_{\underline{B}} \,|\det \underline{\boldsymbol{b}} \big\rangle = \big\langle \mathbf{1} |\, \boldsymbol{Eul}_{\underline{C}} \,|\det \underline{\boldsymbol{c}} \big\rangle \cdot \big\langle \det \underline{\boldsymbol{c}} | \det_{f,g} | \underline{\boldsymbol{a}} \hat{\otimes} \underline{\boldsymbol{b}} \big\rangle.$$

In particular,

$$\mathcal{T}(\underline{C}, \underline{\boldsymbol{c}}) = \pm \mathcal{T}(\underline{A}, \underline{\boldsymbol{a}}) \cdot \mathcal{T}(\underline{B}, \underline{\boldsymbol{b}}).$$

For any chain complex $(\underline{C}, \partial)$ and any $k \in \mathbb{Z}$ we denote by $(\underline{C}[k], \partial)$ the degree shifted complex defined by

$$C[k]_i = C_{i+k}, \quad \forall i \in \mathbb{Z}.$$

Observe that

$$\operatorname{Det}(\underline{C}[k]) =_p \operatorname{Det}(\underline{C})^{(-1)^k},$$

and we have a commutative diagram

$$\begin{array}{ccc} \operatorname{Det}(\underline{C}[k]) & \xrightarrow{\text{Koszul}} & \big(\operatorname{Det}(\underline{C})\big)^{(-1)^k} \\ {\scriptstyle \boldsymbol{Eul}_{\underline{C}[k]}}\Big\downarrow & & \Big\downarrow{\scriptstyle \boldsymbol{Eul}_{\underline{C}}^{(-1)^k}} \\ \operatorname{Det}\big(H_{*+k}(\underline{C})\big) & \xrightarrow{\text{Koszul}} & \big(\operatorname{Det} H_*(\underline{C})\big)^{(-1)^k}. \end{array}$$

Given a chain complex

$$(\underline{C}, \partial) = \bigoplus_{j \in \mathbb{Z}} (C_j, \partial_j)$$

of $\mathbb{K}$-vector spaces we can form its dual $\underline{C}^-$ defined by

$$\underline{C}_j^- := C_{-j}^* := \operatorname{Hom}(C_{-j}, \mathbb{K}),$$

and whose boundary maps are the duals of the boundary maps of $\underline{C}$. We have a commutative diagram

$$\begin{array}{ccccc} \operatorname{Det}(\underline{C}^-) \hat{\otimes} \operatorname{Det}(\underline{C}) & & \xrightarrow{\boldsymbol{Eul}_{\underline{C}^-} \hat{\otimes} \boldsymbol{Eul}_{\underline{C}}} & & \operatorname{Det}\big(H_*(\underline{C}^{-1})\big) \hat{\otimes} \operatorname{Det}\big(H_*(\underline{C})\big) \\ & {\scriptstyle \mathrm{Tr}} \searrow & & \swarrow {\scriptstyle \mathrm{Tr}} & \\ & & \operatorname{Det}(0). & & \end{array}$$

In particular, if $\underline{C}$ is acyclic, and $\underline{\boldsymbol{c}}$ is a basis, then

$$\mathcal{T}(\underline{C}, \underline{\boldsymbol{c}}) \cdot \mathcal{T}(\underline{C}^-, \underline{\boldsymbol{c}}^-) = \pm 1, \tag{1.9}$$

where $\underline{c}_n^- = \underline{c}_{-n}^*$. Suppose now that the field $\mathbb{K}$ is equipped with an involutive automorphism

$$\varepsilon : \mathbb{K} \to \mathbb{K}.$$

Example 1.20. If $\mathbb{K} = \mathbb{C}$ we can take ε to be the complex conjugation. If $\mathbb{K} = \mathbb{Q}(t)$, the field of rational functions in one variable, then the correspondence $t \mapsto t^{-1}$ induces such an involution. □

The ε-conjugate of a $\mathbb{K}$-vector space V is the vector space $\bar{V} = \bar{V}^{\varepsilon}$ which coincides with V as an Abelian group while the scalar multiplication is given by

$$\mathbb{K} \times V \ni (\lambda, v) \mapsto \varepsilon(\lambda)v.$$

We denote by $\varepsilon = \varepsilon_V : V \to \bar{V}^{\varepsilon}$ the tautological bijection. A linear map $A : U \to V$ tautologically induces a linear map $\bar{A}^{\varepsilon} : \bar{U}^{\varepsilon} \to \bar{V}^{\varepsilon}$.

An ε-pairing between the $\mathbb{K}$-vector spaces U, V is a bilinear map

$$\langle \bullet, \bullet \rangle : U \times \bar{V}^{\varepsilon} \to \mathbb{K}.$$

Observe that such a pairing induces a $\mathbb{K}$-linear map

$$T : \bar{V}^{\varepsilon} \to U^*, \quad v \mapsto \langle \bullet, v \rangle.$$

The $\mathbb{K}$-pairing is called *perfect* (or a *duality*) if the induced $\mathbb{K}$-linear map

$$T : \bar{V}^{\varepsilon} \to U^*$$

is an isomorphism. If U, V happen to be $\mathbb{Z}_2$-graded

$$U \cong U_+ \oplus U_-, \quad V = V_+ \oplus V_-$$

then the duality is called *supersymmetric* if the operator T is supersymmetric, i.e. it is either purely odd, $T(\bar{V}^{\varepsilon}_{\pm}) = U^*_{\mp}$, or purely even, $T(\bar{V}^{\varepsilon}_{\pm}) = U^*_{\pm}$. Correspondingly, a supersymmetric duality can be even or odd. We will denote by ν the parity of a supersymmetric duality.

Consider the *length n* chain complexes of $\underline{C} = \oplus_{i=0}^{n} C_i$ and $D = \oplus_{j=0}^{n} D_j$ of $\mathbb{K}$-vector spaces with ambiguities $\mathfrak{A}$. A *chain complex pairing* is a pairing

$$\langle \bullet, \bullet \rangle : \underline{C} \times \underline{\bar{D}}^{\varepsilon} \to \mathbb{K}$$

such that the induced map T is a degree zero morphism between the chain complexes

$$T : \underline{\bar{D}}^{\varepsilon} \to \underline{C}^{-}[n].$$

Observe that such pairings are supersymmetric with respect to the natural $\mathbb{Z}_2$-gradings on the chain complexes. The parity of the pairing is the same as the parity of n- the length of the chain complexes. A pairing is called *perfect* if the induced morphism is an isomorphism. We have the following immediate result.

Proposition 1.21 (Abstract duality theorem). *Suppose*

$$\langle \bullet, \bullet \rangle : \underline{C} \times \underline{\bar{D}}^{\varepsilon} \to \mathbb{K}$$

is a perfect pairing of acyclic complexes of length n. Then the following diagram is commutative:

$$\begin{array}{ccccc}
\operatorname{Det}(\underline{D}) & \xrightarrow{\det \varepsilon} & \operatorname{Det}(\underline{\bar{D}}^{\varepsilon}) & \xrightarrow{\det T} & \operatorname{Det}(\underline{C}^{-})^{(-1)^n} \\
\Big\downarrow \boldsymbol{Eul}_{\underline{D}} & & & & \Big\downarrow \boldsymbol{Eul}_{\underline{C}^{-}}^{(-1)^n} \\
\operatorname{Det}(H_*(\underline{D})) & \xrightarrow{\det \varepsilon_*} & \operatorname{Det}(H_*(\underline{\bar{D}}^{\varepsilon})) & \xrightarrow{\det T_*} & \operatorname{Det}(H_*(\underline{C}^{-}))^{(-1)^n}.
\end{array}$$

In particular, if $\underline{C}$ is acyclic and $\underline{\boldsymbol{c}}$ is a basis then the equality (1.9) *implies*

$$\varepsilon\big(\mathcal{T}(\underline{D}, [\boldsymbol{d}])\big) = \mathcal{T}(\underline{\bar{D}}^{\varepsilon}, [\bar{\boldsymbol{d}}]) = \pm\mathcal{T}(\underline{C}, [\underline{\boldsymbol{c}}])^{(-1)^{n+1}},$$

where $\underline{\boldsymbol{d}} = \varepsilon T^{-1}(\underline{\boldsymbol{c}}^{-})$.

§1.4 Some generalizations

The notion of torsion can be defined in a much more general context than the one discussed above. We refer the reader to [19, 72] for a more in depth study. We will need only a mild generalization of the ideas developed so far.

Often, instead of complexes of vector spaces over a field $\mathbb{K}$ one encounters complexes $\underline{C}$ of free modules over an integral domain R. Denote by $\mathbb{K}$ the quotient field of R. An R-basis of $\underline{C}$ canonically induces a $\mathbb{K}$-basis of $\underline{C} \otimes_R \mathbb{K}$. The torsion of $\underline{C}$ (with respect to some R-basis) is, by definition, the torsion of the complex $\underline{C} \otimes_R \mathbb{K}$ with respect to the induced basis. There is no canonical choice of R-basis and thus we must consider any two of them equivalent. This ambiguity is encoded by the group $\mathfrak{A} = \mathrm{GL}(\underline{C}, R)$ of automorphisms of R-modules. This group acts transitively on the set of R-bases and thus the torsion is well defined as an element of $\mathbb{K}/R^{\times}$, where $R^{\times}$ denotes the multiplicative subgroup of invertible elements of R.

Suppose more generally that R is only a commutative ring with unit, and $\varphi\colon R \to \mathbb{K}$ is a nontrivial morphism from R to a field $\mathbb{K}$. If $\underline{C}$ is a chain complex of free R-modules, then we can form the complex of $\mathbb{K}$-vector spaces

$$\underline{C}^{\varphi} = \underline{C} \otimes_{\varphi} \mathbb{K}.$$

Then a R-basis of $\underline{C}$ defines a $\mathbb{K}$-basis of $\underline{C}^{\varphi}$, and we define

$$\mathcal{T}^{\varphi}(\underline{C}, \bullet) := \mathcal{T}(\underline{C}^{\varphi}, \bullet).$$

Suppose R is a *quasi-field*, i.e. a commutative ring with 1 which decomposes as a finite direct sum of fields of characteristic 0

$$R = \oplus_{j=1}^{m} \mathbb{K}_j.$$

Denote by φ_j the natural projection $R \to \mathbb{K}_j$. Suppose $\underline{C}$ is a chain complex of *free*, R-modules. A R-basis $\underline{c}$ of $\underline{C}$ induces a $\mathbb{K}_j$-basis of $\underline{C}^{\varphi_j}$ and as above we obtain a torsion

$$\mathcal{T}^{\varphi_j}(\underline{C}, [\underline{c}]_{\mathfrak{A}}) \in \mathbb{K}_j^* / \det \mathfrak{A} \cup \{0\}.$$

The direct sum

$$\bigoplus_{j=1}^{m} \mathcal{T}^{\varphi_j}(\underline{C}, [\underline{c}]_{\mathfrak{A}}) \in (\oplus_j \mathbb{K}_j) / \det \mathfrak{A}$$

is an element in $R/\det \mathfrak{A}$ – the space of orbits of the determinant action of $\mathfrak{A}$ on R.

We can further extend the class of coefficient rings to include the *quasi-integral domains*, i.e. the commutative rings R with 1 such that the associated ring of fractions $Q(R)$ (i.e. the localization with respect to the prime ideal of zero divisors) is a quasi-field

$$Q(R) = \oplus_j \mathbb{K}_j.$$

Denote by $\varphi_j \colon R \to \mathbb{K}_j$ the natural morphism. If $\underline{C}$ is a chain complex of free, R-modules then, by definition, its torsion is the direct sum

$$\mathcal{T}(\underline{C}, \bullet) := \bigoplus_j \mathcal{T}^{\varphi_j}(\underline{C}, \bullet) \in Q(R)/\det \mathfrak{A}.$$

Let us observe the following simple fact.

Proposition 1.22. *Suppose R is a quasi-integral domain of characteristic zero, $\mathbb{K}$ is a field of characteristic zero and $\varphi \colon R \to \mathbb{K}$ is a nontrivial morphism. If $\underline{C}$ is a chain complex of free R-modules then*

$$\varphi\big(\mathcal{T}(\underline{C}, \bullet)\big) = \mathcal{T}^{\varphi}(\underline{C}, \bullet).$$

§1.5 Abelian group algebras

In this section we want to describe a few special features of the group algebras of finitely generated Abelian groups since they will play a central role in topological applications.

Suppose H is a finitely generated Abelian group. It can be non-canonically decomposed as

$$H = F_H \oplus \operatorname{Tors}(H),$$

where F_H denotes the free part of H, $F_H \cong H/\operatorname{Tors}(H)$. Denote by $\mathbb{Q}(H)$ the ring of fractions of the group algebra $\mathbb{Z}[H]$.

Proposition 1.23. *$\mathbb{Z}[H]$ is a quasi-integral domain of characteristic zero.*

Proof. Let us first consider the two extremes, $\mathrm{Tors}(H) = 0$, or $F_H = 0$.

• If $\mathrm{Tors}(H) = 0$ then $H = F_H$, and if rank $H = n$, then $\mathbb{Q}(H)$ is the field of rational functions in n variables with rational coefficients.

• If $F_H = 0$, so that $H = \mathrm{Tors}(H)$, then $\mathbb{Q}[H]$ is a semisimple, commutative algebra, and thus decomposes as a sum of fields; see [55]. In particular, $\mathbb{Q}[H] = \mathbb{Q}(H)$.

In general we have

$$\mathbb{Q}[H] = \mathbb{Q}[\mathrm{Tors}(H)][F_H] \cong \bigoplus_i \mathbb{K}_i[F_H],$$

where the summands $\mathbb{K}_i$ are the fields entering into the direct sum decomposition of $\mathbb{Q}[\mathrm{Tors}(H)]$. Thus,

$$Q(\mathbb{Z}[H]) = \mathbb{Q}(H) = \bigoplus_i \mathbb{K}_i(F_H).$$

Each of the above summands is a field of rational functions in $n = \mathrm{rank}(H)$ variables. □

Example 1.24. If H is finite cyclic, then the fields in the decomposition of $\mathbb{Q}[H]$ are all cyclotomic fields. We illustrate this on the special case when H is a finite cyclic group of order $n > 1$,

$$H = \{1, x, \dots, x^{n-1}\}.$$

Then

$$\mathbb{Q}[H] \cong \mathbb{Q}[t]/(t^n - 1).$$

The decomposition in $\mathbb{Q}[t]$ of $t^n - 1$ into irreducible factors is (see [55])

$$t^n - 1 = \prod_{d|n} \Phi_d(t),$$

where Φ_m denotes the m-th cyclotomic polynomial

$$\Phi_m(t) = \prod_{(r,m)=1} (t - \zeta^r), \quad \zeta = \zeta_m := \exp(2\pi \mathbf{i}/m).$$

Thus $\mathbb{Q}[t]/(t^n - 1)$ decomposes as a direct sum of cyclotomic fields

$$\mathbb{Q}[H] \cong \mathbb{Q}[t]/(t^n - 1) \cong \bigoplus_{d|n} \mathbb{Q}[t]/(\Phi_d(t)) =: \bigoplus_{d|n} \mathfrak{F}_d.$$

□

Remark 1.25. Since every finite Abelian group H is a direct sum of cyclic groups we deduce inductively that all the fields $\mathbb{K}_j$ in the decomposition

$$\mathbb{Q}[H] \cong \bigoplus_j \mathbb{K}_j$$

are isomorphic to subfields of $\mathbb{C}$. The natural projections $\pi_j : \mathbb{Q}[H] \to \mathbb{K}_j \subset \mathbb{C}$ induce group morphisms

$$\pi_j : H \to \mathbb{C}^*.$$

These are known as the *characters* of H and determine the harmonic (Fourier) analysis on H. An element $f \in \mathbb{Q}[H]$ can be regarded as a function

$$f : H \to \mathbb{Q}.$$

Its components $\pi_j(f)$ are determined by the Fourier transform of f. We refer to §1.6 for more details. □

The natural morphism $\mathbb{Q}(H) \to \mathbb{Q}(F_H)$ induced by the projection

$$\pi : H \to F_H = H/\operatorname{Tors}(H)$$

is called the *augmentation* map, and we will denote it by $\mathfrak{aug}$. It has a natural right inverse

$$\begin{aligned} \mathfrak{aug}^{-1} &: \mathbb{Q}(F_H) \to \mathbb{Q}(H), \\ &F_H \ni f \mapsto \frac{1}{\nu_H} \sum_{\pi(h)=f} h, \quad (\nu_H := |\pi^{-1}(0)| = |\operatorname{Tors}(H)|). \end{aligned}$$

Set $\mathfrak{I} := \mathfrak{aug}^{-1}(1)$. Observe that

$$\mathfrak{aug}^{-1}\big(\mathfrak{aug}(q)\big) = q\mathfrak{I}, \quad \forall q \in \mathbb{Q}(H).$$

The ideal of $\mathbb{Q}(H)$ generated by $\mathfrak{I}$ will be denoted by $(\mathfrak{I})$. It is the kernel of $\mathfrak{aug}$. The above identity shows that, as a ring, the ideal $(\mathfrak{I})$ is isomorphic to the field $\mathbb{Q}(F_H)$. From the identity $\mathfrak{I}^2 = \mathfrak{I}$ we deduce the following consequence.

Corollary 1.26. *The map*

$$\mathbb{Q}(H) \to \ker \mathfrak{aug}, \quad x \mapsto x - x\mathfrak{I}$$

is a surjective morphism of algebra. Moreover the element

$$\hat{\mathbf{1}} := 1 - \mathfrak{I} \in \ker \mathfrak{aug}$$

is the identity element in the subalgebra $\ker \mathfrak{aug}$.

Following Turaev [111], we define a family of subrings $\mathfrak{N}_k(H) \subset \mathbb{Q}(H)$, $k = 0, 1, 2, \dots$ as follows.

A. If $\operatorname{rank}(H) \geq 1$ then

$$\begin{aligned}\mathfrak{N}_0(H) &:= \mathbb{Z}[H],\\ \mathfrak{N}_k(H) &= \{q \in \mathbb{Q}(H);\ (1-h)q \in \mathfrak{N}_{k-1}(H), \forall h \in H\}.\end{aligned}$$

Roughly speaking, $\mathfrak{N}_k(G)$ consists of all solutions in $x \in \mathbb{Q}(G)$ of the linear system

$$(1-g)^k x \in \mathbb{Z}[G], \quad \forall g \in G.$$

B. If $\operatorname{rank}(H) = 0$ then

$$\mathfrak{N}_k(H) = \ker \mathfrak{aug} \subset \mathbb{Q}(H), \quad \forall k = 0, 1, \dots.$$

Observe that

$$\mathfrak{N}_0(H) \subset \mathfrak{N}_1(H) \subset \cdots \subset \mathfrak{N}_k(H) \subset \cdots.$$

We set

$$\mathfrak{N}(H) = \lim_{k\to\infty} \mathfrak{N}_k(H),$$

$$\mathfrak{S} := \nu_H \mathfrak{I} = \sum_{u \in \operatorname{Tors}(H)} u \in \mathbb{Z}[H], \quad \nu_H = |\operatorname{Tors}(H)|.$$

Proposition 1.27 ([111]). *Let H be a finitely generated Abelian group of rank ≥ 1.*

(a) *If* $\operatorname{rank}(H) \geq 2$ *then* $\mathfrak{N}_k(H) = \mathbb{Z}[H]$, $\forall k = 0, 1, 2, \dots$.

(b) *Suppose* $\operatorname{rank}(H) = 1$. *Denote by t a generator of $F := H/\operatorname{Tors}(H)$ and set $T = \mathfrak{aug}^{-1}(t)$. Then*

$$x \in \mathfrak{N}_k(H) = \mathbb{Z}[H] + \mathfrak{S}\mathbb{Z}[H](1-T)^{-k}.$$

Proof. (a) It suffices to prove $\mathfrak{N}_1(H) = \mathfrak{N}_0(H)$. The equality is obvious if H is torsion free. Suppose now $\operatorname{Tors}(H) \neq 0$. Any $x \in \mathbb{Q}(H)$ decomposes uniquely as

$$x := \bar{x} + x^{\perp}$$

where

$$\bar{x} := \mathfrak{I}x = \mathfrak{aug}^{-1}\mathfrak{aug}(x), \quad x^{\perp} := (1-\mathfrak{I})x.$$

Suppose $x \in \mathbb{Q}(H)$ is such that

$$(1-h)x \in \mathbb{Z}[H], \quad \forall h \in H.$$

Observe that $\mathfrak{aug}(\bar{x}) \in \mathfrak{N}_1(H/\operatorname{Tors}(H)) = \mathbb{Z}[H/\operatorname{Tors}(H)]$ so that

$$\bar{x} = \mathfrak{aug}^{-1}\mathfrak{aug}(\bar{x}) \in \mathfrak{I}\mathbb{Z}[H].$$

By summing over $h \in \mathrm{Tors}(H)$ we deduce

$$\nu_H x^{\perp} = \nu_H(1 - \mathfrak{I})x = \sum_{h \in \mathrm{Tors}(H)} (1-h)x \in \mathbb{Z}[H].$$

We conclude that $x \in \mathbb{Q}[H]$ and thus we can write

$$x = \sum_{h \in S} x_h h,$$

where $S \subset H$ is a *finite set* and $x_h \in \mathbb{Q}$. Since H is infinite we can find $h_0 \in H$ such that S is disjoint from $h_0 + S$. Then

$$\mathbb{Z}[H] \ni (1 - h_0)x = \sum_{h \in S} x_h(h - h_0 h).$$

This shows $x_h \in \mathbb{Z}$.

(b) Again, the conclusion is obvious when H is torsion free. Set

$$\mathfrak{N}'_k(H) = \mathbb{Z}[H] + \mathbb{Z} \cdot \mathfrak{S}\mathbb{Z}[H](1 - T)^{-k}.$$

We will first prove the equality

$$\mathfrak{N}_1 = \mathfrak{N}'_1.$$

Next, using induction, we will establish the general identity

$$\mathfrak{N}_k = \mathfrak{N}'_k, \quad k \geq 2.$$

Pick $\tau \in H$ such that $\mathfrak{aug}(\tau) = t \iff T = \mathfrak{I}\tau$. Since

$$(1 - \tau)\mathfrak{I}(1 - T)^{-1} = \mathfrak{I}(1 - T)(1 - T)^{-1} = \mathfrak{I}$$

we deduce

$$\mathfrak{I}(1 - T)^{-1} = \mathfrak{I}(1 - \tau)^{-1} \iff \mathfrak{S}(1 - T)^{-k} = \mathfrak{S}(1 - \tau)^{-k}, \quad \forall k.$$

We can now prove that $\mathfrak{N}'_k \subset \mathfrak{N}_k$, $\forall k \geq 1$. Indeed, if $x \in \mathfrak{N}'_k$, $h \in \mathrm{Tors}(H)$ and $m \in \mathbb{Z}$ then

$$(1 - h\tau^m)\mathfrak{S} = (\mathfrak{S} - \tau^m \mathfrak{S}) = \mathfrak{S}(1 - T^m),$$

so that

$$(1 - h\tau^m)x \in \mathfrak{N}'_{k-1} = \mathfrak{N}_{k-1}.$$

To prove the reverse inclusion, consider $x \in \mathfrak{N}_k(H)$. Then

$$\mathfrak{aug}(x) \in \mathfrak{N}_k\big(H/\mathrm{Tors}(H)\big)$$

so that

$$\bar{x} = \mathfrak{I}x = \mathfrak{aug}^{-1}\mathfrak{aug}(x) \in \mathfrak{I}\mathbb{Z}[\tau] + \mathfrak{I}\mathbb{Z}[\tau](1-\tau)^{-k} = \frac{1}{\nu_H}\mathfrak{S}\mathbb{Z}[H](1-\tau)^{-k}.$$

Summing the congruences $(1-h)x \in \mathfrak{N}_{k-1}(H)$ over $h \in \mathrm{Tors}(H)$ we deduce

$$\nu_H x^{\perp} \in \mathfrak{N}_{k-1}(H) = \mathbb{Z}[H] + \mathfrak{S}\mathbb{Z}[H](1-T)^{-(k-1)}.$$

Thus

$$x \in \frac{1}{\nu_H}\big(\mathbb{Z}[H] + \mathfrak{S}\mathbb{Z}[H](1-T)^{-k}\big)$$

and

$$(1-h)x \in \mathbb{Z}[H] + \mathfrak{S}\mathbb{Z}[H](1-\tau)^{-(k-1)}, \quad \forall h \in H.$$

We write

$$x^{\perp} = A, \quad \bar{x} = \mathfrak{S}B(1-\tau)^{-k}.$$

We need to consider two cases.

A. $k = 1$. In this case

$$x^{\perp} = A \in \frac{1}{\nu_H}\mathbb{Z}[H], \quad \bar{x} = \mathfrak{S}B(1-\tau)^{-1} \in \frac{1}{\nu_H}\mathfrak{S}\mathbb{Z}[H](1-\tau)^{-1},$$

and we can write

$$A = \sum_{m\in\mathbb{Z}}\Big(\sum_{u\in\mathrm{Tors}(H)} a_{m,u}u\Big)\tau^m, \quad B = \sum_{m\in\mathbb{Z}}\Big(\sum_{u\in\mathrm{Tors}(H)} b_{m,u}u\Big)\tau^m.$$

Set

$$\bar{b}_m = \sum_u b_{m,u}.$$

Then

$$\mathfrak{S}B = \mathfrak{S}\sum_m \bar{b}_m\tau^m.$$

Denote by $\alpha_{m,u}$ (resp. β_m) the image of $a_{m,u}$ (resp. $\bar{b}_m$) in $\mathbb{Q}/\mathbb{Z}$. Observe that

$$\nu_h\beta_m = 0 = \nu_H\alpha_{m,u}.$$

Since $(1-u)\mathfrak{S} = 0$, $\forall u \in \mathrm{Tors}(H)$ we deduce

$$(1-u)x = (1-u)A \in \mathbb{Z}[H], \quad \forall u \in \mathrm{Tors}(H) \iff \alpha_{m,v} = \alpha_{m,u}, \tag{1.10}$$

$\forall u, v \in \mathrm{Tors}(H)$, $\forall m$. Denote by α_m the common value of $\alpha_{m,u}$, $u \in \mathrm{Tors}(H)$ and by k_m the integer $0 \le k_m < \nu_H$ such that

$$\frac{k_m}{\nu_H} = \alpha_m \quad \text{in } \mathbb{Q}/\mathbb{Z}.$$

Define

$$K := \sum_m \frac{k_m}{\nu_H}\tau^m \in \frac{1}{\nu_H}\mathfrak{S}\mathbb{Z}[H].$$

The identities (1.10) can be rephrased in the following compact form,

$$A - \mathfrak{S}K \in \mathbb{Z}[H].$$

On the other hand, the identities

$$(1-\tau)x \in \mathbb{Z} \iff \alpha_{m+1} - \alpha_m + \beta_{m+1} = 0, \quad \forall m$$

can now be rewritten

$$\mathfrak{S}(1-\tau)K + \mathfrak{S}B \in \mathfrak{S}\mathbb{Z}[H]$$

so that

$$x = (A - \mathfrak{S}K) + \mathfrak{S}\big(B + (1-\tau)K\big)(1-\tau)^{-1} \in \mathbb{Z}[H] + \mathfrak{S}\mathbb{Z}[H](1-\tau)^{-1} = \mathfrak{N}'_1.$$

B. $k > 1$. Set

$$S := \{h \in H;\ a_h b_h \neq 0\}.$$

Then if $h_0 \in H$ is such that $S \cap (h_0 + S) = \emptyset$ and we conclude as in part (a). □

The above proposition has the following immediate consequence.

Corollary 1.28. *Suppose that H is a finitely generated Abelian group. Denote by $\imath : H \to \mathfrak{N}(H)$ the natural morphism. If $P, Q \in \mathfrak{N}(H)$ then*

$$P|Q \iff P|Q(\imath(h) - 1), \quad \forall h \in H.$$

Example 1.29. Suppose $H \cong \mathbb{Z} \oplus G$ where G is a finite Abelian group. Denote by t the generator of $\mathbb{Z}$. Then

$$T = \mathfrak{I}t = \frac{1}{N}\mathfrak{S}t = \Big(\frac{1}{N}\sum_{g\in G} g\Big)t, \quad N := |G|$$

The group algebra $\mathbb{Q}[H]$ is isomorphic to the ring of Laurent polynomials

$$\mathbb{Q}[H] \cong \mathbb{Q}[G][t, t^{-1}] \cong \bigoplus \mathbb{K}_i[t, t^{-1}].$$

Then

$$\mathfrak{N}[H] = \mathbb{Z}[H] + \mathfrak{S}\mathbb{Z}[T, T^{-1}, (1-T)^{-1}].$$ □

The correspondence $H \mapsto \mathfrak{N}(H)$ is functorial. More precisely we have the following result due to V. Turaev.

Proposition 1.30 ([111]). *Every* ***epimorphism***

$$f : H_1 \to H_2$$

induces a morphism of $\mathbb{Q}$-algebras $f_\sharp : \mathfrak{N}(H_1) \to \mathfrak{N}(H_2)$ such that the diagram below is commutative.

$$\begin{array}{ccc} H_1 & \xrightarrow{\ f\ } & H_2 \\ {\scriptstyle \iota_1}\downarrow & & \downarrow{\scriptstyle \iota_2} \\ \mathfrak{N}(H_1) & \overset{f_\sharp}{\dashrightarrow} & \mathfrak{N}(H_2). \end{array}$$

Proof. $f_\sharp$ is defined as follows. Observe first that f induces a morphism

$$f_* : \mathbb{Z}[H_1] \to \mathbb{Z}[H_2] \to \mathfrak{N}(H_2).$$

If $\operatorname{rank}(H_1) \geq 2$ then we set $f_\sharp = f_*$. If $\operatorname{rank}(H_1) = 0$ then $f_\sharp$ denotes the restriction of f_* to $\ker \mathfrak{aug} \subset \mathbb{Q}[H_1]$.

When $\operatorname{rank}(H_1) = 1$ the definition is a bit more intricate. Denote by t a generator of $F_{H_1} = H_1/\operatorname{Tors}(H_1)$, choose $\tau \in H_1$ an element projecting to t and set $T := \mathfrak{aug}^{-1}(t)$. We claim that there exists an unique $X = X_f \in \mathfrak{N}(H_2)$ such that

$$X f_*(\tau - 1) = X f_*(T - 1) = f_*(\mathfrak{S}_1), \quad X f_*(h - 1) = 0, \quad \forall h \in \operatorname{Tors}(H_1). \tag{1.11}$$

Uniqueness. If X, X' are two solutions of (1.11) then

$$(X - X') f_*(u - 1) = 0, \quad \forall u \in H_1.$$

Since $f : H_1 \to H_2$ is onto we deduce from Corollary 1.28 that $X - X' = 0$.

Existence. Any element $u \in H$ decomposes uniquely as

$$u = h\tau^k, \quad h \in T_{H_1},\ k \in \mathbb{Z}.$$

Then

$$f_*(\mathfrak{S}_1) f_*(u - 1) = f_*(\mathfrak{S}_1 h\tau - \mathfrak{S}_1) = f_*(\mathfrak{I}_1) f_*(\tau^k - 1).$$

Thus

$$f_*(\tau - 1) | f_*(u - 1) f_*(\mathfrak{S}_1), \quad \forall u \in H_1.$$

Since $f : H_1 \to H_2$ is surjective we deduce from Corollary 1.28

$$f_*(\tau - 1) | f_*(\mathfrak{S}_1) \quad \text{in } \mathfrak{N}(H_2).$$

Thus, there exists $X_1 \in \mathfrak{N}(H_2)$ such that

$$f_*(\mathfrak{S}_1) = X_1 f_*(\tau - 1).$$

We then set

$$X_f := X_1 f_*(\mathfrak{I}_1).$$

Finally, define $f_\sharp = f_*$ on $\mathbb{Z}[H]$ and

$$f_\sharp(\mathfrak{S}_1(T-1)^{-1}) := X_f. \qquad \square$$

Remark 1.31. If $H_1 \stackrel{f_1}{\to} H_2 \stackrel{f_2}{\to} H_3$ is are epimorphisms of Abelian groups then

$$(f_2 \circ f_1)_\sharp = (f_2)_\sharp \circ (f_1)_\sharp.$$

Thus the correspondence $G \mapsto \mathfrak{N}(G)$ defines a covariant functor from the category of finitely generated Abelian groups with epimorphisms as arrows to the category of commutative $\mathbb{Q}$-algebras. We refer to the next section for a more geometric description of the morphism $f_\sharp$ in terms of Fourier transform. $\square$

Example 1.32. Suppose f is the natural projection $\mathbb{Z} \to \mathbb{Z}_n = \mathbb{Z}/(n\mathbb{Z})$. Then

$$\mathfrak{N}(\mathbb{Z}) \cong \mathbb{Z}[t, t^{-1}, (1-t)^{-1}], \quad \mathfrak{N}(\mathbb{Z}_n) = \Big(1 - \frac{1}{n}\sum_{j=1}^{n} s^j\Big) \cdot \mathbb{Z}[s], \quad s^n = 1.$$

Then $f_\sharp$ is determined by

$$1 \mapsto \hat{\mathbf{1}} = (1 - \mathfrak{I}), \quad t \mapsto (1-\mathfrak{I})s, \quad \mathfrak{I} = s\Big(1 - \frac{1}{n}\sum_{j=1}^{n} s^j\Big).$$

Observe that

$$(t-1) \mapsto (s-1)(1-\mathfrak{I}).$$

The inverse of $(1-s)$ in the algebra $\ker \mathfrak{aug}$ with unit $\hat{\mathbf{1}} = 1 - \mathfrak{I}$ is (see [81] or §1.6)

$$(1-s)^{-1} = \Big(\frac{\hat{\mathbf{1}}}{2} - \mathfrak{d}(s)\Big),$$

where

$$\mathfrak{d}(s) := \sum_{k=1}^{n} \Big(\Big(\frac{k}{n}\Big)\Big) s^k,$$

and $((x))$ is denotes Dedekind's symbol

$$((x)) := \begin{cases} 0 & x \in \mathbb{Z} \\ x - \lfloor x \rfloor - \frac{1}{2} & x \in \mathbb{R} \setminus \mathbb{Z}. \end{cases}$$

($\lfloor x \rfloor :=$ the largest integer $\le x$.) $\square$

§1.6 Abelian harmonic analysis

When studying the torsion of a 3-manifold one is often lead to solving linear equations of the form $ax = b$, where a, b belong to the group algebra $\mathbb{C}[G]$ of a finitely generated Abelian group G. When G has torsion elements the ring $\mathbb{Z}[G]$ has zero divisors and thus the above equation may have more than one solution. Finding the annihilator of a given element $a \in \mathbb{C}[H]$ is never an easy job due to the complexity of the multiplication operation in this algebra. This complexity is only artificial and magically disappears if we perform a simple but extremely versatile trick, namely taking the Fourier transform of the above equation. In the Fourier picture the above equation simplifies dramatically to the point that it can be solved explicitly.

The versatility of the Fourier transform can be very clearly seen in the very simple description of the rings $\mathfrak{N}(G)$ and morphisms $f_\sharp$ introduced in the previous section. These rings are essentially described in terms of linear equations in the ring $\mathbb{Z}[G]$. More precisely, $\mathfrak{N}(G)$ is obtained by adjoining to $\mathbb{Z}[G]$ certain solutions $x \in \mathbb{Q}(G)$ of the family of linear equations linear equations

$$(1-g)^k \cdot x = f, \quad g \in G,\, f \in \mathbb{Z}[G].$$

The Fourier transform fits these equations like a glove. The goal of the present section is to explain in detail these claims.

Suppose G is a finitely generated Abelian group. We denote by μ_G the counting measure on G, $\mu_G(\{x\}) = 1$, $\forall x \in G$. The group algebra $\mathbb{C}[G]$ can be thought of as the vector space $C_0(G, \mathbb{C})$ of continuous, compactly supported functions $f\colon G \to \mathbb{C}$ equipped with the convolution product. More precisely, if $\delta_g\colon G \to \mathbb{C}$ denotes the Dirac function concentrated at $g \in G$,

$$\delta_g(x) = \begin{cases} 1 & \text{if } x = g \\ 0 & \text{if } x \neq g, \end{cases}$$

then the correspondence $\mathbb{C}[G] \to C_0(G; \mathbb{C})$ is given by

$$\mathbb{C}[G] \in A := \sum_{g \in G} a_g g \mapsto A(\bullet) := \sum_{g \in G} a_g \delta_g(\bullet) \in C_0(G, \mathbb{C}).$$

The convolution product on $C_0(G, \mathbb{C})$ is given by

$$(f_0 * f_1)(g) = \sum_{h \in G} f(g-h)g(h).$$

We denote by $\hat{G} := \operatorname{Hom}(G, \mathrm{U}(1))$ the *Pontryagin dual of* G, i.e. the group of characters. $\hat{G}$ is a locally compact topological group, and we denote by $\hat{\mu}_G$ the Haar measure on $\hat{G}$ normalized so that $\hat{\mu}_G$ is the counting measure if $\hat{G}$ is finite and $\hat{\mu}_G = d\theta$ if $\hat{G} = S^1$. The *Fourier transform* is a linear isomorphism

$$\mathcal{F}\colon L^2(G, \mu_G) \to L^2(\hat{G}, \hat{\mu}_G)$$

defined by

$$\hat{f}(\chi) := \langle f, \chi \rangle = \int_G f(g)\overline{\chi(g)}\, d\mu_G(g), \quad \forall \chi \in \hat{G}.$$

Its inverse is described by the *Fourier inversion formula*

$$f(g) = \frac{1}{\hat{\mu}_G(\hat{G})} \int_{\hat{G}} \hat{f}(\chi)\chi(g) d\hat{\mu}_G(\chi), \quad \forall f \in C_0(G, \mathbb{C}).$$

If $f \in C_0(G, \mathbb{C})$ then $\hat{f} \in C(\hat{G}, \mathbb{C})$ and

$$\widehat{f * g}(\chi) = \hat{f}(\chi) \cdot \hat{g}(\chi), \quad \forall f, g \in C_0(G, \mathbb{C}),\ \chi \in \hat{G}.$$

The Fourier transform produces a morphism of $\mathbb{C}$-algebras

$$\big(\mathbb{C}[G], +, *\big) \to \big(C(\hat{G}, \mathbb{C}), +, \cdot\big), \quad A \mapsto \hat{A},$$

where " $\cdot$ " denotes the *pointwise* multiplications of functions.

Remark 1.33. In applications it is convenient to consider the holomorphic counterpart of the Pontryagin dual. Thus, if G is a finitely generated Abelian group, we set

$$\tilde{G} := \operatorname{Hom}(G, \mathbb{C}^*).$$

We will refer to the elements of $\tilde{G}$ as *holomorphic characters*. Note that $\hat{G} \subset \tilde{G}$. $\tilde{G}$ is an union of complex tori of dimension $\operatorname{rank}(G)$. Given a function $f \in \mathbb{C}[G]$ we define its *complex Fourier transform* by

$$\hat{f}(\chi) = \sum_{g \in G} f(g)\chi^{-1}(g), \quad \forall \chi \in \tilde{G}.$$

Observe that the restriction of the complex Fourier transform to $\hat{G}$ is the usual Fourier transform. □

We want to discuss in detail a few concrete situations relevant in topological problems.

1. rank$(G) = 0$. We denote the group operation multiplicatively. For any $\chi \in \hat{G}$ we denote by $R_\chi \subset S^1$ the range of χ. R_χ is a finite cyclic group. The integration along the fibers of $\chi : G \to R_\chi$ produces a linear map

$$\chi_* : \mathbb{C}[G] \to \mathbb{C}[R_\chi], \quad f \mapsto f^\chi.$$

More explicitly,

$$f^\chi(\alpha) = \sum_{\chi(g)=\alpha} f(g), \quad \forall f : G \to \mathbb{Q},\ \alpha \in R_\chi.$$

When χ is the trivial character $\mathbf{1}$ then

$$f^{\mathbf{1}} = \mathfrak{aug}(f).$$

Observe the following identities

$$\hat{f}(\chi) = \sum_{\alpha \in R_\chi} f^\chi(\alpha) \cdot \bar{\alpha}, \quad \hat{f}(\mathbf{1}) = \mathfrak{aug}(f),$$

$$\mathfrak{N}(G) = \{f \in \mathbb{Q}[G];\ \hat{f}(\mathbf{1}) = 0\}.$$

We conclude that a function $f : G \to \mathbb{C}$ is completely determined by the functions

$$f^\chi : R_\chi \to \mathbb{C}, \quad \forall \chi \in \hat{G}.$$

In the special case $f \in \mathbb{Q}[G]$, the components of f with respect to the decomposition of $\mathbb{Q}[G]$ as a direct sum of fields are all amongst the elements of $f^\chi \in \mathbb{Q}[R_\chi]$. Thus, in order to understand the components of f we need to understand the Fourier transform of f.

The Fourier transform of δ_1 is the constant function $\mathbf{1}$ on $\hat{G}$. The Fourier transform of the idempotent $\mathfrak{I}$ (with respect to the convolution product) is the Dirac function

$$\delta_1 : \hat{G} \to \mathbb{C}$$

concentrated at the origin. This is an idempotent with respect to the pointwise multiplication. In particular, the function $\mathbf{1} - \hat{\mathfrak{I}}$ can be interpreted as the identity element on the algebra of functions $f : \hat{G} \setminus \{1\} \to \mathbb{C}$.

We have seen that if $\phi : G_0 \to G_1$ is an epimorphism of finite Abelian groups there is an induced morphism

$$\phi_\sharp : \mathfrak{N}(G_0) \to \mathfrak{N}(G_1).$$

We want to present a description of this morphism using Fourier analysis.

Let $\hat{G}_i^* = \hat{G}_i \setminus \{1\}$, $i = 0, 1$. The Fourier transform maps $\mathfrak{N}(G_i)$ isomorphically onto a subring $\mathfrak{N}(\hat{G}_i)$ of the ring of functions $\hat{G}_i \to \mathbb{C}$ consisting of functions vanishing at $\mathbf{1}$. We will identify this subring with a space of continuous functions $\hat{G}_i^* \to \mathbb{C}$. The epimorphism ϕ induces a monomorphism $\hat{\phi} : G_1 \to G_0$ and thus a pull-back map

$$\hat{\phi}^* : C(\hat{G}_0^*, \mathbb{C}) \to C(\hat{G}_1^*, \mathbb{C}).$$

Proposition 1.34. *The following diagram is commutative.*

$$\begin{array}{ccc}
\mathfrak{N}(G_0) & \xrightarrow{\mathcal{F}} & C(\hat{G}_0^*, \mathbb{C}) \\
{\scriptstyle \phi_\sharp}\downarrow & & \downarrow{\scriptstyle \hat{\phi}^*} \\
\mathfrak{N}(G_1) & \xrightarrow{\mathcal{F}} & C(\hat{G}_1^*, \mathbb{C}).
\end{array}$$

Proof. The morphism $\phi_\sharp$ is the restriction of the integration-along-fibers map

$$\phi_*\colon \mathbb{C}[G_0] \to \mathbb{C}[G_1]$$

to the augmentation ideal, ker $\mathfrak{aug}_{G_0}$. Since

$$\mathfrak{aug}_{G_1}(\phi_*(f)) = \mathfrak{aug}_{G_0}(f), \quad \forall f \in \mathbb{C}[G_0]$$

we deduce that $\phi_*(\ker \mathfrak{aug}_{G_0}) \subset \ker \mathfrak{aug}_{G_1}$. The proposition follows from the more general statement

$$\hat{\phi}^* \circ \mathcal{F} = \mathcal{F} \circ \phi_*.$$

Indeed for every $\chi \in \hat{G}_1$ and $f \in \mathbb{C}[G_0]$ we have

$$\hat{\phi}^*(\hat{f})(\chi) = \hat{f}(\hat{\phi}(\chi)) = \langle f, \hat{\phi}(\chi)\rangle = \sum_{g \in G_0} f(g)\overline{\hat{\phi}(\chi)(g)} = \sum_{g \in G_0} f(g)\overline{\chi(\phi(g))}$$

$$= \sum_{g_1 \in G_1} \Big(\sum_{\phi(g)=g_1} f(g) \Big) \bar{\chi}(g_1) = \sum_{g_1 \in G_1} \phi_*(f)(g_1)\bar{\chi}(g_1) = \mathcal{F} \circ \phi_*(f).$$

□

Example 1.35. Suppose that the finite Abelian group G is equipped with a nondegenerate, symmetric, pairing

$$\boldsymbol{q}\colon G \times G \to S^1, \quad (u, v) \mapsto \boldsymbol{q}(u, v) =: u \cdot v.$$

In this case we have a natural isomorphism

$$G \to \hat{G}, \quad g \mapsto g_\sharp = \boldsymbol{q}(g, \bullet).$$

Observe that

$$R_g := R_{g_\sharp} \cong G/g^\perp, \quad g^\perp := \{u \in G;\ \boldsymbol{q}(g, u) = 1 \in S^1\}.$$

The element $f^g := f^{g_\sharp} \in \mathbb{Q}[R_g]$ can be alternatively described by

$$f^g(\alpha) = \sum_{u \cdot g = \alpha} f(u).$$

□

2. rank$(G) = 1$. In this case there exist isomorphisms $G = \mathbb{Z} \oplus H$ where H is finite. Then

$$\hat{G} \cong S^1 \times \hat{H}.$$

More invariantly, H is the torsion subgroup of G, and if $\boldsymbol{\imath}\colon H \to G$ denotes the inclusion map, the subgroup $S^1 \subset \hat{G}$ can be identified with the kernel of the dual map $\hat{\imath}\colon \hat{G} \to \hat{H}$. This kernel is the component of $\mathbf{1} \in \hat{G}$.

To define the Fourier transform we need to have a way of identifying the elements in $\mathfrak{N}(G)$ with functions on G. Such an identification requires a bit of additional data. Fix an *orientation* $\mathfrak{o}$ on $G \otimes \mathbb{R}$. This is equivalent to choosing an isomorphism $G/\operatorname{Tors}(G) \cong \mathbb{Z}$. This induces an epimorphism

$$\deg = \deg_{\mathfrak{o}} : G \to \mathbb{Z}.$$

Fix $t \in G$ such that $\deg t = 1$. This defines a splitting $G \cong \mathbb{Z} \oplus H$, and an identification

$$\mathfrak{N}(G) \cong \mathbb{Z}[G, \mathfrak{S}(1-t)^{-1}].$$

Using the formal equality

$$\frac{1}{1-t} = \sum_{n \geq 0} t^n$$

we can identify the element $\mathfrak{S}(1-t)^{-1}$ with the function

$$\omega_{\mathfrak{o}} : G \to \mathbb{Z}, \quad \omega_{\mathfrak{o}}(g) = \begin{cases} 1 & \text{if } \deg_{\mathfrak{o}}(g) \geq 0 \\ 0 & \text{if } \deg_{\mathfrak{o}}(g) < 0. \end{cases}$$

More generally, we can identify $\mathfrak{S} \cdot (1-t)^{-k}$ with the function

$$G \ni g \mapsto \omega_{\mathfrak{o}}(g) \cdot \binom{-k}{\deg_{\mathfrak{o}}^{+} g} \in \mathbb{Z},$$

where $\deg_{\mathfrak{o}}^{+} = \max(\deg_{\mathfrak{o}}, 0)$. Define the Novikov ring $\Lambda_{\mathfrak{o}}(G)$

$$\Lambda_{\mathfrak{o}}(G) := \{ f : G \to \mathbb{Z};\ \exists C \in \mathbb{R} \text{ such that } f(g) = 0 \text{ if } \deg_{\mathfrak{o}}(g) < C \}.$$

The multiplication in this ring is again the convolution product which is well defined due to the support constraint on the functions in this ring. We have an injective morphism

$$\mathfrak{N}(G) \hookrightarrow \Lambda_{\mathfrak{o}}(G), \quad f \mapsto f_{\mathfrak{o}},$$

uniquely determined by the requirements

$$\mathbb{Z}[G] \ni \sum_{g \in G} P_g g = P \mapsto P_{\mathfrak{o}} \in \Lambda_{\mathfrak{o}}(G), \quad P_{\mathfrak{o}}(g) = P_g,$$

and

$$\mathfrak{S}(1-t)^{-k} \mapsto \mathfrak{S}_{\mathfrak{o}} * \underbrace{\omega_{\mathfrak{o}} * \cdots * \omega_{\mathfrak{o}}}_{k}.$$

This morphism depends on $\mathfrak{o}$, but not on the choice of t such that $\deg_{\mathfrak{o}} t = 1$. We denote by $\mathcal{N}_{\mathfrak{o}}(G)$ the image of this morphism. Note that a function $f \in \mathcal{N}_{\mathfrak{o}}(G)$ need not have a compact support. In fact, the function $\omega_{\mathfrak{o}}$ is not even L^1 with respect to the discrete measure on G.

The characters of G have the form

$$\chi := e_\theta \cdot \varphi, \quad \varphi \in \hat{H},\ e_\theta(t^n) := e^{\mathbf{i}n\theta},\ 0 \le \theta < 2\pi.$$

If $f \in L^1(G, \mu_G)$ then

$$\hat{f}(\chi) = \sum_{g \in G} f(g)\bar{\chi}(g) = \sum_{n \in \mathbb{Z}} \Big(\sum_{h \in H} f(n, h)\bar{\varphi}(h) \Big) e^{-\mathbf{i}n\theta}.$$

In particular, if $\delta_h : \mathbb{Z} \oplus H \to \mathbb{C}$ denotes the Dirac function concentrated at $(0, h) \in G$ then

$$\hat{\delta}_t(\chi) = \bar{\chi}(t), \quad \hat{\delta}_h(\chi) = \bar{\chi}(h), \quad \hat{\mathfrak{I}}(\chi) = \frac{1}{|H|} \sum_{h \in H} \bar{\chi}(h).$$

$\hat{\mathfrak{I}}$ is an idempotent in the algebra of continuous functions $\hat{G} \to \mathbb{C}$. One can check immediately that $\hat{\mathfrak{I}}$ is the characteristic function the identity component of $\hat{G}$. If we set $T = \mathfrak{I} * t$ then

$$\hat{T}(\chi) = \hat{\delta}_t(\chi)\hat{\mathfrak{I}}(\chi).$$

$\hat{T}$ is a function on $\hat{G}$ supported on the identity component $S^1 \hookrightarrow \hat{G}$ where it is equal to

$$T(\theta) = e^{-\mathbf{i}\theta}.$$

The Fourier transform extends in a natural way to the ring $\mathcal{N}_{\mathfrak{o}}(G)$, but its range will contain distributions on $\hat{G}$ of a special kind. We begin with the simplest situation.

A. Tors $G = 0$. Fix an orientation $\mathfrak{o}$ on $G \otimes \mathbb{R}$. In this case there exists an unique $t = t_{\mathfrak{o}} \in G$ such that $\deg_{\mathfrak{o}} t = 1$. We also have an identification

$$\mathbb{C}^* \xrightarrow{\mathfrak{o}} \tilde{G} := \mathrm{Hom}(G, \mathbb{C}^*), \quad z \mapsto \chi_z, \quad \chi_z(t) = z.$$

Denote by $\mathcal{M}(\mathbb{C}^*)$ the field of meromorphic functions on $\mathbb{C}^*$. To each function $f \in C_0(G, \mathbb{C}) \cong \mathbb{C}[G]$ we associate its complex Fourier transform $\tilde{f} : \tilde{G} \to \mathbb{C}$ which can be identified with a Laurent polynomial in $\mathcal{M}(\mathbb{C}^*)$,

$$\tilde{f}(\chi_z) \longleftrightarrow \sum_{n \in \mathbb{Z}} f(t^n) z^{-n}, \quad \chi_z(t) = z.$$

Observe $\hat{f} = \tilde{f}|_{|z|=1}$. The Fourier transform $\mathcal{F} : C_0(G, \mathbb{C}) \to C(S^1, \mathbb{C})$ is completely determined by algebra morphism

$$\mathcal{F}_{\mathfrak{o}} : \mathbb{C}[G] \ni f \mapsto \tilde{f} \in \mathcal{M}(\mathbb{C}^*).$$

To understand the obstacle we face when trying to extend the Fourier transform to $\mathcal{N}_{\mathfrak{o}}$ we only need to look at a simple example. Observe that

$$u(\chi) := \widetilde{(1 - t)} = 1 - z^{-1} \in \mathcal{M}(\mathbb{C}^*).$$

This has an inverse in the ring $\mathcal{M}(\mathbb{C}^*)$. However, its restriction to the circle $|z| = 1$ is not invertible in the ring $C(S^1, \mathbb{C})$ because $u(1) = 0$. This degeneracy can be detected working directly with the Fourier transform.

We have identified $(1-t)^{-1}$ with the function ω_o which does not have compact support, and it does not belong to $L^1(\mathbb{Z})$. Its Fourier transform is no longer a function on S^1, it is a *distribution* $\hat{\omega}_o$ described by

$$\langle \hat{\omega}_o, \varphi \rangle = \sum_{n\geq 0} \int_0^{2\pi} \varphi(\theta) e^{-\mathbf{i}n\theta} d\theta, \quad \forall \varphi \in C_0^\infty(S^1, \mathbb{C}).$$

The above sum is convergent because

$$\left| \int_0^{2\pi} \varphi(\theta) e^{-\mathbf{i}n\theta} d\theta \right| = O(n^{-k}), \quad \text{as } n \to \infty,\ \forall k > 0.$$

Observe that

$$\langle \hat{\omega}_o, e^{\mathbf{i}m\theta} \rangle = 2\pi \omega(t^m).$$

However, this distribution can be suitably identified with the boundary value[4] of the holomorphic function $1/u(z) = (1 - z^{-1})^{-1} = \frac{z}{z-1} \in \mathcal{H}$. More precisely, we have the following result. (For more information on this type of distribution we refer to [37].)

Proposition 1.36.

$$\langle \omega_o, \varphi \rangle = \lim_{r \searrow 1} \int_0^{2\pi} \frac{\varphi(\theta)}{u(re^{\mathbf{i}\theta})} d\theta, \quad \forall \varphi \in C_0^\infty(S^1, \mathbb{C})$$

so that

$$\hat{\omega}_o = \lim_{r \searrow 1} (1/u)|_{|z|=r}$$

in the sense of distributions. Moreover $(1 - e^{-\mathbf{i}\theta}) \cdot \hat{\omega}_o = 1$ *in the sense of distributions, i.e.* $\hat{\omega}$ *is indeed a distributional inverse of the smooth function* $u(\zeta) = (1 - \zeta^{-1})$, $|\zeta| = 1$.

Proof. For simplicity, we write ω instead of ω_o since we will be using the same orientation throughout the proof below. Observe that if φ is constant, $\varphi \equiv c$, then

$$\langle \hat{\omega}, c \rangle = 2\pi c = \lim_{r \searrow 1} \int_0^{2\pi} c u(re^{\mathbf{i}\theta}) d\theta.$$

Thus, it suffices to prove that

$$\langle \hat{\omega}, \varphi \rangle = \lim_{r \searrow 1} \int_0^{2\pi} u(re^{\mathbf{i}\theta}) \varphi(\theta) d\theta, \quad \forall \varphi \colon S^1 \to \mathbb{C},\ \varphi(1) = 0.$$

[4] We are indebted to Brian Hall for this observation

Observe that

$$K_r(\zeta) := u(r\zeta)\varphi(\zeta) = \frac{r\varphi(\zeta)}{r-\zeta}, \quad |\zeta| = 1.$$

Since $\varphi(1) = 0$ we deduce from the dominated convergence theorem that the series of $L^1(S^1)$ functions

$$\sum_{n=0}^{\infty} \varphi(\zeta)\zeta^{-n}, \quad |\zeta| = 1.$$

converges in the L^1-norm to

$$K_0(\zeta) := \frac{\varphi(\zeta)}{1-\zeta^{-1}}, \quad |\zeta| = 1.$$

Thus

$$\langle \hat{\omega}, \varphi \rangle = \int_0^{2\pi} K_0(e^{\mathbf{i}\theta})\varphi(\theta)d\theta$$

and we need to show that

$$\lim_{r \searrow 1} \int_0^{2\pi} \big(K_r(e^{\mathbf{i}\theta}) - K_0(e^{\mathbf{i}\theta})\big)\varphi(\theta)d\theta = 0.$$

This follows easily from the dominated convergence theorem.

To prove that $\hat{\omega}$ is the distributional inverse of $u(\zeta)$ we need to show that for every smooth function $\varphi\colon S^1 \to \mathbb{C}$ we have the identity

$$\langle \hat{\omega}, u\varphi \rangle = \int_0^{2\pi} \varphi(e^{\mathbf{i}\theta})d\theta.$$

Since $u(1) = 0$ the above arguments show that

$$\langle \hat{\omega}, u\varphi \rangle = \int_0^{2\pi} K_0(e^{\mathbf{i}\theta})u(e^{\mathbf{i}\theta})\varphi(e^{\mathbf{i}\theta}) = \int_0^{2\pi} \varphi(e^{\mathbf{i}\theta})d\theta$$

because $K_0(e^{\mathbf{i}\theta})u(e^{\mathbf{i}\theta}) = 1$. □

Every element $f \in \mathfrak{N}(G)$ can be uniquely written as

$$f = \frac{P(t)}{(1-t)^k}, \quad P(t) \in \mathbb{Z}[t, t^{-1}].$$

Arguing as above we deduce that the Fourier transform of $f_\circ$ is the distribution $\hat{f}_\circ \in \mathcal{D}'(S^1)$, with singular support concentrated at $1 \in S^1$ defined by

$$\hat{f}_\circ := \lim_{|z| \searrow 1} \frac{P(z^{-1})}{(1-z^{-1})^k}.$$

Definition 1.37. The complex Fourier transform of a function $f \in \mathcal{N}_\mathfrak{o}(G)$ is by definition the unique meromorphic function $\mathcal{F}_\mathfrak{o}(f)$ on $\mathbb{C}^*$ whose restriction to $S^1 \setminus \{1\}$ coincides with the Fourier transform $\hat{f}_\mathfrak{o}$ of f. □

In view of the above discussion we deduce that

$$\hat{f}_\mathfrak{o} = \lim_{|z| \searrow 1} \mathcal{F}_\mathfrak{o}(f).$$

The complex Fourier transform maps the ring $\mathfrak{N}(G)$ to a space of meromorphic functions on $\tilde{G}$ via the composition

$$\mathfrak{N}(G) \xrightarrow{\mathfrak{o}} \mathcal{N}_\mathfrak{o}(G) \xrightarrow{\mathcal{F}_\mathfrak{o}} \mathcal{M}(\mathbb{C}^*) := \text{meromorphic functions on } \mathbb{C}^*.$$

We denote by $\mathfrak{N}_\mathfrak{o}(\tilde{G})$ the image of $\mathfrak{N}(G)$ via the complex Fourier transform $\mathcal{F}_\mathfrak{o}$. Observe that

$$\mathfrak{N}_\mathfrak{o}(\tilde{G}) \overset{\mathfrak{o}}{\cong} \mathbb{Z}\big[z, z^{-1}, (1-z)^{-1}\big],$$

where z is the function $\chi \mapsto \chi(t_\mathfrak{o})$. Similarly, we can defined the (real) Fourier transform on $\mathfrak{N}(G)$

$$\mathfrak{N}(G) \ni f \mapsto \hat{f}_\mathfrak{o} := \lim_{r \searrow 1} \mathcal{F}_\mathfrak{o}(f)|_{|z|=r} \in \mathcal{D}'(S^1)$$

where the limit is in the sense of distributions as in Proposition 1.36.

We want to describe the dependence of these construction on the choice of orientation $\mathfrak{o}$. Denote by $t_\pm$ the unique element in G such that $\deg_{\pm\mathfrak{o}}(t_\pm) = 1$. Set

$$u := (1 - t_+) = (1 - t_-^{-1}) \in \mathfrak{N}(G).$$

For every $\chi \in \tilde{G}$ we set $z_\pm = \chi(t_\pm)$, so that $z_- = 1/z_+$.

$$\mathcal{F}_\mathfrak{o}(1/u_\mathfrak{o})(\chi) = \frac{1}{1 - z_+^{-1}} = \frac{1}{1 - z_-} = \mathcal{F}_{-\mathfrak{o}}(1/u_{-\mathfrak{o}})(\chi).$$

This shows that the *complex* Fourier transform of $1/u$ is a meromorphic function on $\tilde{G}$, *independent* of the orientation $\mathfrak{o}$. Thus the complex Fourier transform is a morphism

$$\mathfrak{N}(G) \to \mathcal{M}(\tilde{G})$$

independent of the orientation. We denote its range by $\mathfrak{N}(\tilde{G})$.

The situation with the real Fourier transform is a bit more subtle. In this case the range of the real Fourier transform consists of distributions on $\hat{G}$. To be able to identify the space of smooth functions on $\tilde{G}$ with a subspace[5] of the space of distributions on $\hat{G}$ we need to have an integration, i.e. an orientation on $\hat{G}$. This is equivalent to fixing an orientation $\mathfrak{o}$ on G.

The following proposition summarizes the facts established so far.

[5]We do not want to get into a discussion about densities as in [42].

Proposition 1.38. *Suppose* $\operatorname{rank}(G) = 1$, $\operatorname{Tors}(G) = 0$. *Then an orientation* $\mathfrak{o}$ *on* $G \otimes \mathbb{R}$ *defines isomorphisms* $G \longleftrightarrow \mathbb{Z}$, $\hat{G} \longleftrightarrow S^1 \subset \mathbb{C}^*$, $\tilde{G} \longleftarrow \mathbb{C}^*$, *and* $\mathfrak{N}(G) \longleftrightarrow \mathcal{N}_{\mathfrak{o}}(G)$. *The (real) Fourier transform on* $\mathbb{Z}[G] \subset \mathcal{N}_0(G)$ *extends to a map*

$$\mathfrak{N}(G) \to \mathcal{N}_{\mathfrak{o}}(G) \to \mathcal{D}'(S^1), \quad f \mapsto \hat{f}_{\mathfrak{o}}$$

whose range $\mathfrak{N}_{\mathfrak{o}}(\hat{G})$ *is a space of distributions on* S^1 *with singular support concentrated at* $1 \in S^1$ *which are solutions of certain division problems.*

The complex Fourier transform on $\mathbb{Z}[G]$ *extends to a morphism of algebras* $\mathcal{F}\colon \mathfrak{N}(G) \to \mathcal{M}(\tilde{G})$ *independent of* $\mathfrak{o}$ *such that*

$$\hat{f}_{\mathfrak{o}}|_{S^1\setminus\{1\}} = \mathcal{F}(f)|_{S^1\setminus\{1\}}.$$

We denote by $\mathfrak{N}(\tilde{G})$ *the range of the complex Fourier transform.*

Suppose $G = \mathbb{Z}$, and $\mathfrak{o}$ is the natural orientation. Denote by π the natural projection $\pi : \mathbb{Z} \to \mathbb{Z}_n$. We know that it induces in a natural way a morphism

$$\pi_\sharp\colon \mathfrak{N}(\mathbb{Z}) \to \mathfrak{N}(\mathbb{Z}_n).$$

We would like to give a very intuitive definition of this morphism using the Fourier transform. Note that π induces an inclusion

$$\hat{p}\colon \hat{\mathbb{Z}}_n \to \hat{\mathbb{Z}} \cong S^1$$

The Fourier transform of $\mathfrak{N}(\mathbb{Z}_n)$ is a ring $\mathfrak{N}(\hat{\mathbb{Z}}_n)$ of functions h on $\hat{\mathbb{Z}}_n$ such that $h(1) = 0$. This can be naturally identified with the ring of functions on the subset $\hat{\mathbb{Z}}_n^* = \hat{\mathbb{Z}}_n \setminus \{1\}$. We can use $\hat{p}$ to pullback the functions on S^1 to functions on $\hat{\mathbb{Z}}_n$, and more generally, we can pullback to $\hat{\mathbb{Z}}_n^*$ the distributions in $\mathfrak{N}(\hat{\mathbb{Z}})$. We thus have a map

$$\hat{\pi}^*\colon \mathfrak{N}(\hat{\mathbb{Z}}) \to C(\hat{\mathbb{Z}}_n^*, \mathbb{C}), \quad \mathfrak{N}(\hat{\mathbb{Z}}) \ni \Phi \mapsto \Phi|_{\hat{\mathbb{Z}}_n^*}.$$

Note that if $\Phi \in \mathfrak{N}(\hat{\mathbb{Z}})$ is the distributional restriction of the holomorphic function $\tilde{\Phi} \in \mathfrak{N}(\tilde{\mathbb{Z}})$, then

$$\Phi|_{\hat{\mathbb{Z}}_n^*} = \tilde{\Phi}|_{\hat{\mathbb{Z}}_n^*}$$

where the above restriction exists classically, not just as a distribution.

Proposition 1.39. *The diagram below is commutative.*

$$\begin{array}{ccc} \mathfrak{N}(\mathbb{Z}) & \xrightarrow{\ \mathcal{F}\ } & \mathfrak{N}(\hat{\mathbb{Z}}) \\ {\scriptstyle \pi_\sharp}\downarrow & & \downarrow{\scriptstyle \hat{\pi}^*} \\ \mathfrak{N}(\mathbb{Z}_n) & \xrightarrow{\ \mathcal{F}\ } & C(\hat{\mathbb{Z}}_n^*, \mathbb{C}). \end{array}$$

Proof. Note first that

$$\pi_\sharp(f) = (1 - \mathfrak{I}) * \pi_*(f), \quad \forall f \in \mathbb{C}[t, t^{-1}]$$

($* =$ convolution product) while $\pi_\sharp((1-t)^{-1})$ is uniquely determined by the division problem

$$\pi_\sharp((1-t)^{-1}) * (1-t) = 1 - \mathfrak{I}.$$

If $f = \sum_{j\in\mathbb{Z}} f(j)t^j \in \mathbb{C}[t, t^{-1}]$, and $\zeta^n = 1$, $\zeta \neq 1$, then

$$\begin{aligned}\mathfrak{F} \circ \pi_{\sharp(f)}(\zeta) &= \mathfrak{F}\big((1 - \mathfrak{I}) * \pi_*(f)\big) = (1 - \delta_{\mathbf{1}}(\zeta)) \cdot \widehat{\pi_*(f)}(\zeta) \qquad (\delta_{\mathbf{1}}(\zeta) = 0)\\ &= \sum_{k=1}^{n} \pi_*(f)(k)\zeta^{-k} = \sum_{k=1}^{n}\sum_{j\in\mathbb{Z}} f(nj+k)\zeta^{-k} = \tilde{f}(\zeta) = \hat{\pi}^*(\tilde{f})(\zeta).\end{aligned}$$

Set $V := \pi_\sharp((1-t)^{-1}) \in \mathfrak{N}(\mathbb{Z}_n)$. Then $\hat{V}(1) = 0$ and

$$\hat{V}(\zeta)(1 - \zeta^{-1}) = (1 - \delta_1(\zeta)), \quad \forall \zeta^n = 1.$$

We conclude that if $\zeta \neq 1$ we have

$$\hat{V}(\zeta) = 1/u(\zeta) = \hat{\omega}(\zeta).$$

This concludes the proof of Proposition 1.39. □

Example 1.40. The Fourier transform of $\pi_\sharp((1-t)^{-1})$ is the function

$$\hat{V}(\zeta) := \begin{cases} 0 & \zeta = 1 \\ \frac{\zeta}{\zeta - 1} & \zeta \neq 1. \end{cases}$$

On the other hand, the Fourier transform of the Dedekind symbol

$$\Delta_n : \mathbb{Z}_n \to \mathbb{Q}, \quad k \bmod n\mathbb{Z} \mapsto \Delta_n(k) = ((k/n))$$

is (see [88, Chap 2, Sec. C])

$$\hat{\Delta}_n(\zeta) = \begin{cases} 0 & \zeta = 1 \\ \frac{1}{2} - \frac{\zeta}{\zeta - 1} & \zeta \neq 1 \end{cases}$$

We conclude that

$$V = \frac{1}{2}(1 - \mathfrak{I}) - \Delta_n.$$

□

B. rank$(G) = 1$, Tors$(G) \neq 0$. Set $H :=$ Tors(G), $F := G/H$ and

$$\mathfrak{S} = \sum_{h\in H} h \in \mathbb{Z}[G].$$

Fix an orientation $\mathfrak{o}$ on $G \otimes \mathbb{R}$. Then $\tilde{G} = \mathrm{Hom}(G, \mathbb{C}^*)$ is an union of one-dimensional complex tori, and the orientation $\mathfrak{o}$ defines an orientation on $\hat{G}$, and thus identifies the identity component of $\tilde{G}$ with $\mathbb{C}^*$.

A function $f \in \mathcal{N}_{\mathfrak{o}}(G)$ has noncompact support, but has temperate growth, and thus it has a Fourier transform as a temperate distribution. Denote by $\mathfrak{N}_{\mathfrak{o}}(\hat{G}) \subset \mathcal{D}'(\hat{G})$ the Fourier transform of $\mathcal{N}_{\mathfrak{o}}(G)$.

Since $\hat{\delta}_g(\chi) = \bar{\chi}(g)$ we conclude that $\hat{\delta}_1$ is the constant function 1 on $\hat{G}$ and

$$\hat{\mathfrak{S}}(\chi) = \sum_{h \in H} \hat{\delta}_h(\chi) = \sum_{h \in H} \bar{\chi}(h).$$

We set $K_\chi := \ker \chi|_H$ and we deduce

$$\hat{\mathfrak{S}}(\chi) = |K_\chi| \sum_{\alpha \in R_\chi} \alpha = \begin{cases} |H| & \chi|_H = 1 \\ 0 & \text{otherwise.} \end{cases}$$

Fix $t \in G$ such that $\deg_{\mathfrak{o}} t = 1$. $\frac{1}{|H|}\hat{\mathfrak{S}} = \hat{\mathfrak{I}}$ is the characteristic function of the identity component of $\hat{G}$, so that the Fourier transform of $\mathfrak{S}_{\mathfrak{o}} \cdot (1 - t)^{-k} \in \mathcal{N}_{\mathfrak{o}}(G)$ is a distribution supported on the identity component of $\hat{G}$. Via the isomorphism

$$\mathbb{Z} \oplus H \to G, \quad (n, h) \mapsto ht^n,$$

which identifies the identity component of $\hat{G}$ with S^1, this distribution is defined by the limit

$$S^1 \ni z \lim_{r \searrow 1} \frac{|H|}{\left(1 - r^{-1}z^{-1}\right)^k}.$$

We deduce that

$$\mathfrak{N}_{\mathfrak{o}}(\hat{G}) := \widehat{\mathcal{N}_{\mathfrak{o}}(G)} \cong \widehat{\mathbb{Z}[G]} + \mathfrak{N}_{\mathfrak{o}}(\hat{F}).$$

Arguing as in part **A** we obtain a complex Fourier transform $\mathcal{F}\colon \mathfrak{N}(G) \to \mathcal{M}(\tilde{G})$ which is independent of the orientation $\mathfrak{o}$, such that for every $f \in \mathfrak{N}(G)$ we have

$$\hat{f}_{\mathfrak{o}}|_{\hat{G}\setminus\{1\}} = \mathcal{F}(f)|_{\hat{G}\setminus\{1\}}.$$

Observing that we have a diagram

$$\begin{array}{ccccc} & & \hat{G} & & \\ & \overset{\iota}{\nearrow} & & \overset{\pi}{\searrow} & \\ \hat{F} & & & & \hat{H} \end{array}$$

we can represent the range $\mathfrak{N}(\tilde{G})$ of the complex Fourier transform as a sum of the space of Laurent polynomials on $\tilde{G}$ with a space of holomorphic functions on $\tilde{G} \setminus \{1\}$, supported on the identity component of $\tilde{G}$. To simplify the presentation we will denote the Fourier transform of $g \in \mathfrak{N}$ by $\hat{f}$. Suppose now that $\phi\colon G_1 \to G_2$ is an epimorphism of Abelian groups and $\operatorname{rank}(G_1) = 1$. It induces a monomorphism

$$\hat{\phi}\colon \tilde{G}_2 \to \tilde{G}_1.$$

This induces by pullback a morphism

$$\hat{\phi}^*\colon \mathfrak{N}(\tilde{G}_1) \to \mathfrak{N}(\tilde{G}_2).$$

Arguing as in the proofs of Propositions 1.34 and 1.39 we deduce that the following diagram is commutative

$$\begin{array}{ccc} \mathfrak{N}(G_1) & \xrightarrow{\mathcal{F}} & \mathfrak{N}(\tilde{G}_1) \\ {\scriptstyle \phi_\sharp}\downarrow & & \downarrow{\scriptstyle \hat{\phi}^*} \\ \mathfrak{N}(G_2) & \xrightarrow{\mathcal{F}} & \mathfrak{N}(\tilde{G}_2). \end{array}$$

3. rank$(G) \geq 2$. Set $r := \operatorname{rank}(G)$, $H = \operatorname{Tors}(G)$, $F := G/H$. Then $\hat{F}$ is an r-dimensional torus which can be identified with the identity component of $\hat{G}$.

In this case $\mathfrak{N}(G) = \mathbb{Z}[G]$, and thus $\mathfrak{N}(\hat{G}) := \mathcal{F}(\mathfrak{N}(G)) \subset C(\hat{G}, \mathbb{C})$. More precisely, $\mathfrak{N}(\hat{G})$ coincides with the subring generated by the Fourier transforms of the Dirac functions δ_g. Observe that

$$\hat{\delta}_g(\chi) = \bar{\chi}(g), \quad \forall \chi \in \hat{G}.$$

The complex Fourier transform is defined in the obvious way.

Arguing as before we deduce that if $\phi\colon G_1 \to G_2$ is an epimorphism of Abelian groups, $\operatorname{rank}(G_1) \geq 2$, then the diagram below is commutative

$$\begin{array}{ccc} \mathfrak{N}(G_1) & \xrightarrow{\mathcal{F}} & \mathfrak{N}(\tilde{G}_1) \\ {\scriptstyle \phi_\sharp}\downarrow & & \downarrow{\scriptstyle \hat{\phi}^*} \\ \mathfrak{N}(G_2) \otimes \mathbb{C} & \xrightarrow{\mathcal{F}} & \mathfrak{N}(\tilde{G}_2). \end{array}$$

The above analysis has the following elementary consequence.

Corollary 1.41. (a) *Suppose G is a finitely generated Abelian group of positive rank, and $f \in \mathfrak{N}(G)$. Then the complex Fourier transform of f is holomorphic on $\tilde{G} \setminus \{1\}$. If moreover* $\operatorname{rank}(G) > 1$ *then the complex Fourier transform of f is holomorphic on $\tilde{G}$.*

(b) *Suppose that $\alpha\colon G \to H$ is an epimorphism of finitely generated Abelian groups. This induces an injection $\hat{\alpha}\colon \tilde{H} \hookrightarrow \tilde{G}$, and for every $f \in \mathfrak{N}(G)$ we have*

$$\widehat{\alpha_{\sharp} f} = \hat{\alpha}^{*}(\hat{f}) = \hat{f} \circ \hat{\alpha}.$$

Chapter 2
The Reidemeister torsion

We now begin in earnest our topological journey. In this chapter we present the basic definitions and facts concerning the Reidemeister torsion of a CW (or simplicial) complex. We have decided that it would be more profitable to the reader to limit to an acceptable minimum the foundational arguments, and instead present many, and diverse examples which in our view best convey the reason why a particular fact could be true. The reader interested in filling in our deliberate foundational gaps can consult the classical survey [72] of J. Milnor, or the recent monograph [117] by V. Turaev.

§2.1 The Reidemeister torsion of a CW-complex

Suppose X is a compact metric space and $S(X)$ is a finite CW-decomposition of X. Set $H := H_1(X)$.

Remark 2.1. To eliminate any ambiguity, let us mention that for us a CW-decomposition is a filtration of X by closed subsets

$$X^{(0)} \subset X^{(1)} \subset \cdots \subset X^{(n)} \subset \cdots$$

such that there exist homeomorphisms

$$\big(J_n \times D^n \sqcup X^{(n-1)}\big)/ \sim \rightarrow X^{(n)}$$

in which J_n is a finite set, D^n denotes the closed n-dimensional ball, and $\sim$ denotes the equivalence relation defined by an attaching map

$$\varphi_n \colon J_n \times \partial D^n \rightarrow X^{(n-1)}.$$

The set $X^{(n)}$ is called the n-skeleton of X. The components of $X^{(n)} \setminus X^{(n-1)}$ are called the (open) n-cells. We denote by $S_n(X)$ the set of n-cells. An orientation of a cell $\sigma \in S_n(X)$ is a choice of an isomorphism

$$H_n(\sigma, \partial\sigma; \mathbb{Z}) \rightarrow \mathbb{Z}. \qquad \square$$

Consider the maximal Abelian cover $\pi : \hat{X} \rightarrow X$ of X, that is the cover of X defined by the Hurewicz morphism

$$\pi_1(X) \rightarrow H.$$

We can view X as a quotient $X = \hat{X}/H$.

The CW-decomposition of X canonically defines a chain complex $\underline{C}(X)$ of free Abelian groups,

$$\underline{C}(X) := \bigoplus_n \bigoplus_{\sigma \in S_n(X)} H_n(\sigma, \partial\sigma).$$

By orienting the n-cells and ordering $S_n(X)$ we obtain bases of this chain complex. We will refer to these as *geometric bases*. The ambiguities in fixing such bases are encoded by the action of the group

$$\mathfrak{S}_X := \prod_{k \geq 0} \mathfrak{S}_{S_k(X)} \times \mathbb{Z}_2^{S_k(X)}$$

where $\mathfrak{S}_S$ denotes the group of permutations of a set S. The CW-decomposition of X lifts to a CW-decomposition $S(\hat{X})$ of $\hat{X}$ which, as a $\mathbb{Z}$-module, is generated by all the lifts of the cells in $S(X)$. Denote by $\underline{C}(\hat{X})$ the associated chain complex. The group H can be identified with the group of deck transformations of $\hat{X} \to X$ and as such it acts on $\underline{C}(\hat{X})$. Hence, $\underline{C}(\hat{X})$ has a natural structure of *free* $\mathbb{Z}[H]$-module.

We can obtain $\mathbb{Z}[H]$-bases of $\underline{C}(\hat{X})$ as follows. Fix a basis $\underline{\boldsymbol{c}}$ of $\underline{C}(X)$. Choose a lift $\hat{\alpha}$ of each *oriented* cell α of $\underline{\boldsymbol{c}}$. We obtain the following $\mathbb{Z}[H]$-basis $\hat{\underline{\boldsymbol{c}}}$ of $\underline{C}(\hat{X})$.

$$\hat{\underline{\boldsymbol{c}}} := \{\hat{\alpha};\ \alpha \in \underline{\boldsymbol{c}},\}$$

This construction is not unique, for two reasons. Firstly, the choice $\underline{\boldsymbol{c}}$ depends on a re-ordering, and a change in orientations. Secondly, the lifts are not unique. These ambiguities can be gathered in the group

$$\mathfrak{A} = \mathfrak{S}_X \times \prod_{k \geq 0} \prod_{\alpha \in S_k(X)} H.$$

Observe that

$$\det \mathfrak{A} \cong \pm H \hookrightarrow (\mathbb{Q}(H), \cdot), \quad \mathbb{Q}(H) := \mathbb{Q}(\mathbb{Z}[H]).$$

Definition 2.2. The torsion of the chain complex $\underline{C}(\hat{X})$ of free $\mathbb{Z}[H]$-modules with respect to the above $\mathfrak{A}$-orbit of $\mathbb{Z}[H]$-bases is called the *Reidemeister torsion* of $S(X)$ and is denoted by $\mathcal{T}(S(X))$, (or $\mathcal{T}_X$ when the CW-structure is clear from the context). It is well defined as an element of $\mathbb{Q}(H)/\pm H$. □

Notation. If $x, y \in \mathbb{Q}(H)$ then $x \sim y \iff \exists h \in H,\ \exists \epsilon = \pm 1\colon\ x = \epsilon h y$.

If $Y \subset X$ is a subcomplex of X then we can define the relative torsion as follows. First form the chain complex of $\mathbb{Z}[H]$-modules $C(S(\hat{X}, \hat{Y}))$ associated to the CW-pair $(\hat{X}, \hat{Y})$, $\hat{Y} := \pi^{-1}(Y)$. Next, choose a basis $\underline{\boldsymbol{c}}/Y$ of the relative complex $\underline{C}(S(X, Y))$, and then lift it to a $\mathbb{Z}[H]$-basis of $S(\hat{X}, \hat{Y})$ of the form

$$\underline{\boldsymbol{c}}/Y = \big\{\hat{\alpha};\ \alpha \in S(X) \setminus S(Y)\big\}.$$

As before, the ambiguities of this construction form a group $\mathfrak{A}$ and

$$\det \mathfrak{A} \cong \pm H \hookrightarrow (\mathbb{Q}(H), \cdot).$$

The torsion $\mathcal{T}(S(\hat{X}, \hat{Y}), [\underline{c}/Y]_{\mathfrak{A}})$ is well defined as an element of $\mathbb{Q}(H)/\pm H$ which we denote by $\mathcal{T}_{X,Y}$.

If R is an integral domain and φ is a morphism $\mathbb{Z}(H) \to R$ then we can form the complex of free R-modules

$$C^{\varphi}(X, Y) := C(S(\hat{X}, \hat{Y})) \otimes_{\varphi} R.$$

The cell decomposition of X provides us as above with natural choices of bases in this complex. The torsion will be an element

$$\mathcal{T}^{\varphi}_{X,Y} \in \mathbb{K}/\pm\varphi(H),$$

where $\mathbb{K}$ is the field of fractions of R. Using Proposition 1.22 we deduce that if $\mathrm{char}(\mathbb{K}) = 0$ then

$$\mathcal{T}^{\varphi}_{X,Y} = \varphi(\mathcal{T}_{X,Y}).$$

Remark 2.3. A morphism $\varphi\colon \mathbb{Z}[H] \to R$ defines a system of local coefficients R_{φ} on X and the homology of the complex $C(S(\hat{X}, \hat{Y})) \otimes_{\varphi} R$ is canonically isomorphic with the homology of (X, Y) with coefficients in the local system R_{φ} (see [17, Chap. 5], [102] or [121, Chap. VI]).

In dealing with the gluing properties of the torsion it is perhaps more convenient to adopt this new point of view because, as explained in [17, ibid], the homology with local coefficients satisfies the same set of defining axioms the ordinary homology. □

Example 2.4 (The circle). Suppose X is the circle S^1 with the natural CW-decomposition consisting of a single 0-cell α_0, and a single 1-cell α_1. Then $H = H_1(S^1, \mathbb{Z}) = 0$. We write it multiplicatively, and we denote its natural generator by t. The group $\mathbb{Z}$ acts on $\hat{X} = \mathbb{R}$ by translations

$$t^n \cdot x = x + n.$$

Define

$$\hat{\alpha}_0 = 0, \quad \hat{\alpha}_1 = [0, 1].$$

In $\underline{C}(\hat{X})$ we have

$$\partial\hat{\alpha}_1 = \langle 1 \rangle - \langle 0 \rangle = (t-1)\hat{\alpha}_0, \quad \partial\hat{\alpha}_0 = 0.$$

This shows $\underline{C}(\hat{X})$ is acyclic. Define

$$\boldsymbol{c}_0 = \{\hat{\alpha}_0\}, \quad \boldsymbol{c}_1 = \{\hat{\alpha}_1\}, \quad \boldsymbol{b}_0 = \emptyset, \quad \boldsymbol{b}_1 = \{\hat{\alpha}_1\}.$$

Then

$$[\partial \boldsymbol{b}_1 \boldsymbol{b}_0/\boldsymbol{c}_0] = (t-1), \quad [\boldsymbol{b}_1/\boldsymbol{c}_1] = 1$$

so that

$$\mathcal{T}_{S^1} \sim (t-1)^{-1} \sim \pm t^n (t-1)^{-1} \in \mathbb{Q}(t)/(\pm t). \qquad \square$$

Example 2.5 (The two-dimensional torus). Suppose X is the torus T^2 equipped with the CW structure consisting of

- One 0-cell α.
- Two 1-cells β_1, β_2.
- One 2-cell γ.

with attaching maps described by the classical diagram in Figure 2.1.

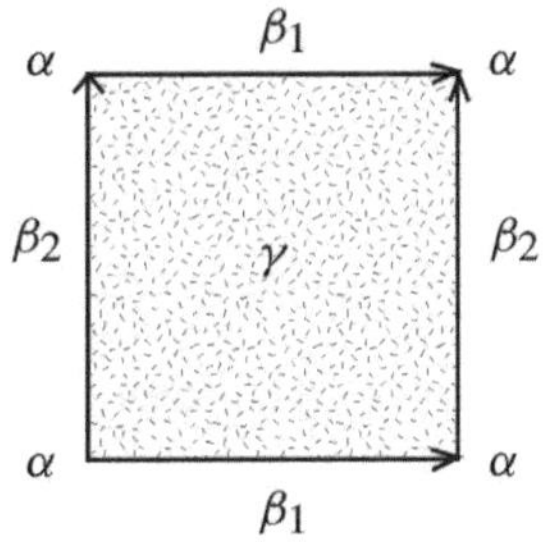

Figure 2.1. The CW-structure of a 2-torus.

Observe that $\hat{X} \cong \mathbb{R}^2$ and $H = \mathbb{Z}^2$ with (multiplicative) generators t_1 and t_2. We choose the bases $\boldsymbol{c}_i$ as follows.

- $\boldsymbol{c}_0 = \{\hat{\alpha} = \langle 0 \rangle \in \mathbb{R}^2\}$.
- $\boldsymbol{c}_1 = \{\hat{\beta}_1 = I \times \{0\} \subset \mathbb{R}^2,\ \hat{\beta}_2 = \{0\} \times I \subset \mathbb{R}^2, I = [0,1]\}$.
- $\boldsymbol{c}_2 = \{\hat{\gamma} = I \times I \subset \mathbb{R}^2\}$.

Inspecting Figure 2.1 we deduce

$$\partial \hat{\alpha} = 0,$$
$$\partial \hat{\beta}_1 = (t_1 - 1)\hat{\alpha}, \quad \partial \hat{\beta}_2 = (t_2 - 1)\hat{\alpha},$$
$$\partial \hat{\gamma} = (1 - t_2)\hat{\beta}_1 - (1 - t_1)\hat{\beta}_2.$$

Now choose

$$\boldsymbol{b}_2 = \{\hat{\gamma}\}, \quad \boldsymbol{b}_1 = \{\hat{\beta}_1\}, \quad \boldsymbol{b}_0 = \emptyset$$

Then

$$[\boldsymbol{b}_2/\boldsymbol{c}_2] = 1, \quad [(\partial \boldsymbol{b}_2)\boldsymbol{b}_1/\boldsymbol{c}_1] = \det \begin{bmatrix} (1-t_2) & 1 \\ -(1-t_1) & 0 \end{bmatrix} = (t_1 - 1),$$
$$[(\partial \boldsymbol{b}_1)\boldsymbol{b}_0/\boldsymbol{c}_0] = (t_1 - 1).$$

We conclude that

$$\mathcal{T}_{T^2} \sim 1 \sim \pm t_1^{n_1} t_2^{n_2}. \qquad \square$$

Example 2.6 (The higher dimensional tori). Suppose X is the n-dimensional torus $X := \mathbb{R}^n/\mathbb{Z}^n$. Denote by t_i the generators of the (multiplicative) group $H := \mathbb{Z}^n \cong \pi_1(X)$. They determine a basis (e_i) of the (additive) group $\mathbb{Z}^n$. For each ordered multi-index $I = 1 \leq i_1 < \cdots < i_k \leq n$ we set

$$e_I := e_{i_1} \wedge \cdots \wedge e_{i_k} \in \Lambda^k \mathbb{Z}^n.$$

The universal Abelian cover is $\mathbb{R}^n \to \mathbb{R}^n/\mathbb{Z}^n$ and

$$C_k(\hat{X}) \cong \mathbb{Z}[H] \otimes_{\mathbb{Z}} \Lambda^k H.$$

The monomials $\{e_I;\ |I| = k\}$ determine a $\mathbb{Z}[t_i, t_j^{-1}]$-basis $(\hat{e}_I)$ of $C_k(\hat{X})$ defined by

$$\hat{e}_I = \big\{(x_1, \dots, x_n) \in [0,1]^n;\ x_j = 0\ \forall j \notin I\big\}.$$

We use the wedge product to introduce a $\mathbb{Z}_2$-graded $\mathbb{Z}[H]$-algebra structure on $\underline{C}(\hat{X})$. The boundary operator

$$\partial\colon \underline{C}(\hat{X}) \to \underline{C}(\hat{X})$$

is then more than just a morphism of $\mathbb{Z}[H]$-modules. It is an odd derivation uniquely determined by the conditions

$$\partial e_\emptyset = 0, \quad \partial e_i := (t_i - 1)e_\emptyset, \quad i = 1, \dots, n.$$

Equivalently, it can be defined as the contraction with the formal vector field

$$T := \sum_k (t_k - 1)e_k.$$

Fix an element $u \in H$, and form the Koszul map

$$\kappa_u\colon C_k(\hat{X}) \to C_{k+1}(\hat{X}), \quad \omega \mapsto u \wedge \omega.$$

For simplicity we let $u := e_1$. Observe that

$$\partial(e_1 \wedge \bullet) + e_1 \wedge \partial\bullet = (1 - t_1)\bullet$$

so that, if we define

$$\eta\colon \underline{C}(\hat{X}) \to \underline{C}(\hat{X}), \quad \omega \mapsto (t_1 - 1)^{-1} e_1 \wedge \omega$$

then $\partial\eta + \eta\partial = 1$. Hence η is a contraction of the complex $\underline{C}(\hat{X})$. As in §1.1 we form the operators

$$D = \partial + \eta.$$

The torsion of $\underline{C}(\hat{X})$ is then

$$\mathcal{T}_X \sim \det\big(D\colon \underline{C}_{\text{even}}(\hat{X}) \to \underline{C}_{\text{odd}}(\hat{X})\big).$$

Set $L_n := \mathbb{Z}^n$. We have produced a sequence of operators

$$D_n : \Lambda^{\text{even}} L_n \otimes_{\mathbb{Z}} \mathbb{Z}[L_n] \to \Lambda^{\text{odd}} L_n \otimes_{\mathbb{Z}} \mathbb{Z}[L_n]$$

such that $\det(D_n) = \tau(T^n)$. If we write $L_{n+1} = L_n \oplus \mathbb{Z} \cdot e_{n+1}$ then we obtain decomposition

$$\Lambda^{\text{even}} L_{n+1} = \Lambda^{\text{even}} L_n \oplus \Lambda^{\text{odd}} L_n \wedge e_{n+1}, \quad \Lambda^{\text{odd}} L_{n+1} = \Lambda^{\text{odd}} L_n \oplus \Lambda^{\text{even}} L_n \wedge e_{n+1}.$$

Using these splittings we obtain the following block decomposition for D_{n+1}

$$D_{n+1} = \begin{bmatrix} D_n & * \\ 0 & \tilde{D}_n \end{bmatrix},$$

where

$$\tilde{D}_n = \partial + \eta : \Lambda^{\text{odd}} L_n \otimes \mathbb{Z}[L_n] \to \Lambda^{\text{even}} L_n \otimes \mathbb{Z}[L_n].$$

Since $D_n \tilde{D}_n = \mathbf{1}$ we deduce

$$\det(D_{n+1}) = 0, \quad \forall n \geq 1.$$

Thus

$$\mathfrak{T}_{T^n} \sim 1, \quad \forall n > 1. \qquad \square$$

The above identity is a consequence the following more general principle.

Example 2.7 (The torsion of $S^1 \times X$, X-finite cell complex). Denote by π both the natural morphism

$$H_1(S^1 \times X) \to H_1(S^1),$$

and the induced map

$$\mathbb{Q}(H_1(S^1 \times X)) \to \mathbb{Q}(t) \cong \mathbb{Q}(H_1(S^1)).$$

Then

$$\mathfrak{T}^{\pi}_{S^1 \times X} \sim (t-1)^{-\chi(X)} \sim \mathfrak{T}^{\chi(X)}_{S^1}.$$

Observe that $\mathfrak{T}^{\pi}_{S^1 \times X}$ can be computed using the cover $\mathbb{R} \times X \to S^1 \times X$. The associated cell complex has a $\mathbb{Z}[t]$-module structure. More precisely

$$\underline{C}(\mathbb{R} \times X) \cong \underline{C}(\mathbb{R}) \otimes_{\mathbb{Z}} \underline{C}(X),$$

where $\underline{C}(\mathbb{R})$ is the chain complex of $\mathbb{Z}[t]$-modules discussed in Example 2.4,

$$0 \to \mathbb{Z}[t] \xrightarrow{(t-1)\cdot} \mathbb{Z}[t] \to 0.$$

Then

$$\underline{C}_k(\mathbb{R} \times X) = \mathbb{Z}[t]e_1 \otimes \underline{C}_{k-1}(X) \oplus \mathbb{Z}[t]e_0 \otimes \underline{C}_k(X),$$

with basis

$$e_1 \otimes S_{k-1}(X) \cup e_0 \otimes S_k(X).$$

The boundary operator $\partial\colon \underline{C}_k(\mathbb{R} \times X) \to \underline{C}_{k-1}(\mathbb{R} \times X)$ acts according to the prescriptions

$$\hat{\partial}(e_1 \otimes \sigma_{k-1}) = (1-t)e_0 \otimes \sigma_{k-1} - e_1 \otimes \partial\sigma_k, \quad \hat{\partial}(e_0 \otimes \sigma_k) = e_0 \otimes \partial\sigma_{k-1}.$$

Define the morphism of $\mathbb{Z}[t]$-module

$$\eta\colon \underline{C}_{k-1}(\mathbb{R} \times X) \to \underline{C}_k(\mathbb{R} \times X),$$

$$e_0 \otimes \sigma_{k-1} \mapsto \frac{1}{t-1} e_1 \otimes \sigma_{k-1}, \quad e_1 \otimes \sigma_{k-2} \mapsto 0.$$

Observe that $\eta^2 = 0$ and $\hat{\partial}\eta + \eta\hat{\partial} = 1$, i.e. η is an algebraic contraction. Then

$$\mathcal{T}^{\pi}_{S^1 \times X} = \det\bigl(\hat{\partial} + \eta\colon \underline{C}_{\mathrm{even}}(\mathbb{R} \times X) \to \underline{C}_{\mathrm{odd}}(\mathbb{R} \otimes X)\bigr).$$

With respect to the bases

$$e_0 \otimes S_0(X) \cup e_1 \otimes S_1(X) \cup e_0 \otimes S_2(X) \cup \cdots$$

of $\underline{C}_{\mathrm{even}}$, and

$$e_1 \otimes S_0(X) \cup e_0 \otimes S_1(X) \cup e_1 \otimes S_2(X) \cup e_0 \otimes S_3(X) \cup \cdots$$

of $\underline{C}_{\mathrm{odd}}$, the operator $\hat{\partial} + \eta$ has the description

	n_0 columns	n_1 columns	n_2 columns	$\cdots$
n_0 rows	$(t-1)^{-1}$	$-\partial$	0	$\cdots$
n_1 rows	0	$(t-1)$	∂	$\cdots$
n_2 rows	0	0	$(t-1)^{-1}$	$\cdots$
$\vdots$	$\vdots$	$\vdots$	$\vdots$	

where $n_k := \#S_k(X)$. We deduce

$$\mathcal{T}^{\pi}_{S^1 \times X} = (t-1)^{-\sum_k (-1)^k n_k} = (t-1)^{-\chi(X)} = \mathcal{T}^{\chi(X)}_{S^1}. \qquad \square$$

Example 2.8 (The torsion of a product). The computations in the above example generalize as follows. Suppose X and Y are compact CW-complexes. Denote by π both the natural map

$$H_1(X \times Y) \to H_1(X),$$

and the induced morphism

$$\mathbb{Q}(H_1(X \times Y)) \to \mathbb{Q}(H_1(X)).$$

Then

$$\mathcal{T}^{\pi}_{X \times Y} = \mathcal{T}^{\chi(Y)}_{X}. \tag{2.1}$$

To prove this consider the Abelian cover of $X \times Y$ induced by π. It coincides with $\tilde{X} \times Y$, where $\tilde{X}$ is the universal Abelian cover of X. Denote by $(\underline{C}(\tilde{X}), \tilde{\partial}_0)$ the chain complex of $\mathbb{Z}[H_1(X)]$-modules generated by the CW-decomposition of $\tilde{X}$. Then the chain complex of $\mathbb{Z}[H_1(X)]$ modules corresponding to the CW-decomposition of $\tilde{X} \times Y$ has the form

$$\big(\underline{C}(\tilde{X} \times Y), \tilde{\partial}\big) \cong \big(\underline{C}(\tilde{X}), \tilde{\partial}_0\big) \hat{\otimes}_{\mathbb{Z}} \big(\underline{C}(Y), \partial_Y\big),$$

where the hat ˆ over the $\otimes$ signs signifies that we are taking a graded tensor product. Suppose for simplicity $(\underline{C}(\hat{X}), \tilde{\partial}_0)$ is acyclic and denote by η an algebraic contraction. Now define

$$\hat{\eta} \colon \underline{C}_{k-1}(\tilde{X} \times Y) \to \underline{C}_k(\tilde{X} \times Y)$$

by the equality

$$\hat{\eta}(\tilde{\sigma} \hat{\otimes} \delta) = (\eta\tilde{\sigma}) \hat{\otimes} \delta,$$

for every cell $\tilde{\sigma}$ of $\tilde{X}$ and every cell δ of Y. This morphism is an algebraic contraction of the complex $\underline{C}(\tilde{X} \times Y)$. Now order the cells of Y

$$\delta_1, \delta_2, \dots$$

as in (1.3) such that

$$\dim \delta_i \geq \dim \delta_{i+1}.$$

Using the decomposition

$$\underline{C}_{\text{even}/\text{odd}}(\tilde{X} \times Y) \cong \bigoplus_i \underline{C}_{\nu_i}(\tilde{X}) \otimes \delta_i,$$

where $\nu_i + \dim \delta_i = \text{even}/\text{odd}$, we deduce that the operator

$$\tilde{\partial} + \hat{\eta} \colon \underline{C}_{\text{even}}(\tilde{X} \times Y) \to \underline{C}_{\text{odd}}(\tilde{X} \times Y)$$

has lower triangular form, and the i-th diagonal element

$$\underline{C}_{\nu_i}(\tilde{X}) \otimes \delta_i \to \underline{C}_{\nu_i - 1}(\tilde{X}) \otimes \delta_i$$

has the form

$$\big(\tilde{\partial}_0 + \eta\big)^{(-1)^{\nu_i}}.$$

Formula (2.1) is now obvious. We refer to [31, 34, 35] for more information about the torsion of a fiber bundle. □

Example 2.9 (The torsion of lens spaces). The lens space $L(p,q)$ is defined as the quotient

$$L(p,q) := S^3/\diamond_{1,q},$$

where for $(r,p)=(s,p)=1$ we denote by $\diamond_{r,s}$ the action of cyclic group $\mathbb{Z}_p$ on

$$S^3 := \{(z_1,z_2)\in\mathbb{C}^2;\ |z_1|^2+|z_2|^2=1\}$$

defined by the rule

$$\zeta\diamond_{r,s}(z_1,z_2)=(\zeta^r z_1,\zeta^s z_2),\quad \forall\zeta^p=1.$$

The maximal Abelian cover of $L(p,q)$ is the sphere S^3. To compute the torsion of the lens space $L(p,q)$ we first need to produce a $\mathbb{Z}_p$-equivariant CW-decomposition of S^3. For $j\in\mathbb{Z}_p$ define

$$E_j^0=\{(\zeta^j,0)\in S^3\},$$

$$E_j^1=\Big\{(e^{\mathbf{i}\theta},0)\in S^3;\ \frac{2\pi j}{p}\le\theta\le\frac{2\pi(j+1)}{p}\Big\},$$

$$E_j^2=\{(z_1,s\zeta^j)\in S^3;\ t\in[0,1]\}=\{(z_1,s\zeta^j)\in\mathbb{C}^2;\ s\in\mathbb{R},\ \sqrt{|z_1|^2+s^2}=1\},$$

$$E_j^3=\Big\{(z_1,z_2)\in S^3;\ \frac{2\pi j}{p}\le\arg z_2\le\frac{2\pi(j+1)}{p}\Big\}.$$

Each E_j^k is homeomorphic to a closed k-ball, and the collection

$$\{E_j^k\}_{0\le k\le 3,\, j\in\mathbb{Z}_p}$$

forms a $\mathbb{Z}_p$-equivariant decomposition of S^3. Set

$$c_j^k := \mathrm{Int}(E_j^k).$$

We orient the cells inductively over k such that, taking into account the orientations, we have

$$\partial c_j^2=\sum_{i=0}^{p-1}c_i^1,$$

and

$$\partial c_j^1=c_{j+1}^0-c_j^0,\quad \partial c_j^3=c_{j+1}^2-c_j^2.$$

Observe that

$$\zeta\cdot c_j^k=c_{j+1}^k,\ \ k=0,1\quad\text{and}\quad \zeta\cdot c_j^k=c_{j+q}^k,\ \ k=2,3.$$

Hence

$$\partial c_0^2=\Big(\sum_{j=0}^{p-1}\zeta^j\Big)c_0^1,$$

and

$$\partial c_j^1 = (\zeta - 1)c_j^0, \quad \partial c_j^3 = (\zeta^r - 1)c_j^2,$$

where $r \cdot q = 1 \bmod p\mathbb{Z}$. Suppose $\mathbb{K}$ is a field and $\varphi\colon \mathbb{Z}[\zeta] = \mathbb{Z}[\mathbb{Z}_p] \to \mathbb{K}$ is a nontrivial ring morphism. Set $t := \varphi(\zeta)$ and $C_*^\varphi := C_*(S^3) \otimes_\varphi \mathbb{K}$. Observe that

$$C_k^\varphi = \mathbb{K} \cdot c_0^k, \quad k = 0, \ldots, 3.$$

This shows the chain complex C_*^φ is acyclic and an elementary computation yields

$$\mathcal{T}^\varphi_{L(p,q)} \sim (1-t)^{-1}(1-t^r)^{-1}.$$

Observe that if we choose a different generator s of $H_1\big(L(p,q)\big)$ defined by $s^q = t$ we have

$$\mathcal{T}^\varphi_{L(p,q)} \sim (1-s^q)^{-1}(1-s)^{-1}. \qquad \square$$

The torsion of a CW-complex *depends on the* CW-*structure*. The following result states this more precisely. For a proof and more details we refer to [16, Chap. IV] or [72, §7].

Theorem 2.10 (Combinatorial invariance of torsion). *The torsion $\mathcal{T}_{X,Y}$ is invariant under subdivision of the* CW-*pair* (X, Y).

It is known that any compact smooth manifold admits C^1-triangulations, and any two have a common finer subdivision (see [72, §9] for more details). This shows that we can define the torsion of a smooth manifold using C^1-triangulations and the result will be independent of triangulations. In other words the following true.

Theorem 2.11. *The torsion of a compact smooth manifold is a diffeomorphism invariant.*

Remark 2.12. The *absolute* torsion $\mathcal{T}_X$ of a simplicial complex X is known to be a *topological* invariant of the cellular complex X; see [13]. However, the *relative* torsion $\mathcal{T}_{X,Y}$ is *not* a topological invariant; see the beautiful paper [69] or Remark 2.62, page 105. $\square$

From the exact homology sequence of a pair and the multiplicativity of the torsion we deduce the following result.

Theorem 2.13. *Suppose (X, Y) is a CW-pair, $\mathbb{K}$ is a field and $\varphi\colon \mathbb{Z}[H_1(X)] \to \mathbb{K}$ is a ring morphism. If j denotes the inclusion induced morphism*

$$j\colon \mathbb{Q}(H_1(Y)) \to \mathbb{Q}(H_1(X))$$

and either $\mathcal{T}_Y^j \neq 0$ or $\mathcal{T}_{X,Y}^{\varphi\circ j} \neq 0$. Then

$$\mathcal{T}_X^\varphi \sim \mathcal{T}_{X,Y}^\varphi \cdot \mathcal{T}_Y^{\varphi\circ j}.$$

If $f: X \to Y$ is a cellular map then we can form its mapping cylinder

$$M_f := (X \times [0,1] \cup Y)/\{(x,1) = f(x)\}.$$

Y is a strong deformation retract of M_f and in particular, $H_1(M_f) \cong H_1(Y)$. We define the torsion of f by the equality[1]

$$\mathcal{T}_f := \mathcal{T}_{M_f,X} \in \mathbb{Q}(H_1(Y)).$$

If f is a homotopy equivalence then $H_1(X) \cong H_1(Y)$ so the torsion of f is also an element of $\mathbb{Q}(H_1(X))$. In general, the torsion of a homotopy equivalence may not be ~ 1. However, we have the following fundamental result of J. H. C. Whitehead ([16, 19, 72]).

Theorem 2.14. *If f is a **simple homotopy equivalence** then*

$$\mathcal{T}_f \sim 1.$$

We will not present a formal, geometric definition of the notion of simple homotopy. We only want to mention a typical example of such homotopy: the collapse of a simplex onto one of its faces (see Figure 2.2). In general, a simple homotopy is a composition of such elementary collapses.

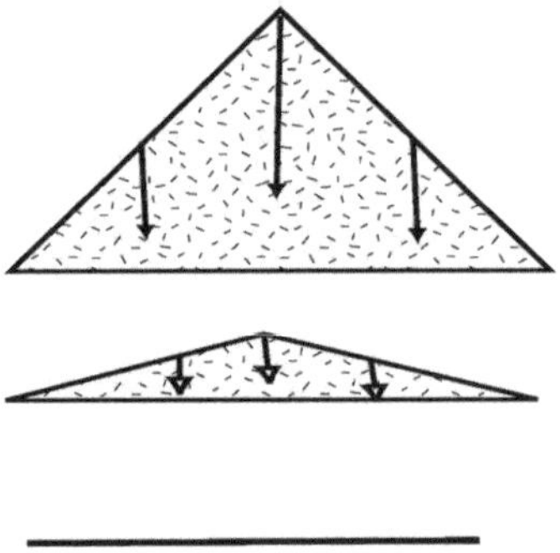

Figure 2.2. Elementary collapse.

For example, a 3-manifold with boundary equipped with a simplicial decomposition is simple homotopy equivalent to a 2-dimensional simplicial complex.

Using Theorem 2.13 and 2.14 we deduce that if two cellular complexes X, Y are simple homotopy equivalent then

$$\mathcal{T}_X \sim \mathcal{T}_Y.$$

[1] The torsion of the pair (M_f, Y) is known to be trivial, ~ 1 (see [72, §7]).

Example 2.15 (The solid torus). The inclusion $S^1 \hookrightarrow D^2 \times S^1$, $p \mapsto (0, p)$ is a simple homotopy equivalence so that the torsion of a solid torsion is identical to the torsion of a circle

$$\mathcal{T}_{D^2\times S^1} \sim (t-1)^{-1}. \qquad \square$$

Using the multiplicativity property of the torsion and the Mayer–Vietoris sequence we deduce the following consequence.

Theorem 2.16 (Gluing Formula). *Suppose X_1, X_2 are subcomplexes of X such that*

$$X = X_1 \cup X_2 \quad \textit{and} \quad X_1 \cap X_2 = Y.$$

Suppose $\mathbb{K}$ is a field and $\varphi\colon \mathbb{Q}[H_1(X)] \to \mathbb{K}$ is a ring morphism. Let

$$j\colon \mathbb{Q}(H_1(Y)) \to \mathbb{Q}(H_1(X)), \quad j_k\colon \mathbb{Q}(H_1(X_r)) \to \mathbb{Q}(H_1(X)), \qquad k = 1, 2$$

denote the inclusion induced morphisms. If $\mathcal{T}_Y^{\varphi\circ j} \neq 0$ then

$$\mathcal{T}_X^{\varphi} \cdot \mathcal{T}_Y^{\varphi\circ j} \sim \mathcal{T}_{X_1}^{\varphi\circ j_1} \cdot \mathcal{T}_{X_2}^{\varphi\circ j_2}.$$

Example 2.17 (The torsion of fibrations over a circle). We follow closely the presentation in [33]. Suppose

$$\begin{array}{ccc} F & \xrightarrow{\ j\ } & X \\ & & \downarrow \pi \\ & & S^1 \end{array}$$

is a smooth fiber bundle over S^1, with compact, closed, connected, oriented fiber F. Fix a Riemann metric on X such that π is a Riemann submersion. π defines a gradient flow which covers the canonical rotational flow on S^1 (with period 2π). We denote by $h\colon X \to X$ the time 2π-map of this flow. It induces a map

$$\mu = h|_F : F = \pi^{-1}(0) \to \pi^{-1}(2\pi) = F$$

known as the *geometric monodromy* of this fibration.

Fix a cellular structure on F and a cellular approximation μ' of the monodromy map μ and form the mapping torus X' of μ'. One can show (see [33, 49]) that the torsion of X is equal to the torsion of X' and we will compute the torsion of X'. The group $\mathbb{Z}$ acts on $C_*(\tilde{\pi}^{-1}(F); \mathbb{Z})$ by

$$t(k, \sigma) = (k + 1, \sigma),$$

so that we have the isomorphism of $\mathbb{Z}[t, t^{-1}]$-modules

$$C_*(\tilde{\pi}^{-1}(F); \mathbb{Z}) \cong C_*(F, \mathbb{Z}) \otimes \mathbb{Z}[t, t^{-1}].$$

The natural projection $\pi : X' \to S^1$ defines an infinite cyclic cover

$$\tilde{\pi} : X'_\pi \to X',$$

the pullback of $\exp : \mathbb{R} \to \mathbb{R}/2\pi\mathbf{i}\mathbb{Z} = S^1$ via π. The space X' has a natural CW-structure which lifts to a CW-structure on the cover X'_π. The cellular chain complex $C_*(X'_\pi; \mathbb{Z})$ has a structure of $\mathbb{Z}[t, t^{-1}]$-module given by the actions of the group of deck transformations of X'_π and can be described as the algebraic mapping torus of the morphism of $\mathbb{Z}[t, t^{-1}]$-cellular complexes

$$\mu'_\pi = t\mu' : C_*(\tilde{\pi}^{-1}(F); \mathbb{Z}) \to C_*(\tilde{\pi}^{-1}(F); \mathbb{Z}),$$
$$(k, \sigma) \mapsto (k+1, \mu'(\sigma)) = t(k, \mu'(\sigma))$$

Using the identity (1.6) in Example 1.6 we conclude that

$$\mathcal{T}\big(C_*(X'_\pi; \mathbb{Z})\big) \sim \zeta_\mu(t),$$

where $\zeta_\mu(t)$ is the s-zeta function of the induced morphism μ_* on the $\mathbb{Z}_2$-graded vector space $H_*(F, \mathbb{R})$. The map π induces a morphism $H := H_1(X) \to \mathbb{Z}$ and thus a morphism

$$\pi : \mathbb{Q}(H) \to \mathbb{Q}(t).$$

The last equality can now be formulated

$$\mathcal{T}_X^\pi \sim \zeta_\mu(t).$$

We refer to [33] for a description of the whole torsion of X. The final result is however not so explicit.

To understand how much information about the torsion of X is contained in the above equality we need to understand $H_1(X, \mathbb{Z})$. This homology can be determined from the Wang exact sequence, [121, Chap. VII], which is a consequence of the fact that the chain complex $C_*(X; \mathbb{Z})$ is the algebraic mapping torus of $C_*(F)$ with respect to μ_*,

$$\cdots \longrightarrow H_k(F) \xrightarrow{1-\mu_*} H_k(F) \xrightarrow{j_*} H_k(X) \xrightarrow{\cap F} H_{k-1}(F) \xrightarrow{1-\mu_*} \cdots.$$

We obtain a short (split) exact sequence

$$0 \to \operatorname{coker}(1-\mu_*) \to H_1(X) \to \mathbb{Z} \to 0.$$

The linear map $H_1(X) \to \mathbb{Z}$ has a simple geometric interpretation: it is given by the integral of the angular form

$$\omega := \frac{1}{2\pi}\pi^*(d\theta)$$

along a cycle $c \in H_1(X)$. Note that if

$$\det\big((1-\mu_*) : H_1(F) \to H_1(F)\big) \neq 0,$$

then $H_1(X)$ has rank 1. In this case we have

$$\mathcal{T}_X^{\mathfrak{aug}} \sim \zeta_\mu(t).$$

If F happens to be a Riemann surface of genus g then, using the computations in §A.2 we deduce

$$\mathcal{T}_X^{\mathfrak{aug}} \sim \frac{\det\big(1 - tH_1(\mu)\big)}{(1-t)^2}. \qquad \square$$

Example 2.18 (The torsion of connected sums). Suppose N_1 and N_2 are two closed, oriented, triangulated, smooth 3-manifolds such that

$$r_i := \operatorname{rank} H_1(N_i) > 0, \quad i = 1, 2.$$

We want to prove that

$$\mathcal{T}_{N_1 \# N_2} = 0,$$

i.e.

$$H_*(N_1 \# N_2; \mathbb{K}_\pi) \neq 0,$$

where $\mathbb{K}$ is an arbitrary field entering into the decomposition of $\mathbb{Q}\big(H_1(N_1 \# N_2)\big)$, π denotes the natural projection

$$\mathbb{Q}\big(H_1(N_1 \# N_2)\big) \to \mathbb{K},$$

and $\mathbb{K}_\pi$ denotes the corresponding local system of coefficients. We will follow an approach we learned from Frank Connolly. We refer to [110, §4.3] for a different proof of a slightly weaker result. Set $N := N_1 \# N_2$,

$$G_i := H_1(N_i), \quad G := H_1(N_1 \# N_2) \cong G_1 \oplus G_1.$$

Let M_i denote N_i with a small open disk D_i removed and $\pi_i : \hat{N}_i \to N_i$ denote the universal Abelian cover of N_i. Set

$$\hat{M}_i := \pi_i^{-1}(M_i).$$

Finally denote by S the 2-sphere $M_1 \cap M_2 \subset N$. The universal Abelian cover $\tilde{N}$ of N is obtained as follows.

• Fix a lift $\hat{D}_i$ of D_i to $\hat{N}_i$, $i = 1, 2$ we can identify $\partial \hat{M}_i \cong G_i \times S^2$ so that

$$\partial(G_1 \times \hat{M}_2) \cong (G_1 \oplus G_2) \times S^2 \cong \partial(G_2 \times \hat{M}_1).$$

For $(g_1, g_2) \in G_1 \times G_2$ we denote by $\partial_{g_i} \hat{M}_i$ the component of $\partial \hat{M}_i$ labelled by g_i.

• Glue $G_1 \times \hat{M}_2$ to $G_2 \times \hat{M}_1$ along the boundary using the identifications (see Figure 2.3).

$$\{g_1\} \times \partial_{g_2} \hat{M}_2 \sim \{g_2\} \times \partial_{g_1} \hat{M}_1.$$

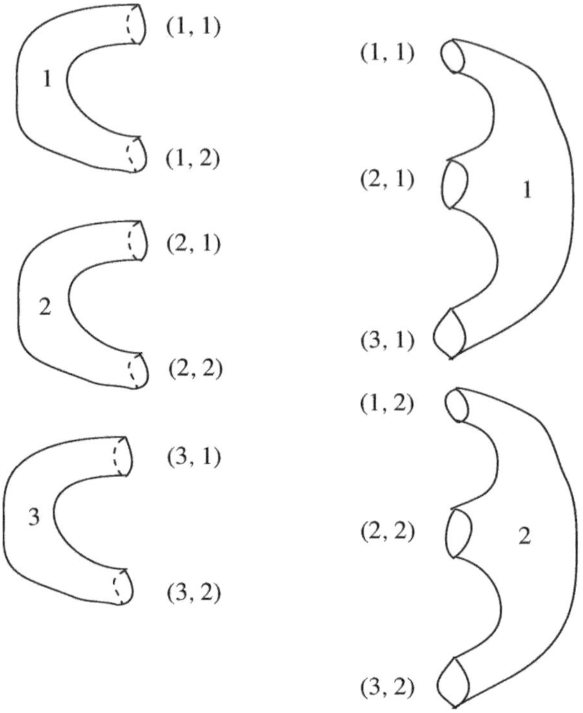

Figure 2.3. Covering a connected sum.

Suppose $\mathbb{K}$ is one of the fields entering into the decomposition of $\mathbb{Q}(G)$. The discussion in §1.5 shows that $\mathbb{K}$ is the field of rational functions in $r_1 + r_2$ variables with coefficients in some finite extension of $\mathbb{Q}$. In particular, since $r_i > 0$ this shows that the monodromy groups

$$\mathfrak{M}_i = \operatorname{Range} \pi_i, \quad \pi_i := G_i \hookrightarrow G \to \mathbb{Q}(G) \xrightarrow{\pi} \mathbb{K}$$

are infinite, so that the coverings of M_i and N_i defined by the morphisms π_i are infinite, thus noncompact. Since M_i is an *open* 3-dimensional manifold we deduce

$$H_3(M_i; \mathbb{K}_{\pi_i}) = H_3(N_i; \mathbb{K}_{\pi_i}) = 0. \tag{$*$}$$

We deduce similarly that

$$H_3(N; \mathbb{K}_{\pi}) = 0.$$

The Mayer–Vietoris sequence for the homology with local coefficients (see [17, Chap. 5]) now implies

$$0 \to H_2(S; \mathbb{K}) \to H_2(M_1; \mathbb{K}_{\pi_1}) \oplus H_2(M_2; \mathbb{K}_{\pi_2}) \to H_2(N; \mathbb{K}_{\pi}) \to H_1(S; \mathbb{K}).$$

Since S is simply connected we deduce that the last group is trivial and the first is isomorphic to $\mathbb{K}$. We thus have a short exact sequence of $\mathbb{K}$-*vector spaces*

$$0 \to \mathbb{K} \to H_2(M_1; \mathbb{K}_{\pi_1}) \oplus H_2(M_2; \mathbb{K}_{\pi_2}) \to H_2(N; \mathbb{K}_\pi) \to 0. \qquad (**)$$

The middle part of $(**)$ can be determined from the long exact sequence of the pair (N_i, M_i):

$$0 \to H_3(N_i, M_i; \mathbb{K}_\bullet) \to H_2(M_i; \mathbb{K}_{\pi_i}) \to H_2(N_i; \mathbb{K}_{\pi_i}) \to H_2(N_i, M_i; \mathbb{K}_\bullet) \to .$$

Using excision we deduce

$$0 \to H_3(D_i, S; \mathbb{K}) \to H_2(M_i; \mathbb{K}_{\pi_i}) \to H_2(N_i; \mathbb{K}_{\pi_i}) \to H_2(D_i, S; \mathbb{K}) \to 0. \quad (***)$$

The sequence $(**)$ implies

$$\dim_{\mathbb{K}} H_2(N; \mathbb{K}_\pi) = \dim_{\mathbb{K}} H_2(M_1; \mathbb{K}_{\pi_1}) + \dim_{\mathbb{K}} H_2(M_2; \mathbb{K}_{\pi_2}) - 1,$$

while $(***)$ implies

$$\dim_{\mathbb{K}} H_2(M_i; \mathbb{K}_{\pi_i}) \geq 1,$$

so that $\dim_{\mathbb{K}} H_2(N; \mathbb{K}_\pi) \geq 1$. □

§2.2 Fitting ideals

We interrupt for a little while the flow of topological arguments to discuss some basic algebraic notions needed to go deeper inside the structure of torsion.

Let R be a Noetherian integral domain and denote by $\mathbb{K}$ its field of fractions. Suppose $T: U \to V$ is a morphism of free R-modules

$$U \cong R^p, \quad V \cong R^q.$$

Choosing bases in U and V we can represent T by a $q \times p$ matrix with entries in R, and we denote by $I_1(T) \subset R$ the ideal generated by the entries of T. Clearly $I_1(T)$ is independent of the chosen bases of U and V. Equivalently, this means that $I_1(T)$ is invariant under elementary row and column operations on T. T induces morphisms

$$\Lambda^k T: \Lambda^k U \to \Lambda^k V,$$

and we set

$$I_k(T) := I_1(\Lambda^k T) \subset R.$$

In more concrete terms, $I_k(T)$ is the ideal generated by all the $k \times k$ minors of T. Observe that

$$R := I_0(A) \supset I_1(T) \supset I_2(T) \supset \cdots$$

and

$$I_k(T_1 \oplus T_2) \supset I_k(T_1) + I_k(T_2), \quad I_k(T_1 \cdot T_2) \subset I_k(T_1) \cdot I_k(T_2), \quad \forall k, T_1, T_2.$$

Suppose

$$U \xrightarrow{T} V \to M \to 0$$

is a presentation of the finitely generated R-module M. Let $g = \operatorname{rank} V$, $r = \operatorname{rank} U$. Then g is the number of generators, while r is the number of relations of this presentation. Assume $r \geq g$. Define

$$F_k(T) := I_{g-k}(T).$$

Any other presentation can be obtained from the above by performing a succession of elementary transformations described below, [92].

1. Change bases in U and V.

2. Replace U, V by $R \oplus U$ and $R \oplus V$ and T by $0 \oplus T$, where 0 denotes the trivial map $R \to R$.

3. The reverse of **2.**

4. Replace U by $R \oplus U$ T by $T \circ \pi_U$, where π_U denotes the natural projection $R \oplus U \to U$.

5. The reverse of **4.**

Clearly the ideal $F_k(T)$ is invariant under these elementary transformations. This shows it is an invariant of the module M. It is called the *k-th Fitting ideal*, of M and is denoted by $F_k(M)$. Observe that

$$F_0(M) \subset F_1(M) \subset \cdots \subset F_k(M) \subset \cdots.$$

$F_0(M)$ is called the *order ideal* of M and is denoted by $\mathfrak{O}(M)$. In case M admits a presentation in which there are fewer relations than generators then we set $\mathfrak{O}(M) = (0)$.

If R happens to be factorial then we define the *order* of M, $\operatorname{ord}(M) \in R/R^\times$, as the greatest common divisor of the elements in $F_0(M)$.

Example 2.19.

$$F_k(R^q) = \begin{cases} 0 & \text{if } 0 \leq k < q \\ R & \text{if } k \geq q. \end{cases}$$

□

Example 2.20. Suppose G is the Abelian group generated by e_1, e_2, e_3 subject to the relations

$$e_1 + e_2 + e_3 = 0, \quad 2e_1 - e_2 + 3e_3 = 0, \quad e_1 = 3e_3.$$

Then it admits the presentation

$$\mathbb{Z}^3 \xrightarrow{A} \mathbb{Z}^3 \to G \to 0$$

where A is the 3×3 matrix

$$\begin{bmatrix} 1 & 2 & 1 \\ 1 & -1 & 0 \\ 1 & 3 & -3 \end{bmatrix}.$$

The order ideal is the ideal generated by $|\det(A)| = 13$. It coincides with the order of the group. In particular, $\operatorname{ord}(G) = \pm 13 \in \mathbb{Z}/\pm 1$. □

Example 2.21. Suppose $M = R/I$, $I \subset R$ is an ideal of R. Then $\mathfrak{O}(M) \cong I$. Indeed, if I is generated by n elements, then M admits a presentation

$$R^n \xrightarrow{L} R \to M \to 0$$

so that

$$\mathfrak{O}(M) \cong I_1(L) = I.$$ □

Example 2.22. Suppose $M = \mathbb{Z}_n$. Then

$$F_0(M) = n\mathbb{Z}, \quad F_1(M) = \mathbb{Z}.$$ □

Proposition 2.23. *Suppose M can be generated by q elements. Then*

$$\big(\operatorname{ann}_R(M)\big)^q \subset \mathfrak{O}(M) \subset \operatorname{ann}_R(M) := \{r \in R;\ r \cdot M = 0\}.$$

Proof. Let $x_1, \dots, x_q$ be generators of M. If $a_1, \dots, r_q \in \operatorname{ann}_R(M)$ then we can form a presentation of M containing the relations

$$a_i x_j = 0, \quad i, j = 1, \dots, q$$

which proves the first inclusion. To prove the second inclusion, consider a $q \times q$-matrix of relations between the x_i's. Then, $\det(A)x_i = 0$, thus proving the second inclusion. □

Corollary 2.24. *Suppose R is a Noetherian integral domain. If M is a submodule of a free R module then $\mathfrak{O}(M) \subset \operatorname{ann}_R(M) = 0$.*

For a proof of the next result we refer to [55, XIX,§2].

Proposition 2.25.

$$F_n(M' \oplus M'') = \sum_{r+s=n} F_r(M')F_s(M'').$$

Corollary 2.26.

$$F_k(M \oplus R^q) = 0, \ \ \forall k < q, \quad F_q(M \oplus R^q) = \mathfrak{O}(M).$$

Example 2.27. Suppose $R = \mathbb{Z}$ and

$$M = \mathbb{Z}_{n_1} \oplus \cdots \oplus \mathbb{Z}_{n_k}, \quad n_k | n_{k-1} | \cdots | n_1.$$

Set $N := n_1 \ldots n_k$. Then

$$F_0(M) = N\mathbb{Z}, \ \ F_1(M) = \frac{N}{n_1}\mathbb{Z}, \ldots, F_{k-1}(M) = \frac{N}{n_1 \ldots n_{k-1}}\mathbb{Z}, \ \ F_k(M) = \mathbb{Z}.$$

□

Example 2.28 (The elementary invariants of a matrix). Suppose $R = \mathbb{Z}[t, t^{-1}]$. Any matrix $A \in \mathrm{SL}_n(\mathbb{Z})$ defines a R-module structure on $\mathbb{Z}^n$

$$p(t, t^{-1}) \cdot u := p(A, A^{-1})u, \quad \forall p(t, t^{-1}) \in R, \ u \in \mathbb{Z}^n.$$

We denote this is R-module by $(\mathbb{Z}^n, A)$. It admits the presentation

$$R^n \stackrel{t-A}{\longrightarrow} R^n \stackrel{f_A}{\longrightarrow} (\mathbb{Z}^n, A) \to 0,$$

where f_A is the map

$$R^n \cong \mathbb{Z}^n \otimes_{\mathbb{Z}} R \to \mathbb{Z}^n, \quad \sum_j \vec{u}_j t^j \mapsto A^j \vec{u}_j.$$

Denote by $q_i \in \mathbb{Z}[t]$ the elementary invariants of A (see [55, XIV, §2]), i.e. the monic polynomials uniquely determined by the conditions

$$q_k | q_{k-1} | \cdots | q_1, \quad p_A(t) := \det(t - A) = q_1 \ldots q_k,$$

such that, as a $\mathbb{Q}[t]$-module, $(\mathbb{Q}^n, A)$ is isomorphic to the direct sum of cyclic modules

$$(\mathbb{Q}^n, A) \cong \bigoplus_{j=1}^{k} \mathbb{Q}[t]/(q_j).$$

(Observe that q_k is the minimal polynomial of A.) We deduce

$$F_0(\mathbb{Z}^n, A) = (\, p_A(t)\,), \quad F_1(\mathbb{Z}^n, A) = \Big(\frac{p_A(t)}{q_1}\Big), \quad F_2(\mathbb{Z}^n, A) = \Big(\frac{p_A(t)}{q_1 q_2}\Big), \ldots.$$

□

§2.3 The Alexander function and the Reidemeister torsion

Assume R is a Noetherian, unique factorization domain with 1 and

$$(\underline{C}, \partial)\colon 0 \to C_\ell \to C_{\ell-1} \to \cdots \to C_1 \to C_0 \to 0$$

is a chain complex of finitely generated R-modules. We define the *Alexander function* $A(\underline{C})$ to be zero if $\operatorname{ord}\big(H_i(\underline{C})\big) = 0$ for some i. Otherwise, we set

$$A(\underline{C}) := \prod_{j=0}^{\ell} \operatorname{ord}\big(H_i(\underline{C})\big)^{(-1)^{i+1}}.$$

Example 2.29 (The circle revisited). The ring $\mathbb{Z}[t, t^{-1}]$ is the localization of a factorial Noetherian ring and so itself must be factorial and Noetherian. Consider the complex of $\mathbb{Z}[t, t^{-1}]$-modules discussed in Example 2.4,

$$\underline{C}\colon\ \ 0 \to C_1 = \mathbb{Z}[t, t^{-1}] \xrightarrow{\partial} C_0 = \mathbb{Z}[t, t^{-1}] \to 0,$$

where δ is the multiplication by $(t - 1)$. Then

$$H_1(\underline{C}) \cong \mathbb{Z}[t, t^{-1}], \quad \operatorname{ord}\big(H_1(\underline{C})\big) \sim 1,$$

$$H_0(\underline{C}) \cong \mathbb{Z}[t, t^{-1}]/(1 - t),$$

so that (see Example 2.21)

$$\operatorname{ord}\big(H_0(\underline{C})\big) \sim (t - 1).$$

Thus $A(\underline{C}) \sim (t - 1)^{-1}$. □

Example 2.30. Suppose U is a torsion module over $R = \mathbb{Z}[t, t^{-1}]$ which is free as a $\mathbb{Z}$-module. We deduce that the rank of U over $\mathbb{Z}$ is finite, say r. The polynomial t defines an automorphism

$$A \in \operatorname{Aut}_{\mathbb{Z}}(U) \cong \operatorname{SL}(U) \cong \operatorname{SL}(r, \mathbb{Z}), \quad u \mapsto t \cdot u.$$

Then, according to Example 2.28 we have

$$\operatorname{ord}_{\mathbb{Z}[t,t^{-1}]} U \sim p_A(t) := \det(t - A).$$ □

Suppose (X, Y) is a compact CW-pair. Set $H := H_1(X)$, $F = H/\operatorname{Tors}(H)$, denote by $\hat{X} \to X$ the universal (maximal) Abelian cover of X, and by $\mathfrak{aug}\colon H \to F$ the natural projection. We can form the complex of $\mathbb{Z}[F]$-modules

$$C^{\mathfrak{aug}}(X, Y) := C(S(\hat{X}), S(\hat{Y})) \otimes_{\mathfrak{aug}} \mathbb{Z}[F]. \tag{2.2}$$

$\mathbb{Z}[F]$ is Noetherian and factorial so the Alexander function of this complex is an element in $\mathbb{Q}(F)$, well defined up to multiplication by a unit in the ring $\mathbb{Z}[F]$. We denote this Alexander function by $A(X, Y)$ and we will refer to it as the *Alexander function* of the pair (X, Y). The next result, due to Turaev, generalizes the computation in Example 2.29.

Theorem 2.31 ([33, 73, 111]). *If (X, Y) is a compact* CW*-pair then*

$$\mathcal{T}^{\mathfrak{aug}}_{X,Y} \sim A(X, Y) \in \mathbb{Q}(F).$$

Remark 2.32. The above result is similar in spirit with the classical Euler–Poincaré theorem which states that the Euler characteristic of a simplicial complex is equal to the Euler characteristic of its homology. In the above theorem, $\mathcal{T}^{\mathfrak{aug}}$ is defined in terms of a simplicial (CW) decomposition while the Alexander function is defined entirely in homological terms. □

Proof. We follow the approach in [111]. This theorem is a consequence of the following abstract result.

Lemma 2.33. *Suppose R is a Noetherian, factorial ring of characteristic zero, $\mathbb{K}$ is its field of fractions and*

$$\underline{C}\colon\ 0 \to C_m \to C_{m-1} \to \cdots \to C_1 \to C_0 \to 0$$

is a chain complex of free R-modules, equipped with a basis $\underline{\mathbf{c}}$ such that

$$\operatorname{rank}_{\mathbb{K}} H_*(\underline{C}) = 0, \quad i.e.\ H_*(\underline{C}) = \operatorname{Tors} H_*(\underline{C}).$$

Then

$$\mathcal{T}(\underline{C}, [\underline{\mathbf{c}}]) = \zeta \prod_{j=0}^{m} \operatorname{ord}\big(H_j(\underline{C})\big)^{(-1)^{j+1}},$$

where ζ is a unit of R.

Proof. Denote by J_i the cokernel of $\partial_{i+1}\colon C_{i+1} \to C_i$, and by A_i the matrix describing

$$\partial_{i+1}\colon C_{i+1} \to C_i$$

with respect to the basis $\underline{\mathbf{c}}$. Set $n_i := \dim_{\mathbb{K}} C_i \otimes \mathbb{K}$ and $r_i := \dim_{\mathbb{K}} A_i \otimes \mathbb{K}$. We have an exact sequence

$$0 \to H_i(\underline{C}) \to J_i \to C_{i-1},$$

and since C_{i-1} is free we deduce that $\operatorname{Tors} H_i(\underline{C}) = \operatorname{Tors}(J_i)$. Hence,

$$\mathfrak{O}\big(H_i(\underline{C})\big) = \mathfrak{O}\big(\operatorname{Tors}(J_i)\big) \Longrightarrow \operatorname{ord}(H_i(\underline{C})) = \operatorname{ord}\big(\operatorname{Tors}(J_i)\big). \tag{2.3}$$

On the other hand, observe that if B_{2i} is *any* nonsingular square submatrix of A_{2i} of order r_{2i}, $0 \leq i \leq \lfloor (m-1)/2 \rfloor$, then there exists an unique τ-chain

$$\{(S_0, \tilde{D}_0), \ldots, (S_{m-1}, \tilde{D}_{m-1})\}$$

such that $\tilde{D}_{2i} = B_{2i}$. Thus

$$\det \tilde{D}_{2i} = \det(B_{2i}) \in \mathfrak{O}\big(\operatorname{Tors}(J_{2i})\big). \tag{2.4}$$

If we write $\mathfrak{T}(\underline{C}, \underline{c})$ as an irreducible fraction x/y, we deduce from Proposition 1.5 that

$$x/y = \prod_{i \geq 0} \det \tilde{D}_{2i+1} / \det \tilde{D}_{2i}$$

Since the $r_{2i} \times r_{2i}$ sub-matrix B_{2i} in the equality (2.4) is *arbitrary*, we deduce that the denominator y of $\mathfrak{T}$ divides *all* the generators of $\prod_{i \geq 0} \mathfrak{O}\big(\operatorname{Tors}(J_{2i})\big)$ because they are all of the form $\prod_{i \geq 0} \det B_{2i}$. Hence

$$\mathfrak{T}(\underline{C}, \underline{c}) \cdot \prod_{i \geq 0} \mathfrak{O}\big(\operatorname{Tors}(J_{2i})\big) \subset \prod_{i \geq 0} \mathfrak{O}\big(\operatorname{Tors}(J_{2i+1})\big).$$

The opposite inclusion

$$\mathfrak{T}(\underline{C}, \underline{c})^{-1} \cdot \prod_{i \geq 0} \mathfrak{O}\big(\operatorname{Tors}(J_{2i+1})\big) \subset \prod_{i \geq 0} \mathfrak{O}\big(\operatorname{Tors}(J_{2i})\big)$$

is proved in a similar fashion. By writing $\mathfrak{T}(\underline{C}) = x/y$, $x, y \in R$ we deduce

$$x \cdot \prod_{i \geq 0} \mathfrak{O}\big(\operatorname{Tors}(J_{2i})\big) = y \cdot \prod_{i \geq 0} \mathfrak{O}\big(\operatorname{Tors}(J_{2i+1})\big).$$

Replacing $\mathfrak{O} \to \operatorname{ord}$, and using (2.3) we deduce

$$x \cdot \prod_{i \geq 0} \operatorname{ord}\big(H_{2i}(\underline{C})\big) \sim y \cdot \prod_{i \geq 0} \operatorname{ord}\big(H_{2i+1}(\underline{C})\big),$$

so that

$$\mathfrak{T}(\underline{C}) \sim \prod_{j=0}^{m} \operatorname{ord}(H_j(\underline{C}))^{(-1)^{j+1}},$$

where we recall that $\sim$ denotes the equality up to multiplication by a unit of R. □

Definition 2.34. Suppose X is a finite CW-complex of dimension ≤ 2 and denote by F the free part of $H_1(X)$, $F := H_1(X)/\operatorname{Tors}(H_1(X))$. The *Alexander polynomial* of X is by defined by the equality

$$\Delta(X) \sim \operatorname{ord}\big(H_1\big(C^{\mathfrak{aug}}(\hat{X})\big)\big) \in \mathbb{Z}[F]$$

where $C^{\mathfrak{aug}}(\hat{X})$ is defined as in (2.2). □

Theorem 2.35 ([110, 111]). *Let X be a finite* CW*-complex with $\chi(X) = 0$ having the simple homotopy type of a finite two-dimensional cell complex. Then*

$$\mathcal{T}_X \in \mathfrak{N}_1(H_1(X))/\pm H_1(X),$$

and

$$\mathcal{T}_X^{\mathrm{aug}} \sim \begin{cases} \Delta(X) & \text{if } b_1(X) \geq 2 \\ \frac{\Delta(X)}{(t-1)} & \text{if } b_1(X) = 1, \end{cases}$$

where in the second line t denotes a generator of the free part of $H_1(X)$.

Proof. We can consider X to be a finite two-dimensional cell complex with a single zero-cell. Denote by m the number of 2-cells which we label by

$$t_1, \dots, t_m.$$

Since $\chi(X) = 0$ the number of 1-cells is $m + 1$ and we label them by

$$\ell_0, \ell_1, \dots, \ell_m.$$

Set $H := H_1(X)$ and denote the maximal Abelian cover by $\hat{X} \to X$. The $1 \times (m+1)$-matrix with entries in $\mathbb{Z}[H]$ representing the boundary operator

$$\partial : C_1(\hat{X}) \to C_0(\hat{X})$$

has the form

$$\left[\hat{\ell}_0 - 1 \;\; \hat{\ell}_1 - 1 \;\; \cdots \;\; \hat{\ell}_m - 1\right],$$

where $\hat{\ell}_i$ denotes the element of H determined by the cell ℓ_i.

The boundary operator $\partial : C_2(\hat{X}) \to C_1(X)$ is represented by a $(m + 1) \times m$, matrix D. Now choose

$$\underline{\boldsymbol{b}}_2 = \{t_1, \dots, t_m\}, \quad \underline{\boldsymbol{b}}_1 = \{\ell_0, \dots, \ell_m\} \setminus \{\ell_k\}$$

where ℓ_k is such that $\hat{\ell}_k \neq 0 \in H$. Then

$$\Delta(X) \cdot (\hat{\ell}_k - 1) \sim \det D_{(k)},$$

where $D_{(k)}$ is the matrix obtained from D by deleting the row corresponding to the cell ℓ_k. Thus

$$(1 - h)\mathcal{T}_X \in \mathbb{Z}(H), \quad \forall h \in H.$$

Using Proposition 1.27 we deduce that $\mathcal{T}_X \in \mathfrak{N}_1(H)$. The second part follows immediately from Theorem 2.31. We leave the details to the reader. □

§2.4 The Reidemeister torsion of 3-manifolds

The smooth, oriented manifolds are very special topological spaces and this special structure is reflected in their Reidemeister torsion as well. We will discuss in some detail the simplest nontrivial situation, that of 3-manifolds.

One important distinguishing characteristic of an oriented manifold is the *Poincaré duality*. J. Milnor has shown that this phenomenon has a Reidemeister torsion counterpart; see [70].

Consider a smooth, compact, oriented n-dimensional manifold M. (We do not exclude the possibility that $\partial M \neq \emptyset$.) We assume M is triangulated and we denote by $\underline{C}(M)$ the corresponding simplicial chain complex. Denote by $\underline{C}^\flat(M, \partial M)$ the dual cellular chain complex [67, §5.3], and by $\langle \bullet, \bullet \rangle$ the natural Poincaré pairing

$$\langle \bullet, \bullet \rangle : C^\flat_{n-k}(M, \partial M) \times C_k(M) \to \mathbb{Z}, \quad k = 0, \dots, n.$$

More precisely, if σ is a k-simplex of M and $D(\sigma)$ denotes is dual $(n-k)$-polyhedron then

$$\langle D(\sigma), \eta \rangle = \begin{cases} 1 & \text{if } \sigma = \eta \\ 0 & \text{if } \sigma \neq \eta. \end{cases}$$

Set $H := H_1(M)$, and denote as usual by $\hat{M} \to M$ the universal Abelian cover. The simplicial decomposition of M induces a simplicial decomposition $(\hat{\sigma})$ on $\hat{M}$. This produces a dual cellular decomposition $\big(D(\hat{\sigma})\big)$. This is equivariant in the sense that

$$D(h \cdot \hat{\sigma}) = h \cdot D(\hat{\sigma}), \quad \forall \hat{\sigma},\ h \in H.$$

$\underline{C}(\hat{M})$ has a natural structure of $\mathbb{Z}[H]$-module. Using the involution ε of $\mathbb{Z}[H]$ induced by the automorphism $h \to -h$ of the additive group H we obtain a conjugate $\mathbb{Z}[H]$-module $\underline{C}^\varepsilon(\hat{M})$. The equivariance of $\hat{\sigma} \mapsto D(\hat{\sigma})$ shows that the Poincaré pairing extends to a $\mathbb{Z}[H]$- bilinear pairing

$$\langle \bullet, \bullet \rangle : \underline{C}^\flat(\hat{M}) \times \underline{C}^\varepsilon(\hat{M}) \to \mathbb{Z}[H].$$

Set $F := H/\mathrm{Tors}\,(H)$. $\mathbb{Q}(F)$ is the field of rational functions in the variables $t_1, \dots, t_b$ $(b := b_1(M))$, and is equipped with the involution ε defined by $t_i \mapsto t_i^{-1}$. We get complexes

$$\underline{C}_{\mathfrak{aug}}(\hat{M}) := \underline{C}(\hat{M}) \otimes_{\mathbb{Z}[H]} \mathbb{Q}(F), \quad \underline{C}^\flat_{\mathfrak{aug}}(\hat{M}) := \underline{C}^\flat(\hat{M}, \partial\hat{M}) \otimes_{\mathbb{Z}[H]} \mathbb{Q}(F),$$

and a $\mathbb{Q}(F)$-bilinear ε-pairing

$$\underline{C}^\flat_{\mathfrak{aug}}(\hat{M}, \partial\hat{M}) \times \underline{C}^\varepsilon_{\mathfrak{aug}}(\hat{M}) \to \mathbb{Q}(F).$$

The Poincaré duality implies that the complex $\underline{C}^\flat$ computes the homology of the pair $(M, \partial M)$ (with various twisted coefficient systems) and the above pairing is perfect. Using the abstract duality result in Proposition 1.21 we deduce

$$\varepsilon\big(\mathcal{T}^{\mathfrak{aug}}_M\big) \cdot \big(\mathcal{T}^{\mathfrak{aug}}_{M,\partial M}\big)^{(-1)^{\dim M+1}} \sim 1. \tag{2.5}$$

In particular, if M is a 3-manifold then

$$\varepsilon\big(\mathcal{T}_M^{\mathfrak{aug}}\big) \sim \mathcal{T}_{M,\partial M}^{\mathfrak{aug}}. \tag{2.6}$$

If moreover $\chi(M) = 0$ (which means that either ∂M is empty or it is an union of tori) then we deduce from the short exact sequence of the pair $(M, \partial M)$ that

$$\mathcal{T}_M^{\mathfrak{aug}} \sim \mathcal{T}_{M,\partial M}^{\mathfrak{aug}},$$

because $\mathcal{T}_{\partial M} \sim 1$. Thus, if $\chi(M) = 0$ we have

$$\varepsilon\big(\mathcal{T}_M^{\mathfrak{aug}}\big) \sim \mathcal{T}_M^{\mathfrak{aug}}. \tag{2.7}$$

More generally, if $\mathbb{K}$ is one of the fields in the decomposition of $\mathbb{Q}(H)$ and $\pi : \mathbb{Q}(H) \to \mathbb{K}$ denotes the natural projection, then the involution ε on $\mathbb{Q}[H]$ descends to an involution on $\mathbb{K}$ we deduce that if $\mathcal{T}_M^{\pi} \in \mathbb{K} \setminus \{0\}$ then

$$\varepsilon\big(\mathcal{T}_M^{\pi}\big) \sim \mathcal{T}_{M,\partial M}^{\pi}.$$

In particular, if we (non-canonically) regard $\mathcal{T}_M$ as a function $\mathcal{T}_M : H \to \mathbb{Q}$, then the above duality statements can be rephrased as

$$\mathcal{T}_M(h) \sim \mathcal{T}_M(h^{-1})$$

meaning there exist $\epsilon = \epsilon(M, \partial M) = \pm 1$ and $h_0 \in H$ such that

$$\mathcal{T}_M(h^{-1}) = \epsilon \mathcal{T}_M(hh_0), \quad \forall h \in H. \tag{2.8}$$

Example 2.36. The smooth 3-manifolds with boundary admit cell decompositions which are simple homotopic to 2-dimensional cell complexes and thus they are covered by Theorem 2.35. Suppose M is the complement of a knot $\mathcal{K} \hookrightarrow S^3$, $M \cong S^3 \setminus \mathcal{K}$. Then $H_1(M) \cong \mathbb{Z}$ and

$$\mathcal{T}_M \sim \mathcal{T}_M^{\mathfrak{aug}} \sim \Delta_{\mathcal{K}}(t)/(1-t),$$

where $\Delta_{\mathcal{K}}(t)$ is the Alexander polynomial of $S^3 \setminus \mathcal{K}$. The duality (2.7) is equivalent to

$$\Delta_{\mathcal{K}}(t) \sim \Delta_{\mathcal{K}}(t^{-1}).$$

The polynomial $\Delta_{\mathcal{K}}(t)$ is usually referred to as the *Alexander polynomial of the knot K*.

More generally, if $\mathcal{K} \hookrightarrow S^3$ is a link with $n \geq 2$ components, then the torsion of its exterior is an element in the group algebra $\mathbb{Z}[\mathbb{Z}^n]$ called the Alexander polynomial of the link. We refer to Appendix §B.1 on methods of computing the Alexander polynomial of a knot or link in S^3. □

Exercise 2.1. Suppose $\mathcal{K} \hookrightarrow S^3$ is a *split link*, i.e. there exists an embedded $S^2 \hookrightarrow S^3$, disjoint from $\mathcal{K}$, such that each component of $S^3 \setminus S^2$ contains at least one component of $\mathcal{K}$. Prove that the Alexander polynomial of $\mathcal{K}$ is trivial. (*Hint:* Use the same strategy as in Example 2.18.) □

The Reidemeister torsion of a closed 3-manifold has special arithmetic properties. The next result, due to V. Turaev, summarizes some of them. Later on in §3.6 we will discuss more refined versions of these arithmetic properties.

Theorem 2.37 ([110, 111, 114]). *Let M be a smooth, closed, oriented, three-manifold. Then*

$$\mathcal{T}_M \in \mathfrak{N}_2(H_1(M))/\pm H_1(M)),$$

and

$$\mathcal{T}_M^{\mathfrak{aug}} \sim \begin{cases} \Delta(M) & \text{if } b_1(M) \geq 2 \\ \frac{\Delta(M)}{(t-1)^2} & \text{if } b_1(M) = 1 \\ 0 & \text{if } b_1(M) = 0. \end{cases}$$

The original proof of this theorem can be found in [110] and is based on a clever use of the gluing formula (see also [112]). For a more elementary approach, based on the definition of torsion, we refer to [114]. Corollary 1.41 implies the following result.

Corollary 2.38. *Suppose M is a 3-manifold, possibly with boundary, and let $G := H_1(M)$. If $b_1(M) > 0$ then the complex Fourier transform of the torsion of M is a holomorphic on $\tilde{G} \setminus \{1\}$. If $b_1(M) > 1$ then the complex Fourier transform is a holomorphic function on $\tilde{G}$.*

The following result generalizes the classical fact stating that the sum of the coefficients of the Alexander polynomial of a knot in S^3 is ± 1.

Theorem 2.39 (Alexander formula). *Suppose M is an oriented 3-manifold such that $b_1(M) = 1$, $r := |\mathrm{Tors}\, H_1(M)|$. Then $|\Delta_M(1)| = r$.*

Proof. Set $H := H_1(M)$, $F\colon H/\mathrm{Tors}(H) \cong \mathbb{Z}$. We will consider only the special case when $\partial M \neq \emptyset$. (For example, M is the complement of a knot in a rational homology sphere.) For the general case we refer to [112].

In this case M is simple homotopy equivalent to a 2-dimensional CW-complex X with a single 0-cell. We can assume that the closure of a one-dimensional cell is a circle which describes a generator t of F. We denote this circle by Y. From the multiplicativity properties of the torsion we deduce

$$\Delta_M(t) = \mathcal{T}_X^{\mathfrak{aug}} \cdot (t-1) = \mathcal{T}_{X,Y}^{\mathfrak{aug}}.$$

The chain complex $C^{\mathfrak{aug}}(X, Y)$ of $\mathbb{Z}[H]$ modules is very simple. It has no cells in dimensions other that 1, 2 and the torsion is given by the determinant of the boundary map

$$\partial\colon C_2^{\mathfrak{aug}}(X, Y) \to C_1^{\mathfrak{aug}}(X, Y).$$

If m denotes the number of 2-cells of (X, Y) then we can regard ∂ as a $m\times m$ matrix with entries in $\mathbb{Z}[t, t^{-1}]$. We will write $\partial(t)$ to emphasize this. The matrix $\partial(1) := \partial(t)|_{t=1}$ is the boundary map

$$C_2(X, Y) \to C_1(X, Y),$$

and

$$|\Delta_M(1)| = |\det\partial(1)| = |H_1(X, Y)| = |\operatorname{Tors} H_1(M)|. \qquad \square$$

Remark 2.40. The above argument can be significantly strengthened. More precisely, suppose M is a closed oriented 3-manifold without boundary such that $b_1(M) = 1$ which is equipped with a CW decomposition. Fix an orientation of $H_1(M, \mathbb{R})$ and choose a generator t of the free part of $H_1(M)$ compatible with the above orientation. Fix ordered bases $\boldsymbol{c}_k$ of the cellular complex $C_k(M)$, $k = 0, \dots, 3$. The Alexander polynomial of M depends on these bases and we will denote this dependence by $\Delta_M(t; \underline{\mathbf{c}})$.

The Poincaré duality on M induces a canonical orientation on $H_*(M, \mathbb{R})$ so that the canonical Euler isomorphism

$$\boldsymbol{Eul}\colon \operatorname{Det} C_*(M) \otimes \mathbb{R} \to \operatorname{Det} H_*(M, \mathbb{R})$$

can be identified with an real number

$$\epsilon(\underline{\boldsymbol{c}}) \in \{\pm 1\}.$$

Then one can show (see [114, Thm. 4.2.3]) that

$$\epsilon(\underline{\boldsymbol{c}})\Delta_M(1; \underline{\boldsymbol{c}}) = |\operatorname{Tors} H_1(M, \mathbb{Z})|.$$

We will have more to say about his issue later on in §3.5. □

§2.5 Computing the torsion of 3-manifolds using surgery presentations

One of the most efficient methods for computing the torsion of a 3-manifold is based on the following surgery formula due V. Turaev.

Theorem 2.41 (Surgery Formula, [111]). *Suppose $\mathcal{K} = \mathcal{K}_1 \cup \dots \cup \mathcal{K}_n$ is an oriented link in the interior of a compact, oriented 3-manifold X such that $\chi(X) = 0$. Denote by $U_{\mathcal{K}}$ a tubular neighborhood of $\mathcal{K}$ in X. Then the natural morphism*

$$\alpha\colon H_1(X \setminus U_{\mathcal{K}}) \to H_1(X)$$

is an epimorphism and

$$\alpha_\sharp(\mathcal{T}_{X\setminus U_{\mathcal{K}}}) = \mathcal{T}_X \cdot \prod_{i=1}^{n}(1 - [\mathcal{K}_i]), \tag{2.9}$$

where $\alpha_\sharp : \mathfrak{N}(H_1(X \setminus U_{\mathcal{K}})) \to \mathfrak{N}(H_1(X))$ *is the morphism induced by* α *described in* §1.5.

This result is an application of the multiplicativity property of the homological exact sequence of the pair $(X, X \setminus U_K)$. For more details about surgery we refer to Appendix §B.2.

Exercise 2.2. Prove Theorem 2.41. □

In applications it is much more convenient to use the Fourier transform of this formula. Set $G = H_1(X \setminus U_K)$ and $H = H_1(X)$. The morphism α induces an injection

$$\hat{\alpha} : \hat{H} \hookrightarrow \hat{G}$$

while the element $[\mathcal{K}_i] \in H$ defines via the Fourier transform a function

$$\widehat{[\mathcal{K}_i]} : \hat{H} \to \mathbb{C}, \quad \widehat{[\mathcal{K}_i]}(\chi) = \bar{\chi}([\mathcal{K}_i]), \quad \forall \chi \in \hat{H}.$$

The surgery formula can now be rewritten as

$$\hat{\alpha}^*(\hat{\mathcal{T}}_{X\setminus U_{\mathcal{K}}}) = \hat{\mathcal{T}}_X \cdot \prod_{i=1}^{n}(1 - \widehat{[\mathcal{K}_i]}). \tag{2.10}$$

The formulation (2.10) has one major advantage over the formulation in (2.9). More precisely, the product in (2.10) is the pointwise product in the algebra of complex valued functions on $\hat{H}$ and the zero divisors of are given by the functions which are zero at some point of $\hat{H}$. On the other hand, the zero divisors of the group algebra $\mathbb{C}[H]$ are much harder to detect. From (2.10) we deduce

$$\hat{\mathcal{T}}_X(\chi) = \begin{cases} \dfrac{\hat{\alpha}^*\big(\hat{\mathcal{T}}_{X\setminus U_{\mathcal{K}}}(\chi)\big)}{\prod_{i=1}^n\big(1-\bar{\chi}([\mathcal{K}_i])\big)} & \text{if } \prod_{i=1}^n(1 - \bar{\chi}([K_i])) \neq 0 \\ ? & \text{if } \prod_{i=1}^n\big(1 - \bar{\chi}([\mathcal{K}_i])\big) = 0. \end{cases} \tag{2.11}$$

We can sometime fill in the question mark above using the following simple observation.

Lemma 2.42. *Suppose* $b_1(X) > 0$ *and the homology classes* $[\mathcal{K}_1], \dots, [\mathcal{K}_n]$ *have infinite orders in* $H_1(X)$. *Then* (2.11) *uniquely determines* $\hat{\mathcal{T}}_X$.

Proof. Set $\tilde{H} = \mathrm{Hom}(H, \mathbb{C}^*)$. The *complex* Fourier transform $\tilde{\mathcal{T}}_X$ of $\mathcal{T}_X$ is holomorphic on $\tilde{H} \setminus \{1\}$ if $b_1(X) = 1$, and in fact holomorphic on $\tilde{H}$ if $b_1(X) > 1$. If the homology classes $[\mathcal{K}_i]$ have infinite orders in H for all $i = 1, \dots, n$ then the zero set

$$Z := \Big\{\chi \in \hat{H} \setminus \{1\};\ \prod_{i=1}^{n}\big(1 - \chi^{-1}([\mathcal{K}_i])\big) = 0\Big\}$$

is an analytic subvariety of $\tilde{H} \setminus \{1\}$ of *positive codimension*. This implies that $\tilde{H} \setminus Z$ is *dense* in $\tilde{H}$. The function $\tilde{\mathcal{T}}_X$ is unambiguously defined on $\tilde{H} \setminus Z$ by (2.11), and admits an unique holomorphic extension to $\tilde{H} \setminus \{1\}$. □

In the remainder of this section we will describe through examples how the above results work in concrete applications.

Example 2.43 (Trivial circle bundles over Riemann surfaces). Suppose $X = S^1 \times \Sigma_g$, where Σ_g is an oriented Riemann surface of genus g. We will consider separately three cases.

1. $g = 0$, $X = S^1 \times S^2$. Then X is obtained by gluing two solid tori along their boundaries via the tautological identification. This shows that $H_1(X) = \mathbb{Z}$ and using the surgery formula we deduce

$$\mathcal{T}_X = (1 - t)^{-2}.$$

2. $g = 1$, $X = T^3$. In this case we have

$$\mathcal{T}_X \sim 1.$$

3. $g \geq 2$. For simplicity we consider only the case $g = 2$. The manifold $S^1 \times \Sigma$ can be obtained from two copies of $S^1 \times T^2$ using the fiber connect sum operation. More precisely, consider two copies Y_1, Y_2 of the complement of a tiny open disk $D \subset T^2$ and set

$$X_i := S^1 \times Y_i.$$

Then

$$T^3 = D \times S^1 \cup X_i,$$

and the inclusion induced morphism $\alpha\colon H_1(X_i) \to H_1(T^3)$ is an isomorphism. From the surgery formula (2.10) we deduce that

$$\mathcal{T}_{X_i} \sim \mathcal{T}_{T^3} \cdot (1 - t) \sim (1 - t),$$

where t denotes the homology class carried by the fiber. We have the decomposition

$$X := S^1 \times \Sigma_2 = X_1 \cup X_2, \quad X_0 := X_1 \cap X_2 \cong T^2.$$

Denote by $\boldsymbol{j}_k \colon H_1(X_k) \to H_1(X)$, $k = 0, 1, 2$, the inclusion induced morphisms. The Mayer–Vietoris gluing formula implies

$$\mathcal{T}_X^{\boldsymbol{j}_0} \cdot \mathcal{T}_{X_0} \sim \mathcal{T}_{X_1}^{\boldsymbol{j}_1} \cdot \mathcal{T}_{X_2}^{\boldsymbol{j}_2} = (1-t)^2. \tag{2.12}$$

This shows $\boldsymbol{j}_0 \neq 0$. We conclude that $\boldsymbol{j}_0 \mathcal{T}_{X_0} = 1$ and

$$\mathcal{T}_X = (1-t)^2.$$

For Riemann surfaces of genus g we have

$$\mathcal{T}_{S^1 \times \Sigma_g} \sim (\varphi - 1)^{2g-2}.$$

This is in perfect agreement with the computations in Example 2.7. □

Example 2.44 (Nontrivial circle bundles over Riemann surfaces). Consider a degree ℓ circle bundle

$$S^1 \hookrightarrow N_\ell \to \Sigma$$

over a Riemann surface of genus $g \geq 0$. N_ℓ can be obtained form N_0 using the following procedure.

• Remove a tubular neighborhood U of a fiber of N_0 so that $U \cong S^1 \times D^2$ and set

$$V := N_0 \setminus U \cong S^1 \times (\Sigma \setminus D^2).$$

• Orient ∂U using the obvious diffeomorphism

$$\partial U \cong \partial D^2 \times S^1.$$

Observe that the above diffeomorphism produces a canonical basis of $H_1(\partial U)$ and hence an identification with $\mathbb{Z}^2$. Similarly, orient ∂V using the obvious diffeomorphism

$$\partial V \cong \partial(\Sigma \setminus D^2) \times S^1.$$

This diffeomorphism produces a natural basis of $H_1(\partial V)$ and thus an identification with $\mathbb{Z}^2$. As in the previous example we deduce

$$\mathcal{T}_V \sim (\varphi - 1)^{2g-3}$$

where φ denotes the homology class carried by the fiber.

• Glue back U to V using the attaching map

$$\Gamma_\ell := \begin{bmatrix} -1 & 0 \\ -\ell & 1 \end{bmatrix}.$$

To obtain more explicit results we need to rely on the Mayer–Vietoris sequence.

Denote by α the natural morphism $\alpha\colon H_1(V)\to H_1(N_\ell)$, and set $G=H_1(N_\ell)$. Denote by β the natural morphism $H_1(U)\to H_1(N_\ell)$. First, notice that $H_2(N_\ell)$ is torsion free, of rank $2g$ and $H_2(V)$ is torsion free of rank $2g$, so that we can split the following short exact portion off the Mayer–Vietoris sequence

$$0\to H_1(T^2)\xrightarrow{j_U\oplus j_V}\begin{matrix}H_1(U)\\ \oplus\\ H_1(V)\end{matrix}\xrightarrow{\beta-\alpha}H_1(N_\ell)\to 0.$$

Denote by c_1, c_2 the natural generators of T^2, and by $\varphi, x_1,\dots,x_{2g}$ the natural generators of $H_1(V)$ (so that φ corresponds to the fiber and the x_i's generate $H_1(\Sigma\setminus D^2)$). Also, denote by y the natural generator of $H_1(U)$. Finally, denote by I_U and respectively I_V the morphisms

$$I_U\colon H_1(\partial U)\to H_1(U),\quad I_V\colon H_1(\partial V)\to H_1(V).$$

In terms of the above generators we have

$$I_V(c_1)=0,\quad I_V(c_2)=\varphi,\quad I_U(c_1)=0,\quad I_U(c_2)=y.$$

Then $\boldsymbol{j}_U=I_U\circ\Gamma_\ell^{-1}$, $\boldsymbol{j}_V=I_V$, so that

$$\boldsymbol{j}_V\colon c_1\mapsto 0,\quad c_2\mapsto y.$$

Since $\Gamma_\ell^{-1}=\Gamma_\ell$ we deduce

$$\boldsymbol{j}_U\colon c_1\mapsto -c_1-\ell c_2\mapsto -\ell\varphi,\quad c_2\mapsto c_2\mapsto\varphi.$$

Using the bases $\{c_1,c_2\}$ in $H_1(T^2)$ and $\{y;\varphi,x_1,\dots,x_{2g}\}$ in $H_1(U)\oplus H_1(V)$ we deduce that $\boldsymbol{j}_U\oplus\boldsymbol{j}_V$ has the $(2g+2)\times 2$-matrix description

$$A:=\begin{bmatrix}0&1\\ -\ell&1\\ 0&0\\ \vdots&\vdots\\ 0&0\end{bmatrix}. \tag{2.13}$$

Denote by Λ the sublattice of $\mathbb{Z}^{2g+2}$ generated by $\ell\varphi$, $y+\varphi$. Since $\ker(\beta-\alpha)=\Lambda$ we deduce

$$\alpha(\varphi^\ell)=0\quad\text{and}\quad \boldsymbol{k}:=\beta(y)=\alpha(\varphi)^{-1}.$$

Re-label $\varphi:=\alpha(\varphi)$. φ generates the torsion part of $H_1(N_\ell)$ which is a cyclic subgroup of order $|\ell|$. Moreover, $\boldsymbol{k}=\varphi^{-1}$ and using (2.11) we deduce

$$\hat{\mathfrak{T}}_{N_\ell}(\chi)\sim\begin{cases}\left(\bar{\chi}(\varphi)-1\right)^{2g-2} & \text{if } \chi(\varphi)\neq 1\\ ? & \text{if } \chi(\varphi)=1.\end{cases}$$

Observe that Lemma 2.42 is unapplicable in this case since φ is a torsion class. When $g = 0$, then N_ℓ can be identified with the lens space $L(1, -\ell)$, and the only character χ such that $\chi(\varphi) = 1$ is the trivial character. In this case $H_1(N_\ell)$ is a torsion group, and we have (see Theorem 2.37 of §2.4)

$$\hat{\mathfrak{T}}_{N_\ell}(1) = 0.$$

To complete the determination of $\mathfrak{T}_{N_\ell}$ for $g > 0$ we will rely on a twisted version of the Gysin sequence.

Consider a nontrivial character $\chi\colon H_1(N_\ell) \to S^1$ such that $\chi(\varphi) = 1$. The usual Gysin sequence of the fibration $N_\ell \to \Sigma$ implies that χ factors through the morphism $\pi_*\colon H_1(N_\ell) \to H_1(\Sigma)$, i.e. there exists a nontrivial character $\tilde{\chi}\colon H_1(\Sigma) \to S^1$ such that the diagram below is commutative:

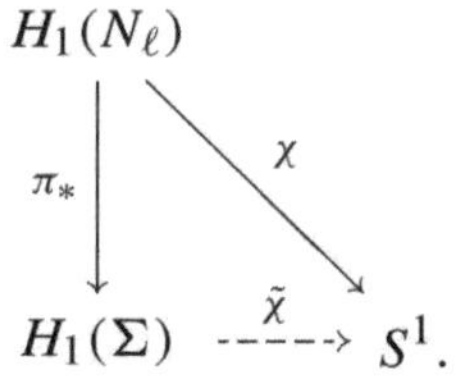

The induced map $\chi\colon \mathbb{Z}[H_1(N_\ell)] \to \mathbb{C}$ defines a system of local coefficients on N_ℓ which we denote by $\mathbb{C}_\chi$. Since χ factors through the morphism π_* we deduce that it defines a system of local coefficients on the total space $\tilde{X}$ of the associated disk bundle. We denote this induced local system by $\mathbb{C}_{\tilde{\chi}}$. Using the Poincaré duality we deduce that we have isomorphisms

$$H_k(\tilde{X}, N_\ell; \mathbb{C}_\chi) \cong H^{4-k}(\tilde{X}, \mathbb{C}_{\tilde{\chi}}) \cong H^{4-k}(\Sigma, \mathbb{C}_{\tilde{\chi}}) \cong H_{k-2}(\Sigma, \mathbb{C}_{\tilde{\chi}}).$$

The homological long exact sequence of the pair $(\tilde{X}, N_\ell)$ can now be rewritten

$$\cdots \to H_k(N_\ell, \mathbb{C}_\chi) \to H_k(\Sigma, \mathbb{C}_{\tilde{\chi}}) \to H_{k-2}(\Sigma, \mathbb{C}_{\tilde{\chi}}) \to \cdots . \tag{2.14}$$

Set $b_k(\chi) := \dim_{\mathbb{C}} H_k(\Sigma, \mathbb{C}_{\tilde{\chi}})$ and $e(\chi) = b_0(\chi) - b_1(\chi) + b_2(\chi)$. Then $e(\chi)$ is independent of χ and we have

$$e(\chi) = e(1) = 2 - 2g, \quad \forall \chi.$$

On the other hand, when $\chi \neq 1$ then $b_0(\chi) = b_2(\chi) = 0$. This can be seen as follows. The 0-th cohomology space is naturally identified with the space global sections of the locally constant sheaf $\mathbb{C}_{\tilde{\chi}}$. Since χ is nontrivial there are no such sections. Hence

$$b_0(\chi) = 0, \quad \forall \chi \neq 1.$$

On the other hand we have a Hodge–DeRham duality

$$\wedge\colon H^2(\Sigma, \mathbb{C}_{\tilde{\chi}}) \times H^0(\Sigma, \mathbb{C}_{\tilde{\chi}^{-1}}) \to H^2(\Sigma, \mathbb{C})$$

so that

$$b_2(\tilde{\chi}) = b_0(\tilde{\chi}^{-1}) = 0, \quad \forall \chi \neq 1.$$

Thus $b_0(\chi) = b_2(\chi) = 0, \forall \chi \neq 1$ and since $e(\chi) = 2 - 2g$ we deduce

$$b_1(\chi) = 2 - 2g, \quad \forall \chi \neq 1.$$

Using this information in the fragment $k = 1$ of the sequence (2.14) we deduce that we have a surjection

$$H_1(N_\ell, \mathbb{C}_\chi) \twoheadrightarrow H_1(\Sigma, \mathbb{C}_{\tilde{\chi}}).$$

In particular, if $g > 1$ we deduce $H_1(N_\ell, \mathbb{K}_\chi) \neq 0$ so that in this case

$$\hat{\mathfrak{T}}_{N_\ell}(\chi) = 0, \quad \forall \chi \neq 1, \; \chi(\varphi) = 1.$$

When $g = 1$ then $H_*(N_\ell, \mathbb{C}_\chi) = 0$ and from the sequence (2.14) we deduce

$$\hat{\mathfrak{T}}_{N_\ell}(\chi) \sim 1.$$

Later in Example 2.57 we will explain how to extend the above technique to the more general case of Seifert manifolds. For a different approach we refer to [31, 35]. We also want to refer to [34, §1,2] where it is explained how to obtain information about the torsion of the total space of an arbitrary fiber bundle. □

It is known (see [41, 92, 96]) that any 3-manifold can be obtained from S^3 by an integral Dehn surgery on an oriented link in S^3. A description of a 3-manifolds as a Dehn surgery on a link is known as a *surgery presentation* of a 3-manifold and it is a very convenient way of operating with 3-manifolds. Many topological invariants can be algorithmically read off a surgery presentation. We will spend the remainder of this section explaining how to obtain almost complete information about the Reidemeister torsion using surgery presentations. For the very basics concerning Dehn surgery we refer to Appendix §B.2 which we follow closely as far as the terminology and the orientation conventions are concerned. For a more in depth look at this important topological operation we refer to [41, 92, 96].

Suppose $\mathcal{K} = \mathcal{K}_1 \cup \cdots \cup \mathcal{K}_n$ is an n-component oriented link in S^3. We denote by E the complement of a tubular neighborhood U of $\mathcal{K} \hookrightarrow S^3$. Then E is an oriented 3-manifold with boundary. Moreover, $H_1(E)$ is a free Abelian group of rank n, and the collection of oriented meridians

$$\{\mu_i; \; i = 1, \ldots, n\}$$

defines an integral basis. We denote by λ_i the canonical longitude of $\mathcal{K}_i \hookrightarrow S^3$ oriented such that

$$\lambda_i \cdot \mu_i = 1,$$

where the above intersection pairing is defined using the canonical orientation of ∂E as boundary of E. $H_1(\partial E)$ is a free Abelian group of rank $2n$ and the collection

$$\{\lambda_i, \mu_j;\ i, j = 1, \dots n\}$$

is an integral basis. Set

$$\ell_{ij} = \boldsymbol{Lk}(\mathcal{K}_i, \mathcal{K}_j) = \boldsymbol{Lk}(\mathcal{K}_j, \mathcal{K}_i),$$

where $\boldsymbol{Lk}$ denotes the $\mathbb{Z}$-valued linking number of two disjoint knots in S^3. Then the inclusion induced morphism

$$j\colon H_1(\partial E) \to H_1(E)$$

is described in the above bases by

$$\mu_j \mapsto \mu_j, \quad \lambda_i \mapsto \sum_{k \neq i} \ell_{kj}\mu_k.$$

We perform an integral Dehn surgery on this link with coefficients $\vec{d} = (d_1, \dots, d_n) \in \mathbb{Z}^n$. The attaching curves of this surgery are

$$c_i = d_i\mu_i + \lambda_i, \quad i = 1, \dots n.$$

Denote resulting manifold by $M_{\vec{d}}$. The natural morphism

$$\alpha\colon H_1(E) \to H_1(M_{\vec{d}}, \mathbb{Z})$$

is onto, and leads to the following presentation of $H_1(N, \mathbb{Z})$

$$H_1(\partial E) \supset \operatorname{span}_{\mathbb{Z}}\{c_i;\ i = 1, \dots, n\} \xrightarrow{j} H_1(E) \twoheadrightarrow H_1(M_{\vec{d}}) \to 0,$$

$$c_i \mapsto d_i\mu_i + \sum_{k \neq i} \ell_{ki}\mu_k.$$

We denote by $P = P_{\vec{d}}$ the symmetric $n \times n$ matrix with entries

$$p_{ij} = \begin{cases} \ell_{ij} & \text{if } i \neq j \\ d_i & \text{if } i = j. \end{cases}$$

The above presentation can be rewritten in the computationally friendly from

$$\mathbb{Z}^n \xrightarrow{P_{\vec{d}}} \mathbb{Z}^n \twoheadrightarrow H_1(M_{\vec{d}}) \to 0. \tag{2.15}$$

The cores of the attaching solid tori define homology classes in $M_{\vec{d}}$ which for simplicity we denote by $\boldsymbol{k}_i$. Algebraically, these homology classes are the images via α of

$K_i \in H_1(E)$, $i = 1, \dots, n$, where $K_i = \boldsymbol{j}\lambda'_i$, and $\lambda'_i \in H_1(E)$ are homology classes determined by the conditions

$$\lambda'_j \cdot c_i = \delta_{ij}, \quad i, j = 1, \dots, n.$$

For example, we can pick

$$\lambda'_j = -\mu_j.$$

To compute the torsion of $M_{\vec{d}}$ we use the following consequence of Turaev's surgery formula Theorem 2.41.

Corollary 2.45.

$$\alpha_\sharp(\mathfrak{T}_E) \sim \mathfrak{T}_{M_{\vec{d}}} \cdot \prod_{i=1}^{n} (1 - \boldsymbol{k}_i), \tag{2.16}$$

where $\alpha_\sharp \colon \mathfrak{N}(H_1(E)) \to \mathfrak{N}(H_1(M_{\vec{d}}))$ *is the morphism induced by* α *described in* §1.5.

As explained before, it is convenient to use the Fourier transform trick described in §1.6. To ease the presentation we set

$$G := H_1(E), \quad H = H_1(M_{\vec{d}}).$$

Then $\alpha \colon G \to H$ is an epimorphism, and by passing to duals we get an injection

$$\hat{\alpha} \colon \hat{H} \hookrightarrow \hat{G}.$$

We can thus view $\hat{H}$ as a subgroup of $\hat{G}$. The Fourier transform of $\mathfrak{T}_E$ is a (generalized) function $\hat{\mathfrak{T}}_E$ on $\hat{G}$, and the Fourier transform of $\mathfrak{T}_{M_{\vec{d}}}$ is a (generalized) function $\hat{\mathfrak{T}}_{M_{\vec{d}}}$ on $\hat{H}$. Then (2.16) becomes the linear equation

$$\hat{\mathfrak{T}}_E(\chi) = \hat{\mathfrak{T}}_{M_{\vec{d}}}(\chi) \cdot \prod_{i=1}^{n} \big(1 - \bar{\chi}(K_i)\big), \quad \forall \chi \in \hat{H} \hookrightarrow \hat{G}. \tag{2.17}$$

The homology classes $\boldsymbol{k}_i \in H$ are represented by the vectors

$$K_i = -\mu_i \in G \cong \mathbb{Z}^n.$$

For each $\chi \in \hat{H}$ we set $\chi_i := \chi([\mu_i])$, where $[\mu_i] := \alpha(\mu_i) \in H$. We can now rewrite (2.16) as

$$\hat{\mathfrak{T}}_E(\chi_1, \dots, \chi_n) = \hat{\mathfrak{T}}_{M_{\vec{d}}}(\chi_1, \dots, \chi_n) \cdot \prod_{i=1}^{n} \big(1 - \chi_i\big). \tag{2.18}$$

The next example will illustrate the strengths and limitations of the above surgery formula.

Example 2.46 (Surgery on the Borromean rings). Consider the Borromean rings depicted in Figure 2.4. Denote by E the complement of this link. This link has the property that any two of its components are unlinked unknots. Hence $\ell_{ij} = 0$, $\forall i, j = 1, 2, 3$. However, this link is nontrivial since its Alexander polynomial is (see [8])

$$\mathcal{T}_E \sim (\mu_1 - 1)(\mu_2 - 1)(\mu_3 - 1).$$

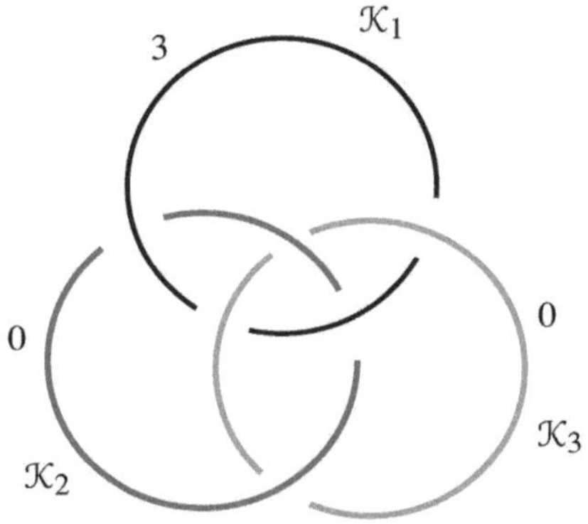

Figure 2.4. Surgery on the Borromean rings.

Set $M = M_{(3,0,0)}$. Note that $\boldsymbol{j}$ has the simple form

$$\mu_i \mapsto \mu_i, \quad \lambda_i \mapsto 0.$$

The attaching curves of this surgery are

$$c_1 = 3\mu_1 + \lambda_1, \quad c_2 = \lambda_2, \quad c_3 = \lambda_3.$$

We can pick

$$K_1 = -\mu_1, \quad K_2 = -\mu_2, \quad K_3 = -\mu_3.$$

The first homology group of M has the presentation

$$\mathbb{Z}^3 \xrightarrow{P} \mathbb{Z}^3 \twoheadrightarrow H_1(M) \to 0$$

where P is the 3×3 matrix

$$P = \begin{bmatrix} 3 & 0 & 0 \\ 0 & 0 & 0 \\ 0 & 0 & 0 \end{bmatrix}.$$

Thus $H = H_1(M) = \mathbb{Z}_3 \oplus \mathbb{Z}^2$ with generators $[\mu_1]$, $[\mu_2]$, $[\mu_3]$. We deduce

$$\hat{H} = U_3 \times (S^1)^2, \quad U_m := \{z \in \mathbb{C}^*;\ z^m = 1\},$$

and (2.18) becomes

$$(\chi_1 - 1)(\chi_2 - 1)(\chi_3 - 1) \sim \hat{\mathfrak{T}}_M(\chi_1, \chi_2, \chi_3)(1 - \chi_1)(1 - \chi_2)(1 - \chi_3),$$

$\forall \chi_1, \chi_2, \chi_3 \in S^1$, $\chi_1^3 = 1$. We deduce

$$\hat{\mathfrak{T}}_M(\chi_1, \chi_2, \chi_3) \sim 1, \quad \forall \chi_1 \neq 1,\ \chi_2 \neq 1,\ \chi_3 \neq 1.$$

Fix $\chi_1 \neq 1$. Then $\hat{\mathfrak{T}}_M(\chi_1, \chi_2, \chi_3)$ is a Laurent polynomial in χ_2, χ_3 so that we deduce

$$\hat{\mathfrak{T}}_M(\chi_1, \chi_2, \chi_3) \sim 1, \quad \forall \chi_1, \chi_2, \chi_3,\ \chi_1 \neq 1.$$

We notice that the surgery formula contains no information about $\hat{\mathfrak{T}}_M(1, \chi_2, \chi_3)$. This is the Fourier transform of $\mathfrak{T}_M^{\mathfrak{aug}}$, which according to Theorem 2.37 is the Alexander polynomial of M. □

Motivated by Lemma 2.42 we isolate a special class of surgeries.

Definition 2.47. A closed 3-manifold M satisfying $b_1(M) > 0$ is said to admit a *nondegenerate surgery presentation* if there exists an oriented link $\mathcal{K} = \mathcal{K}_1 \cup \cdots \cup \mathcal{K}_n \subset S^3$, and surgery coefficients $\vec{d} \in \mathbb{Z}^n$ such that if $M \cong M_{\vec{d}}$, and all the homology classes $\boldsymbol{k}_j$ have infinite orders in $H_1(M_{\vec{d}})$. □

The nondegenerate surgeries can be easily recognized using the following elementary algebraic result.

Proposition 2.48. *Consider an n-component oriented link $\mathcal{K} \hookrightarrow S^3$, and a vector $\vec{d} \in \mathbb{Z}^n$. Set $G := H_1(S^3 \setminus \mathcal{K})$. The surgery defined by the coefficients $\vec{d}$ is nondegenerate if and only if for every $i = 1, \ldots, n$ there exists $\vec{w}_i \in \mathrm{Hom}(G, \mathbb{Z})$ such that*

$$\langle \vec{w}_i, P_{\vec{d}}\mu_j \rangle = 0, \quad \forall j = 1, \ldots, n, \text{ and } \langle \vec{w}_i, \mu_i \rangle \neq 0.$$

Exercise 2.3. Prove Proposition 2.48. □

The proof of the following result is a simple exercise in Kirby calculus ([115, §3.8]).

Proposition 2.49. *Any 3-manifold M with $b_1(M) > 0$ admits a nondegenerate surgery presentation.*

Example 2.50. To fix the "deficiency" of the surgery described in Example 2.46 we slide $\mathcal{K}_2$ over $\mathcal{K}_1$. The link and the surgery coefficients change as indicated in Figure 2.5. This surgery is nondegenerate and produces the same 3-manifold as the surgery in Example 2.46. The new problem we are facing is the computation of the Alexander polynomial of the new link. We leave the quite unpleasant computation to the reader.

□

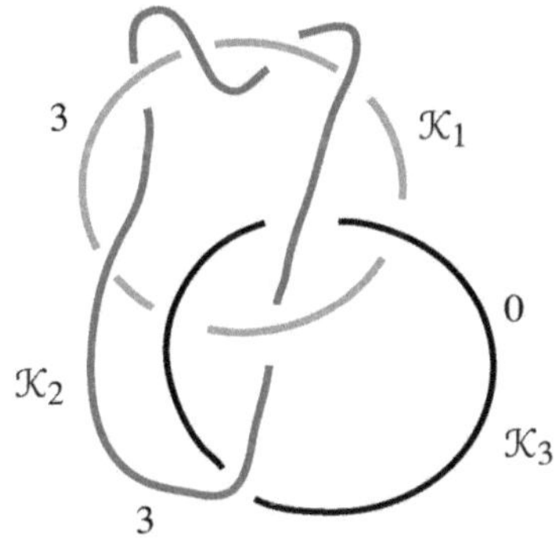

Figure 2.5. The effect of a Kirby move.

The above discussion shows that we can compute the torsion of any 3-manifold with $b_1 > 0$ provided we have a way to compute the Alexander polynomials of links in S^3. There are algorithms for computing Alexander polynomials of links (see e.g. §B.1 and the references therein), but this may not always be a pleasant task. We will take up this issue again in the next chapter and explain how to extend the above consideration to rational homology 3-spheres.

Example 2.51. We want to illustrate the above observations by computing the Reidemeister torsion of the 3-manifold obtained by Dehn surgery on the two component link depicted in Figure 2.6. The torsion of the complement E of the link is the Alexander polynomial of this link which was computed in [26] and is

$$\Delta_{\mathcal{K}}(\mu_1, \mu_2) = (1 - \mu_1 + \mu_1^2)\big((1 - 2\mu_1) - \mu_1^2\mu_2(2 - \mu_1)\big).$$

The linking number of these two knots with the orientations indicated in the figure is $\boldsymbol{Lk}(\mathcal{K}_1, \mathcal{K}_2) = 2$.

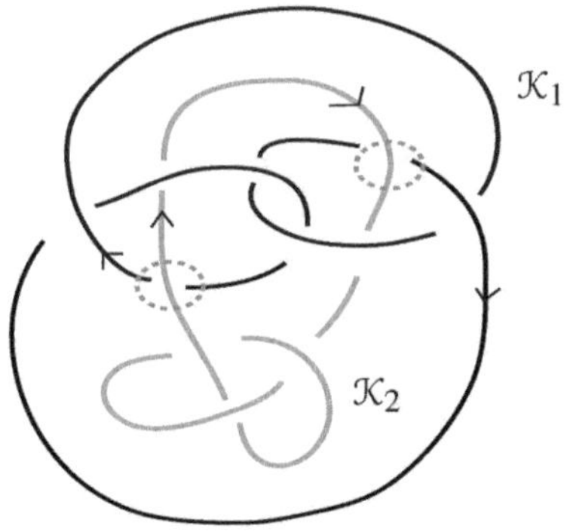

Figure 2.6. A two-component link.

We perform a (4, 1) surgery on this link, and we denote by M the resulting 3-manifold. The group $H = H_1(M)$ admits the presentation

$$F_1 = \operatorname{span}_{\mathbb{Z}}(c_1, c_2) \xrightarrow{P} F_0 = \operatorname{span}_{\mathbb{Z}}(\mu_1, \mu_2) \to H \to 0,$$

where

$$P = \begin{bmatrix} 4 & 2 \\ 2 & 1 \end{bmatrix}$$

Also,

$$K_1 = -\mu \in F_1, \quad K_2 = -\mu_2 \in F_1.$$

Using the `MAPLE` procedure ismith we obtain

$$\begin{bmatrix} 1 & 0 \\ 0 & 0 \end{bmatrix} = UPV, \quad U = \begin{bmatrix} 0 & 1 \\ 1 & -2 \end{bmatrix}, \quad V = \begin{bmatrix} 0 & -1 \\ 1 & 2 \end{bmatrix}$$

This means that the bases $\{Vc_1, Vc_2\}$ in F_0 and $\{U^{-1}\mu_1, U^{-1}\mu_2\}$ diagonalize the presentation matrix P. The coordinates of μ_1 and μ_2 in the new basis are

$$\mu_1 \to \begin{bmatrix} 0 \\ 1 \end{bmatrix}, \quad \mu_2 \to \begin{bmatrix} 1 \\ 2 \end{bmatrix}.$$

We denote by g the generator in H. We deduce that

$$\boldsymbol{k}_1 = -g, \quad \boldsymbol{k}_2 = 2g,$$

so that both these homology classes have infinite orders. In other words, this surgery is nondegenerate. Let $\chi \in \hat{H} \setminus \{1\}$. The surgery formula (2.17) becomes

$$(1 - \chi + \chi^2) \cdot \big((1 - 2\chi) - \chi^2\chi^{-2}(2 - \chi)\big) \sim \hat{\mathfrak{T}}_M(\chi) \cdot (1 - \chi^{-1})(1 - \chi^2)$$

so that

$$(1 - \chi + \chi^2)(1 + \chi) \sim \hat{\mathfrak{T}}_M(\chi)(1 - \chi)^2(1 + \chi)$$

which implies

$$\hat{\mathfrak{T}}_M(\chi) \sim \frac{1 - \chi + \chi^2}{(1 - \chi)^2}.$$

□

Exercise 2.4. Compute the torsion of the 3-manifold obtained from the Dehn surgery depicted in Figure 2.5. □

§2.6 Plumbings

In this section we want to describe a special yet large class of 3-manifolds, and then outline a method for computing their torsions. These 3-manifolds, known as *graph manifolds*, or *plumbings*, play an important role in the study of isolated singularities of complex surfaces. They are all obtained by gluing elementary pieces of the form $S^1 \times \Sigma$, where Σ is a surface with boundary. We begin by describing a combinatorial method of cataloging these manifolds.

Start with a graph G. We denote by V the set of vertices and by E the set of edges. The edges are oriented. We do not exclude the possibility that the graph has tails[2]. We denote by T the set of tails, so that $T \cap E = \emptyset$. Multiple edges connecting the same pair of vertices or loops are also allowed. For each edge e we denote by $v_\pm(e)$ the final/initial point. For each vertex we denote by $E_v^\pm$ the set of outgoing/incoming edges, and by T_v the set of tails at v. Set $E_v = E_v^+ \cup E_v^-$, $\deg_\pm v = |E_v^\pm|$, $t_v = |T_v|$, and $\deg v = \deg_+ v + \deg_- v + t_v$.

A *decoration* of G is a function

$$\Gamma : E \to \mathrm{SL}_2(\mathbb{Z}), \quad e \mapsto \Gamma(e).$$

A *weight* on G is a function $g : V \to \mathbb{Z}_{\geq 0}$. Denote by C the 2×2 matrix

$$C = \begin{bmatrix} -1 & 0 \\ 0 & 1 \end{bmatrix}.$$

Suppose (G, V, E, T, g, Γ) is a weighted decorated graph. We construct a 3-manifold with boundary as follows.

• Associate to each $v \in V$ a Riemann surface Σ_v with $\deg v$ boundary components. Fix a bijection between the components of $\partial \Sigma_v$ and $E_v \cup T_v$. Set $M_v := S^1 \times \Sigma_v$, and denote by Θ_v the fiber of this trivial fibration over Σ_v.

• For each $v \in V$ and each $e \in E_v$ fix an orientation preserving diffeomorphism between the component $\partial_e M_v$ of M_v and the oriented standard torus $\Theta_v \times \partial_e \Sigma_v$.

• For each edge $e \in E$ glue the torus $\partial_e M_{v_-(e)}$ to the torus $\partial_e M_{v_+(e)}$ using the *orientation reversing* diffeomorphism which is described by the matrix $C \circ \Gamma(e)$ with respect to the oriented bases

$$\big([\partial_e \Sigma_{v_-(e)}], [\Theta_{v_-(e)}]\big), \quad \big([\partial_e \Sigma_{v_+(e)}], [\Theta_{v_+(e)}]\big)$$

of $H_1(\partial_e M_{v_\pm(e)})$.

After the above identifications we obtain an oriented 3-manifold with $|T|$ boundary components. We denote it by $M(G)$ and we will say that M is a *generalized plumbing* described by the weighted decorated graph G. G is also known as the *plumbing graph*.

[2]A tail is an arc with one boundary component a vertex of the graph, while the other is free.

This is related to the traditional plumbing construction described in [46, §8] where the decorations have the special form

$$\Gamma(e) = \pm P_{m,n} := \pm \begin{bmatrix} m & 1 \\ mn-1 & n \end{bmatrix} = \pm T_n \Phi T_m,$$

where

$$\Phi = \begin{bmatrix} 0 & 1 \\ -1 & 0 \end{bmatrix}, \quad T_k = \begin{bmatrix} 1 & 0 \\ k & 1 \end{bmatrix}, \quad \forall k \in \mathbb{Z}.$$

To see this, note that

$$CP_{m,n} = \begin{bmatrix} -1 & 0 \\ -n & 1 \end{bmatrix} \cdot \begin{bmatrix} 0 & 1 \\ 1 & 0 \end{bmatrix} \begin{bmatrix} -1 & 0 \\ -m & 1 \end{bmatrix},$$

and the expression in the right hand side is exactly the description of the attaching map in [46, p.67]. Observe also that $\Phi^2 = -1$ and $T_m \circ T_n = T_{m+n}, \forall m, n \in \mathbb{Z}$.

Despite its name, the generalized plumbing construction does not produce more manifolds that the usual plumbing. To see this we will describe a few simple methods of simplifying the combinatorics of a decorated graph G which do not affect the topology of $M(G)$. Assume for simplicity that there are no tails.

If $G(V, E, \{M_v\}, \Gamma)$ is a weighted, decorated graph we define its *conjugate* with respect to a subset $S \subset E$ to be the graph $\bar{G}_S(V, \bar{E}_S, S, \bar{\Gamma}_S)$ such $\bar{G}_S$ has the same edges as G but the ones in S have opposite orientations while the others are unchanged. Moreover

$$\bar{\Gamma}_S(e) = \begin{cases} \Gamma_e & \text{if } e \in E \setminus S \\ C\Gamma_e^{-1}C & \text{if } e \in S \end{cases}$$

Observe that $CT_m^{-1}C = T_m$ and $CP_{m,n}^{-1}C = P_{n,m}$.

Proposition 2.52. *For any weighted decorated graph $G(V, E, \{M_v\}, \Gamma)$ and any subset $S \subset E$ the generalized plumbings $M(G)$ and $M(\bar{G}_S)$ are diffeomorphic.*

Exercise 2.5. Prove the above proposition. □

Figures 2.7 and 2.8 represent pairs of conjugate weighted decorated graphs describing in one instance a circle bundle over a Riemann surface and in the second instance a simple plumbing.

Figure 2.7. A degree d circle bundle over a Riemann surface of genus g.

Figure 2.8: Plumbing a degree m bundle over a Riemann surface of genus g_0 with a degree n bundle over a genus g_1 Riemann surface.

Suppose (G, V, E, g, Γ) is a decorated graph and v_0 is a vertex of G of genus 0 as depicted in Figure 2.9. The *concatenation* of G at v_0 is the decorated graph G_{v_0} obtained via the transformation of the graph G depicted in Figure 2.9. It consists of replacing the two edges at v_0 decorated by $\Gamma_\pm$ by a single edge connecting the two neighbors of v_0 by a single edge decorated by $\Gamma_+ \cdot \Gamma_-$. The following result is now obvious.

Proposition 2.53. *The manifolds $M(G)$ and $M(G_{v_0})$ are diffeomorphic.*

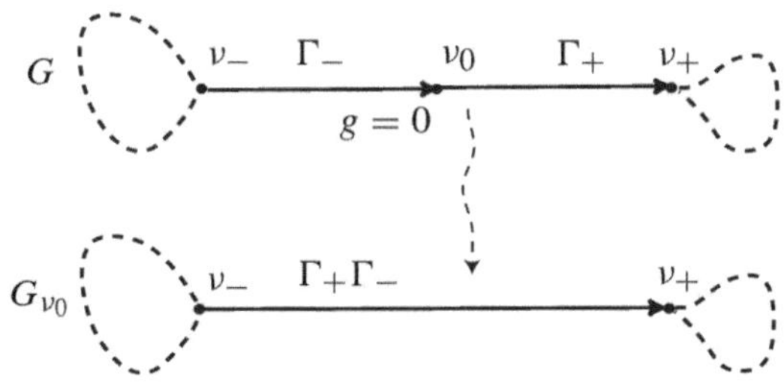

Figure 2.9. Concatenation.

To proceed further we need the following algebraic result.

Proposition 2.54. *Denote by $\mathfrak{G} \subset \mathrm{SL}_2(\mathbb{Z})$ the semigroup generated by the matrices $P^+_{n,m}$, $n, m \in \mathbb{Z}$. Then*

$$\mathfrak{G} = \mathrm{SL}_2(\mathbb{Z}).$$

Proof. Since $\Phi^2 = -1$ we deduce that $-1 = \Phi^2 \in \mathfrak{G}$. In particular

$$\Phi^{-1} = -\Phi \in \mathfrak{G}.$$

Observe next that

$$S = \begin{bmatrix} 1 & 1 \\ 0 & 1 \end{bmatrix} = P^+_{1,1} \in \mathfrak{G} \quad \text{and} \quad S^{-1} = \begin{bmatrix} 1 & -1 \\ 0 & 1 \end{bmatrix} = \Phi^2 P^+_{-1,-1} \in \mathfrak{G}.$$

Thus $\mathfrak{G}$ contains the semigroup generated by $\{\pm 1, S, S^{-1}, \Phi, \Phi^{-1}\}$ which coincides with $\mathrm{SL}_2(\mathbb{Z})$ (see [100, Chap. VIII]). □

The above proposition implies that by applying the concatenation trick several times we can transform any decorated graph into one in which all the decorations are of the form $P^{\pm}_{m,n}$. Thus any manifold which can be obtained by a generalized plumbing can also by obtained by a traditional plumbing.

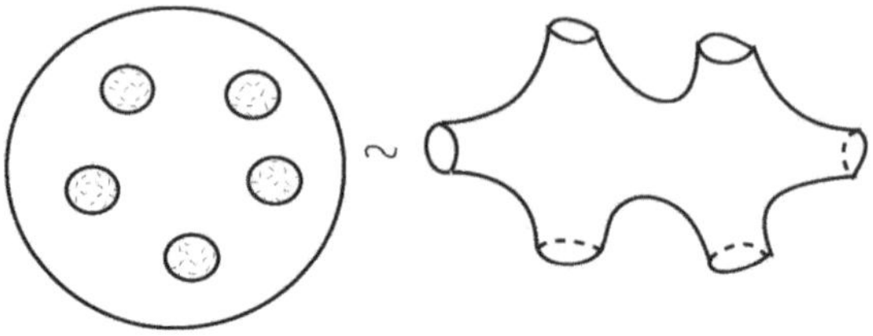

Figure 2.10. A sphere with several holes.

Any compact, oriented Riemann surface, possibly with boundary, can be decomposed into Riemann surfaces of genus zero of the type depicted in Figure 2.10. We have thus proved the following result.

Corollary 2.55. *Any generalized plumbing is diffeomorphic with a regular plumbing of circle bundles over Riemann surfaces of genus zero.*

When the plumbing graph has no loops[3] the combinatorics of the problem simplifies somewhat in the case of usual plumbings of circle bundles over spheres. First, we no longer have to keep track of the genera of the vertices, since $g(v) = 0$, for all v. A decoration can now be identified with a pair of integers $m_{\pm}(e)$, and a sign $\epsilon(e) = \pm 1$ so that

$$\Gamma(e) = \epsilon(e) T_{m_+} \Phi T_{m_-}$$

Due to the equality $CP^{-1}_{m,n}C = P_{n,m}$ we deduce from the conjugation trick that the orientation of the edge e is irrelevant. The decorated graph of a plumbing can be simplified by performing the changes indicated in Figure 2.11.

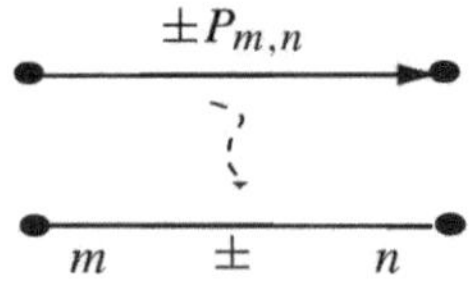

Figure 2.11. Describing a regular plumbing.

Define the *Chern number* of a vertex to be

$$c_1(v) = \sum_{e \in E_v^-} m_-(e) + \sum_{e \in E_v^+} m_+(e).$$

[3]These correspond to selfplumbings and lead to quite subtle phenomena; see [37].

The decorated graph then describe plumbings of circle bundles of degree $c_1(v)$ over spheres, according to the undecorated graph. In the left-hand-side Figure 2.12 we

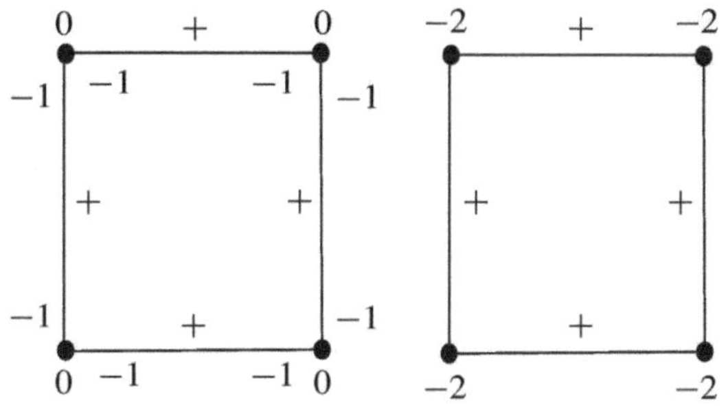

Figure 2.12. Plumbing −2-bundles over spheres.

have the decorated graph describing a plumbing of bundles over spheres. Each vertex has Chern number −2. The usual graphical representation of this plumbing (defined in [46, §8]) is shown in the right-hand side of Figure 2.12. In general, we will drop the numbers $m_{\pm}(e)$ attached to the edges of the graph, and we replace them with numbers $c_1(v)$ attached to the vertices. From now own, we will use only this description of plumbings over spheres. We will refer to this plumbing description as the *usual, regular, traditional* etc. If additionally, all the edges have the same sign + we will no longer indicated it on the plumbing graph.

A usual plumbing diagram can be transformed so that the resulting manifold does not change. For more details on this calculus with plumbings we refer to [78].

To compute the torsion of a plumbing we need to produce a surgery description of such a 3-manifold. Fortunately there is a simple way to do this. We follow closely the prescriptions in [37, 78]. Assume for simplicity that there are no loops[4].

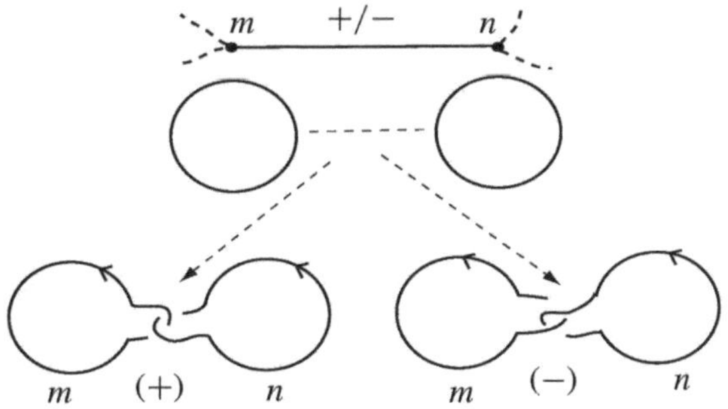

Figure 2.13. Surgery description of plumbings.

First, mark edges $e_1, \ldots, e_m$ of the graph G so that the graph obtained by removing

[4]We can eliminate them by concatenation.

the marked edges is a *connected* tree. Now replace each vertex v of the graph G with an *oriented* unknot K_v with surgery coefficient $c_1(v)$. If two vertices v_1, v_2 are connected by a (marked or unmarked) edge of G, then *locally* link K_{v_1} to K_{v_2} away from the other components as in Figure 2.13 so that the local linking between these to unknots is equal to the sign along the edge connecting them. If the vertices v_1 and v_2 are connected by several edges we have to perform this local linking procedure several times. The unknots K_{v_i} will be transformed into two unknots with linking number equal the signed number of edges between v_1 and v_2.

Next, for every marked edge e_i introduce an unknot K_i with surgery coefficient 0 which links the unknots corresponding to the vertices of e_i as in Figure 2.14. For example, the plumbing described in Figure 2.12 has the surgery description in Figure 2.15.

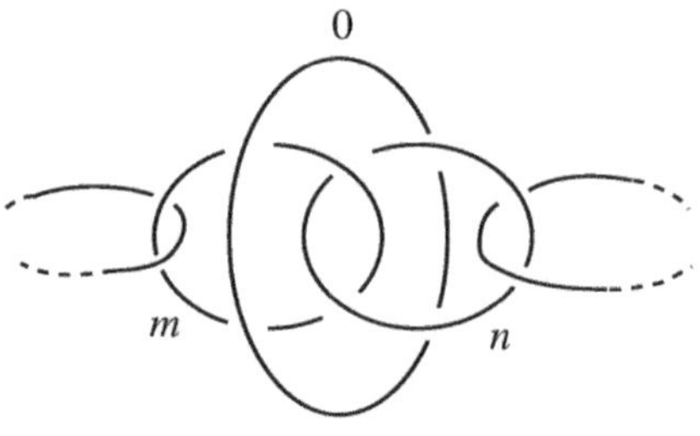

Figure 2.14. Simulating the cycles of the plumbing graph.

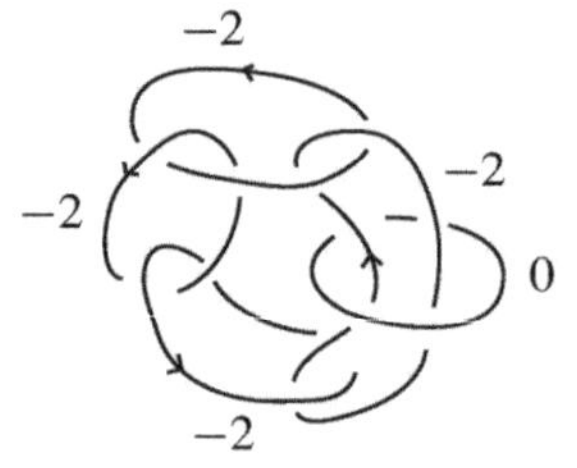

Figure 2.15. A circular plumbing of −2-spheres.

Exercise 2.6. Show that the surgery presentation in Figure 2.15 is nondegenerate in the sense of Definition 2.47. □

The determination of torsion via surgery descriptions has one computational limitation: it requires the computation of Alexander polynomials of links with many components and crossings which often can be a very challenging task. However, the links involved in plumbings are quite special, and if the combinatorics of the plumbing graph is not too sophisticated they can be obtained quite easily using the following simple facts.

Proposition 2.56. (a) *If the link $\mathcal{K}'$ is obtained from the link $\mathcal{K}$ by adding an additional component C which is the meridian of a component $\mathcal{K}_i$ of $\mathcal{K}$ then*

$$\Delta_{\mathcal{K}'} \sim (\mu_i - 1) \cdot \Delta_{\mathcal{K}}. \tag{2.19}$$

(b) (Seifert–Torres formula, [98, 109].) *Denote by $\mathcal{K}'$ the link obtained from the link $\mathcal{K} = \mathcal{K}_1 \cup \cdots \cup \mathcal{K}_n$ by adding a component $\mathcal{K}_{n+1}$ which is a simple closed curve on the boundary a thin tubular neighborhood of $\mathcal{K}_n$ of the form $p\lambda_n + q\mu_n$, $p \neq 0$. Then*

$$\Delta_{\mathcal{K}'}(\mu_1, \dots, \mu_{n+1}) = \Delta_{\mathcal{K}}(\mu_1, \dots, \mu_n^p \mu_{n+1})(T^p \mu_{n+1} - 1), \tag{2.20}$$

where $T = \prod_{i=1}^n \mu_i^{\ell_i}$, $\ell_i = \boldsymbol{Lk}(\mathcal{K}_i, \mathcal{K}_{n+1})$. Moreover, if $\mathcal{K}''$ denotes the link obtained from $\mathcal{K}'$ by removing the component $\mathcal{K}_n$ then

$$\Delta_{\mathcal{K}''}(\mu_1, \dots, \mu_{n-1}, \mu_{n+1}) = \Delta_{\mathcal{K}}(\mu_1, \dots, \mu_{n-1}, \mu_n^p) \cdot \frac{T^p - 1}{T - 1}. \tag{2.21}$$

Exercise 2.7. Prove, *without relying on* (2.19), that the Alexander polynomial of the link depicted in Figure 2.16 is $\sim (1 - \mu)$, where μ denotes the meridian of the middle component. (*Hint:* Find a simple CW-decomposition of the exterior of this link, or use the Fox free calculus in §B.1.) □

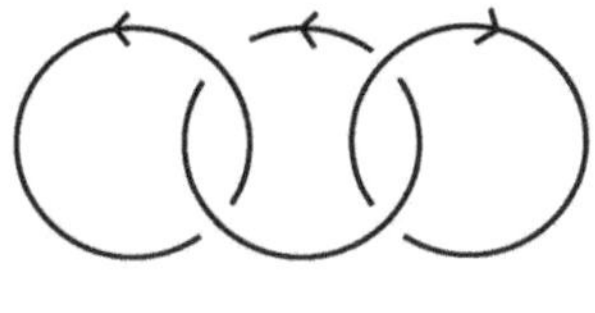

Figure 2.16. A simple link.

Exercise 2.8. Prove the identity (2.19). (*Hint:* Fix a tubular neighborhood U_i of $\mathcal{K}_i$ containing C. This allows us to regard the exterior of $\mathcal{K}$ as a subset of the exterior of $\mathcal{K}'$. Now use the Mayer–Vietoris theorem coupled with the computation in the previous exercise.) □

Exercise 2.9. Prove (2.20) and (2.21). □

Exercise 2.10. Use (2.21) to compute the Alexander polynomial of the (3, 5)-torus knot depicted in Figure 2.17. □

If the graph G of a plumbing is a (connected) tree, all its edges are positive, and $\mathcal{K} = (\mathcal{K}_v)_{v \in V}$ is its associated link, then an iterated application of the trick (2.19)

Figure 2.17. A MAPLE rendition of the (3, 5)-torus knot.

produces the equality

$$\mathcal{T}_{S^3\setminus\mathcal{K}} \sim \prod_{i=1}^{n}(\mu_v - 1)^{\deg v - 1}. \tag{2.22}$$

Things get much more complicated when the plumbing graph has cycles. In our next examples we want to describe how to compute the torsion for simply connected plumbing graphs and for the simplest non simply connected plumbing graphs.

Example 2.57 (The torsion of Seifert manifolds. The case $b_1 > 0$.). Consider the star-shaped generalized plumbing graph depicted in Figure 2.18. The center of the star has genus g. All the other vertices have genus zero. All Seifert manifolds can be described by such star-shaped plumbing graphs, with possibly more than three rays (see [46]).

Applying the concatenation trick we obtain the simpler generalized plumbing description at the bottom of Figure 2.18 where $\Gamma_1, \Gamma_2, \Gamma_3 \in SL_2(\mathbb{Z})$. Denote the resulting manifold by $M = M(g, \Gamma_1, \Gamma_2, \Gamma_3)$. Let $H := H_1(M)$. To compute the torsion of M we use the surgery formula in Theorem 2.41. First we need to find a presentation of H.

Denote by Σ an oriented Riemann surface of genus g with 3 boundary components. Then $G := H_1(S^1 \times \Sigma)$ has a presentation

$$G = \{\varphi, b_1, b_2, b_3, c_1, \dots, c_{2g};\ b_1 + b_2 + b_3 = 0\}$$

where φ denotes the homology class carried by the fiber S^1, b_1, b_2, b_3 denote the cycles carried by the boundary components of Σ, and $c_1, \dots, c_{2g}$ form a symplectic basis of 1-cycles obtained by capping the boundary components of Σ. The boundary of $S^1 \times \Sigma$ consists of three tori, and the manifold M is obtained by filling them with solid tori U_i, $i = 1, 2, 3$, attached according to the prescription given by $\Gamma_1, \Gamma_2, \Gamma_3$. Suppose

$$\Gamma_i = \begin{bmatrix} p_i & x_i \\ q_i & y_i \end{bmatrix}, \quad i = 1, 2, 3.$$

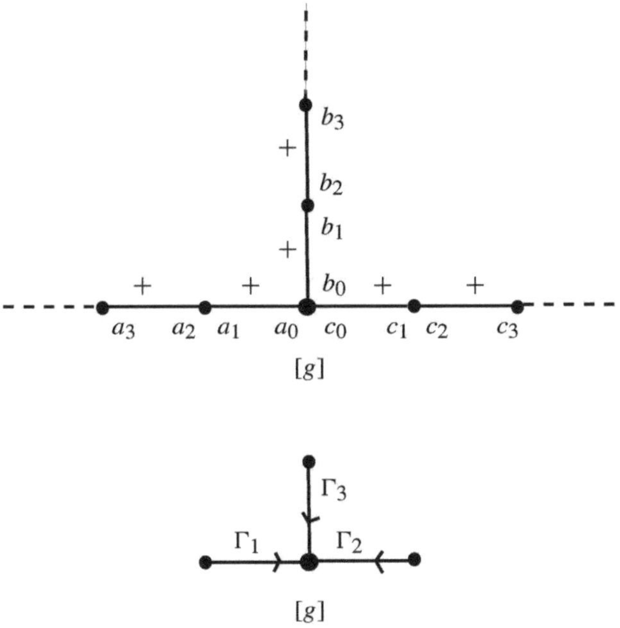

Figure 2.18. A star-shaped plumbing.

We obtain the following presentation of H

$$H = \{\varphi, b_1, b_2, b_3, c_1, \ldots, c_{2g};\ b_1 + b_2 + b_3 = 0,\ -p_i b_i + q_i \varphi = 0\}.$$

The cores of the attaching solid tori U_i define the homology classes K_i represented by $-x_i b_i + y_i \varphi$. If we denote by π the natural projection $G \to H$ we deduce

$$\pi_\sharp(\mathfrak{T}_{S^1 \times \Sigma}) = \mathfrak{T}_M \cdot (1 - K_1)(1 - K_2)(1 - K_2).$$

This implies

$$\pi_\sharp(1 - \varphi)^{2g+1} = \mathfrak{T}_M \cdot \pi_\sharp\Big(\prod_{i=1}^{3}(1 - \varphi^{y_i} b_i^{-x_i})\Big).$$

To see how this works in practice we consider the special case $g = 1$ and

$$\Gamma_1 = \Gamma_2 = \Gamma_3 = \begin{bmatrix} 4 & 5 \\ 3 & 4 \end{bmatrix}.$$

Identify the free Abelian group generated by b_1, b_2, b_3, φ (in this order) with $\mathbb{Z}^4$, and the free Abelian group generated by $b_1, b_2, b_3, \varphi, c_1, c_2$ with $\mathbb{Z}^6$. Then H admits the presentation

$$\mathbb{Z}^4 \xrightarrow{P} \mathbb{Z}^6 \to H \to 0,$$

where

$$P = \begin{bmatrix} 1 & -4 & 0 & 0 \\ 1 & 0 & -4 & 0 \\ 1 & 0 & 0 & -4 \\ 0 & 3 & 3 & 3 \\ 0 & 0 & 0 & 0 \\ 0 & 0 & 0 & 0 \end{bmatrix}.$$

We now use MAPLE's ismith procedure to reduce P to the Smith normal form

$$P = USV,$$

where

$$S := \begin{bmatrix} 1 & 0 & 0 & 0 \\ 0 & 1 & 0 & 0 \\ 0 & 0 & 4 & 0 \\ 0 & 0 & 0 & 36 \\ 0 & 0 & 0 & 0 \\ 0 & 0 & 0 & 0 \end{bmatrix}, \quad U = \begin{bmatrix} 3 & 3 & -5 & 4 & 0 & 0 \\ 2 & 2 & -4 & 3 & 0 & 0 \\ 3 & 2 & -5 & 4 & 0 & 0 \\ -24 & -15 & 39 & -32 & 0 & 0 \\ 0 & 0 & 0 & 0 & 1 & 0 \\ 0 & 0 & 0 & 0 & 0 & 1 \end{bmatrix},$$

and

$$V := \begin{bmatrix} 1 & 0 & -32 & -32 \\ 0 & 1 & -18 & -17 \\ 0 & 0 & -7 & -8 \\ 0 & 0 & 1 & 1 \end{bmatrix}.$$

We deduce that

$$H := \mathbb{Z}^2 \oplus \mathbb{Z}_4 \oplus \mathbb{Z}_{36}.$$

We denote by $e_1, \dots, e_6$ the new basis of $\mathbb{Z}$ given by the matrix U^{-1}. Then e_3 generates the $\mathbb{Z}_4$-summand, e_4 generates the $\mathbb{Z}_{36}$-summand. The vectors e_5, e_6 determine an integral basis of the free summand of H. They are images of the basis vectors $c_1, c_2 \in H_1(T^2)$. The coordinates of φ in the new basis are given by the fourth column of U and we see that $\varphi = 4e_4$ in H. From the first three columns of U we read the following equalities in H.

$$b_1 = e_3 + 12e_4, \quad b_2 = 2e_3 - 15e_4, \quad b_3 = e_3 + 3e_4.$$

Thus $K_i = -5b_i + 4\varphi = -5b_i + 16e_4$ and we deduce

$$K_1 = -e_3 + 28e_4 = -e_3 - 8e_4, \quad K_2 = -2e_3 + 19e_4, \quad K_3 = -e_3 + e_4.$$

We obtain the following equation in $\mathbb{Z}[H]$

$$(1 - e_4^4)^3 = \mathfrak{T}_M(1 - e_3^{-1}e_4^{28})(1 - e_3^2 e_4^{19})(1 - e_3^{-1}e_4).$$

The characters of H have the form $\chi = \rho\zeta z_1 z_2$, $\rho^4 = \zeta^{36} = 1$, $z_1, z_2 \in S^1$. When we Fourier transform the above equation we obtain

$$(1-\zeta^{-4})^3 = \hat{\mathfrak{T}}_M(\rho,\zeta,z_1,z_2)(1-\rho\zeta^8)(1-\rho^2\zeta^{17})(1-\rho\bar{\zeta}),$$

$\forall \rho^4 = \zeta^{36} = 1$, $\forall z_1, z_2 \in S^1$. Observe that

$$4\cdot 17 \equiv -4 \bmod 36, \quad 4\cdot 8 \equiv -4 \bmod 36,$$

so that if we set

$$\omega_0 = \omega_0(\chi) = \zeta^{-1}, \qquad \omega_1 = \omega_1(\chi) = \rho\zeta^8,$$
$$\omega_2 = \omega_2(\chi) = \rho^2\zeta^{17}, \quad \omega_3 = \omega_3(\chi) = \rho\zeta^{-1},$$

then $\omega_0^4 = \omega_1^4 = \omega_2^4 = \omega_3^4$ and

$$(1-\omega_0^4)^3 = \prod_{k=1}^{3}(1-\omega_k^4).$$

Note that the functions ω_0^4, ω_i, $i = 1,2,3$ are precisely the Fourier transforms of $\varphi, [K_i] \in \mathbb{Z}[H]$, $i = 1,2,3$. We conclude

$$\hat{\mathfrak{T}}_M(\rho,\zeta,z_1,z_2) \sim \begin{cases} \frac{(1-\omega_0^4)^3}{(1-\omega_1)(1-\omega_2)(1-\omega_3)} & \text{if } \prod_{i=1}^3(1-\omega_i) \neq 0 \\ ? & \text{if } \prod_{i=1}^3(1-\omega_i) = 0. \end{cases} \tag{2.23}$$

To resolve the ambiguity in the last equality we will analyze in greater detail the gluing process. Set $X := S^1 \times \Sigma$. We have an inclusion

$$\hat{\pi} : \hat{H} \hookrightarrow \hat{G}, \quad G = H_1(X).$$

The functions $\omega_0^4, \omega_i : \hat{H} \to S^1, i = 1,2,3$ are restrictions of functions on $\hat{G}$, namely the Fourier transforms of φ, $[K_i] \in \mathbb{Z}[G]$. We will continue to denote these functions on $\hat{G}$ by the same symbols as their restrictions to $\hat{H}$.

Suppose $\chi \in \hat{H}$ is a character such that $\omega_i(\chi) = 1$, for some $i = 1,2,3$. Then $\omega_0^4(\chi) = 1$. We now regard χ as a character of G with the property $\chi(\varphi) = 1$. In other words, χ factors through a character $\tilde{\chi}$ of $H_1(\Sigma)$. As in Example 2.44, the character $\tilde{\chi}$ determines a local coefficient system on Σ which we denote by $\mathbb{K}_{\tilde{\chi}}$. We can now use the Künneth formula for homology with local coefficients, [6], to conclude that

$$H_*(X, \mathbb{K}_\chi) = H_*(S^1, \mathbb{C}) \otimes H_*(\Sigma, \mathbb{K}_{\tilde{\chi}}).$$

The groups $H_*(\Sigma, \mathbb{K}_{\tilde{\chi}})$ can be easily determined since Σ is simple homotopy equivalent to a wedge of circles (see Figure 2.19).

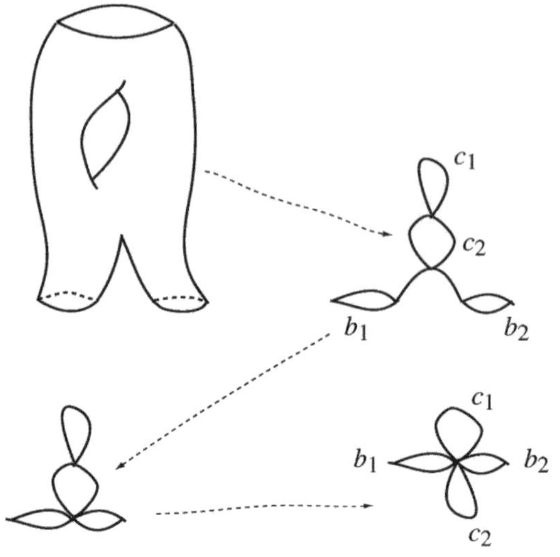

Figure 2.19. A cellular Morse decomposition of the torus with three holes.

More precisely, $H_*(\Sigma, \mathbb{K}_{\bar{\chi}})$ is determined by the based chain complex

$$C_1(c_1, c_2, b_1, b_2) \xrightarrow{\partial} C_0(v) \to 0, \quad \partial c_i = (z_i - 1)v, \ \partial b_i = (\beta_i - 1)v, \tag{2.24}$$

where $z_i = \chi(c_i)$, $\beta_i = \chi(b_i)$, $i = 1, 2$. If we set $b_k(\chi) = \dim_{\mathbb{C}} H_k(\Sigma, \mathbb{K}_{\bar{\chi}})$ and $e(\chi) = b_0(\chi) - b_1(\chi) + b_2(\chi)$ then

$$e(\chi) = 3, \quad b_2(\chi) = 0,$$

$$b_0(\chi) = \begin{cases} 1 & \text{if } \chi = 1 \\ 0 & \text{if } \chi \neq 1. \end{cases}$$

We deduce that if $\chi \neq 1$ then

$$\dim_{\mathbb{C}} H_0(X, \mathbb{K}_\chi) = \dim_{\mathbb{C}} H_3(X, \mathbb{K}_\chi) = 0,$$

$$\dim_{\mathbb{C}} H_1(X, \mathbb{K}_\chi) = \dim_{\mathbb{C}} H_2(X, \mathbb{K}_\chi) = 3.$$

The manifold M decomposes as an union $X \cup U$, where U denotes the union of the attached solid tori U_i, $i = 1, 2, 3$. Denote by V the overlap of these two parts, $V = X \cap U$. V is an union of three tori, T_1, T_2, T_3. Fix $\chi \in \hat{H} \setminus \{1\}$ such that

$$\big(1 - \omega_1(\chi)\big)\big(1 - \omega_2(\chi)\big)\big(1 - \omega_3(\chi)\big) = 0.$$

Set

$$I_\chi := \{1 \le i \le 3;\ \omega_i(\chi) = 1\}.$$

We distinguish two cases.

1. $|I_\chi| < 3$. The character χ induces a local coefficient systems on each of the parts X, U_i, T_j, $i, j = 1, 2, 3$, in the above decomposition of M. Moreover, the local coefficient system induced by χ on T_i is trivial if and only if $i \in I_\chi$. This follows from the fact that along both K_i, and the attaching curve of the Dehn gluing of U_i the character χ is equal to one. These two curves form a basis of $H_1(T_i)$. The Mayer–Vietoris sequence has the form

$$0 \to H_3(M, \mathbb{K}_\chi) \to H_2(V, \mathbb{K}_\chi) \to H_2(X, \mathbb{K}_\chi) \to H_2(M, \mathbb{K}_\chi) \to \cdots$$

Now observe that

$$\dim_{\mathbb{C}} H_2(V, \mathbb{K}_\chi) = |I_\chi| < 3 = \dim H_2(X, \mathbb{K}_\chi).$$

This forces $H_2(M, \mathbb{K}_\chi) \neq 0$ so that $\hat{\mathfrak{T}}_M(\bar{\chi}) = 0$ whenever $|I_\chi| < 3$.

2. $|I_\chi| = 3$. In this case the Mayer–Vietoris sequence implies that $H_*(M, \mathbb{K}_\chi) = 0$ so that $\hat{\mathfrak{T}}_M(\chi) \neq 0$. To compute the torsion we use the more refined version (1.8) of the multiplicative property of the Mayer–Vietoris sequence. The Mayer–Vietoris sequence for the homology with local coefficients defined by χ reduces to three isomorphisms

$$0 \to H_k(V, \mathbb{K}_\chi) \xrightarrow{\phi} H_k(U, \mathbb{K}_\chi) \oplus H_k(X, \mathbb{K}_\chi) \to 0, \quad k = 0, 1, 2, \tag{2.25}$$

where ϕ is the direct sum $\Gamma_1^{-1} \oplus \Gamma_2^{-1} \oplus \Gamma_3^{-1} \oplus \boldsymbol{j}$, and

$$\boldsymbol{j} \colon H_k(V, \mathbb{K}_\chi) \to H_k(X, \mathbb{K}_\chi)$$

is the morphism induced by the inclusion $V \hookrightarrow X$. We need to fix cellular structures on U, V, X, M such that the attaching maps Γ_i are cellular. On the other hand, as explained in [72] on smooth manifolds the choice of cellular structure is irrelevant as far as torsion computations are concerned, and we may as well work with cellular complexes simple homotopic to the original choices. Next, we need to pick bases in $H_k(U, \mathbb{K}_\chi)$, $H_k(V, \mathbb{K}_\chi)$, $H_k(X, \mathbb{K}_\chi)$. Denote by $d_k(\chi)$, $k = 0, 1, 2$, the determinants of the isomorphisms (2.25) with respect to these bases. The cellular structures on X, U, V, and the bases in the corresponding homologies produce via the Euler isomorphisms scalars

$$\boldsymbol{Eul}_V, \quad \boldsymbol{Eul}_U, \quad \boldsymbol{Eul}_X \in \mathbb{K}_\chi^*.$$

The generalized multiplicative formula now implies that

$$\boldsymbol{Eul}_U \cdot \boldsymbol{Eul}_X = \pm \boldsymbol{Eul}_V \cdot \hat{\mathfrak{T}}_M(\bar{\chi}) \cdot \frac{d_2(\chi) d_0(\chi)}{d_1(\chi)}.$$

Let us now explain how to carry out the computations. Observe first that the restrictions of the local coefficients system to V and U are trivial. The spaces $H_0(U)$, $H_0(V)$ have canonical bases and since $H_0(X, \mathbb{K}_\chi) = 0$ we deduce $d_0(\chi) = 1$. Next observe that $H_2(U) = 0$ and $H_2(V)$ has a canonical basis induced by the orientation.

We denote by $\underline{C}$ the complex defined in (2.24) and by $\underline{D}$ the trivial complex

$$0 \to D_1 = \mathbb{C} \stackrel{0}{\to} D_0 = \mathbb{C} \to 0$$

describing the homology of the fiber of X. We use the based complex $\underline{C} \hat{\otimes} \underline{D}$ to compute the twisted homology of X. As basis of $H_1(\Sigma, \mathbb{K}_\chi)$ we choose

$$\{b_1,\ b_2,\ (1 - z_2)c_1 + (z_1 - 1)z_2\}.$$

A quick look at Figure 2.20 shows that $(1 - z_2)c_1 + (z_1 - 1)c_2 = b_1 + b_2 + b_3$. As basis of $H_2(X, \mathbb{K}_\chi)$ we choose

$$\{b_1 \times \varphi,\ b_2 \times \varphi,\ b_3 \times \varphi\}.$$

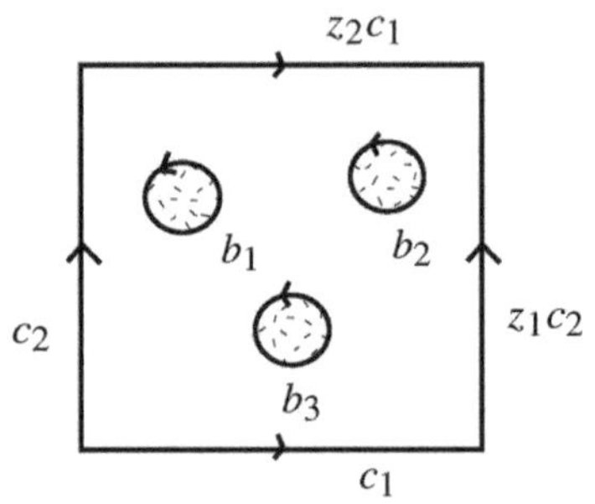

Figure 2.20. The twisted homology of the torus with three holes.

We deduce that $d_2(\chi) = 1$. We now choose $\{b_i, \varphi\}$ as basis of $H_1(T_i)$, $i = 1, 2, 3$. Since

$$\Gamma_i^{-1} = \begin{bmatrix} 4 & -5 \\ -3 & 4 \end{bmatrix}$$

we deduce that

$$\phi(b_i) = -3\varphi \oplus b_i \in H_1(U_i) \oplus H_1(X, \mathbb{K}_\chi),$$
$$\phi(\varphi) = 4\varphi \oplus 0 \in H_1(U_i) \oplus H_1(X, \mathbb{K}_\chi).$$

We conclude that $d_1(\chi) = \pm 4^3$. It is easy to check that $\boldsymbol{Eul}_U = \pm 1$ and $\boldsymbol{Eul}_V = \pm 1$. It remains to compute $\boldsymbol{Eul}_X$. This can be done using the based complex

$$0 \to A_2 = C_1 \otimes D_1 \to A_1 = (C_0 \otimes D_1) \oplus A_0 = (C_1 \otimes D_0) \to C_0 \otimes D_0 \to 0.$$

Concretely, this is the torsion of the *acyclic based complex*

$$0 \to H_2(X, \mathbb{K}_\chi) \to A_2 \oplus H_1(X, \mathbb{K}_\chi) \to A_1 \oplus H_0(X, \mathbb{K}_\chi) \to A_0 \to 0,$$

where $H_k(X, \mathbb{K}_\chi)$ is identified with a subspace in A_k. A simple computation shows that this torsion is ± 1. Hence

$$\hat{\mathfrak{T}}_M(\chi) = \pm 4^3.$$

To determine the torsion we need to use the inverse Fourier transform,

$$\mathfrak{T}_M = \sum_{h \in \mathrm{Tors}(H)} \mathfrak{T}_M(h)h, \quad \mathfrak{T}_M(h) = \frac{1}{4 \cdot 36} \sum_{\rho^4 = \zeta^{36} = 1} \hat{\mathfrak{T}}_M(\rho, \zeta, z_1, z_2)\rho(h)\zeta(h).$$

The surprising thing about the above formula is that a priori it is not immediately clear that the inverse Fourier transform will produce an integer valued function. We will present below an elementary argument proving this arithmetic fact directly. In the process we will shed additional light on the algebraic structure of the torsion of a 3-dimensional Seifert manifolds.

The correspondences $\chi \mapsto \omega_i(\chi)$ define morphisms $\hat{H}_i \to S^1$, $i = 1, 2, 3$. We denote by G_i the range of the morphism ω_i, and by Z_i its kernel. Observe that $Z_0 := Z_1 \cap Z_2 \cap Z_3$ coincides with the identity component of $\hat{H}$ Denote by $\mathbb{I}_{Z_i} : \hat{H} \to \mathbb{C}$ the characteristic function of the subset $Z_i \hookrightarrow \hat{H}$. Set

$$f_i : \hat{H} \to \mathbb{C}, \quad f_i(\rho, \zeta, z_1, z_2) = \sum_{k=0}^{3} \omega_i^k - 4\mathbb{I}_{Z_i}, \quad i = 1, 2, 3.$$

Observe that

$$f_i(\chi) = \begin{cases} \frac{1-\omega_i^4}{1-\omega_i} & \text{if } \omega_i \neq 1 \\ 0 & \text{if } \omega_i = 1. \end{cases}$$

This shows that

$$\hat{\mathfrak{T}}_M(\chi) = f_1(\chi) \cdot f_2(\chi) \cdot f_3(\chi) \pm 4^3 \mathbb{I}_{Z_0}, \quad \forall \chi \in \hat{H}.$$

Denote by $\mathcal{F}^{-1}[\bullet]$ the inverse Fourier transform. We get

$$\mathfrak{T}_M = \prod_{k=1}^{3} \mathcal{F}^{-1}[f_k]$$

where the above product is the convolution product on $\mathbb{C}[H]$. Now observe that the Pontryagin dual of G_i can be naturally identified with a torsion subgroup of H, more precisely the cyclic subgroup generated by $[K_i]$.

$$\mathcal{F}^{-1}[\mathbb{I}_{Z_i}] = \mathbb{I}_{\langle [K_i] \rangle}.$$

Next observe that

$$\mathcal{F}^{-1}[\omega_1] = e_3 e_4^{-8} = [K_1] \in \mathbb{Z}[H],$$
$$\mathcal{F}^{-1}[\omega_2] = e_3^2 e_4^{19} = [K_2] \in \mathbb{Z}[H],$$
$$\mathcal{F}^{-1}[\omega_3] = e_3 \cdot e_4 = [K_3] \in \mathbb{Z}[H].$$

Hence

$$\mathcal{T}_M = \prod_{j=1}^{3}\Big([K_j] - 4\sum_{\ell=0}^{3}[K_j]^{\ell}\Big) \pm \prod_{j=1}^{3} 4\Big(\sum_{\ell=0}^{3}[K_j]^{\ell}\Big).$$

The sign ambiguity can be resolved using the Casson–Walker–Lescop invariant of M, [58], but we will not get into details. □

Consider a generalized plumbing given by the circular decorated graph at the top of Figure 2.21. Such plumbings arise naturally in the study of cusp singularities (see [45]). Using the concatenation trick we see that this graph is equivalent with the one-loop graph at the bottom of Figure 2.21. This 3-manifold fibers over S^1, and the monodromy is $S = \Gamma_\nu \dots \Gamma_1$. Equivalently, this is the mapping torus of the diffeomorphism $S\colon T^2 \to T^2$. We will denote it by M_S. Given this very explicit description of M_S we will adopt a direct approach.

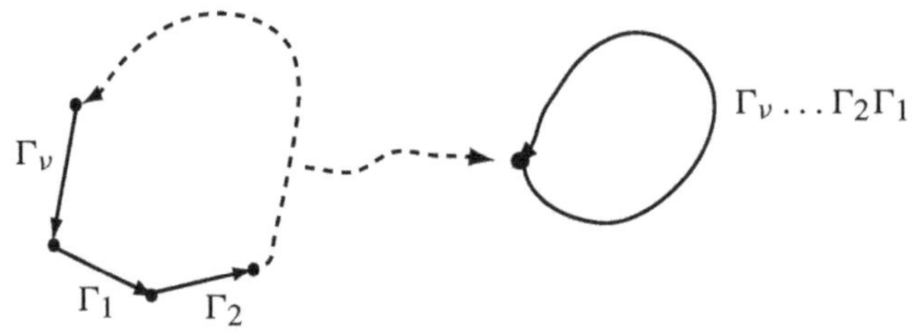

Figure 2.21. An arbitrary circular plumbing.

Example 2.58 (The torsion of circular plumbings). Denote by Λ_0 the standard lattice in $\mathbb{R}^2$. We view S as an automorphism of Λ_0 and we denote by Λ_S the sublattice

$$\Lambda_S := (1 - S)(\Lambda_0).$$

From the Wang exact sequence we deduce that we have a short exact sequence

$$0 \to \Lambda_0/\Lambda_S \to H_1(M_S) \to \mathbb{Z} \to 0.$$

Now we need to distinguish three cases.

1. S is elliptic, i.e. $|\operatorname{Tr} S| \le 1$. There are very few such elements in $SL_2(\mathbb{Z})$ and the manifold M_S is very special. More precisely M_S is finitely covered by a 3-dimensional torus so that it admits a flat metric. Moreover its diffeomorphism type belongs to a very short list of Seifert fibrations over S^2 (see [97, p. 443]). In particular, it can be alternatively represented by a simply connected plumbing graph and the computation is an iterated application of the Mayer–Vietoris principle which we leave to the reader.

2. S is parabolic, i.e. $|\operatorname{Tr} S| = 2$. For example the plumbing in Figure 2.12 leads to

$$S = P^4_{1,1} = \begin{bmatrix} 1 & 4 \\ 0 & 1 \end{bmatrix}.$$

In this case M_S is an S^1-bundle over a torus (see [97, p. 470]) and its Reidemeister torsion is 1.

3. S is hyperbolic, i.e. $|\operatorname{Tr} S| \geq 3$. In this case $\det(1 - S) \neq 0$, i.e. $\operatorname{rank} \Lambda_S = \operatorname{rank}(\Lambda_0) = 2$. Fix a splitting of $H = H_1(M_S)$

$$H = G_S \oplus \mathbb{Z}, \quad G_S = \Lambda_0/\Lambda_S.$$

Observe that $S(\Lambda_S) \subset \Lambda_S$ so that S induces an action on $\mathbb{R}^2/\Lambda_S$ which commutes with the action of the deck group Λ_0/Λ_S of the cover

$$\Sigma_S := \mathbb{R}^2/\Lambda_S \xrightarrow{\pi} \mathbb{R}^2/\Lambda_0 =: \Sigma_0. \tag{2.26}$$

We have an explicit description of the universal Abelian cover $\tilde{M}_S$ of M_S. More precisely

$$\tilde{M}_S = \mathbb{R}^2/\Lambda_S \times \mathbb{R}.$$

The action of $(g, n) \in G_S \oplus \mathbb{Z} = H$ on $\tilde{M}_S$ is given by

$$(u, n) \cdot (x, t) = (S^n(g \cdot x), t + n) = (g \cdot S^n x, t + n), \quad \forall (x, t) \in \mathbb{R}^2/\Lambda_S \times \mathbb{R}.$$

Denote by $0 < d_1 | d_2$, $d_1 d_2 = |\det(1 - S)|$, the elementary divisors of the sublattice Λ_S. Fix a $\mathbb{Z}$-basis $\{\boldsymbol{e}_1, \boldsymbol{e}_2\}$ of Λ_0 such that $\{d_1\boldsymbol{e}_1, d_2\boldsymbol{e}_2\}$ is a basis of Λ_S. We denote by S_0 the matrix representing S with respect to this basis,

$$S_0 := \begin{bmatrix} a & b \\ c & d \end{bmatrix},$$

and set $B := 1 - S_0$. Then

$$d_1 d_2 = |\det B| = |\operatorname{Tr} B| = |2 - \operatorname{Tr} S_0| = |(a - 1) + (d - 1)|,$$

and $d_1 = \gcd((a - 1), b, c, (d - 1))$. Denote by D the 2×2 matrix $\operatorname{diag}(d_1, d_2)$ and set

$$\tilde{S}_0 := D^{-1} S_0 D = \begin{bmatrix} a & \frac{d_2}{d_1} b \\ \frac{d_1}{d_2} c & d \end{bmatrix} \in \mathrm{SL}_2(\mathbb{Z}).$$

The diffeomorphism S of Σ_0 is described by the matrix S_0. It is covered by a diffeomorphism $\tilde{S}$ of Σ_S described by the matrix $\tilde{S}_0$. $\tilde{S}$ commutes with the action of the deck group G_S. Note that we can identify G_S with a subgroup of $\mathbb{R}^2/\Lambda_S$ and as such it acts on Σ_S by translations.

Fix $n \in \mathbb{Z}_{>0}$ and denote by Fix_n the fixed point set of $S^n : \Sigma_0 \to \Sigma_0$. For any $x \in \mathrm{Fix}_n$ the diffeomorphism $\tilde{S}$ defines a permutation of the fiber $\pi^{-1}(x)$ of the cover (2.26). This fiber is an orbit of the action of G_S on Σ_S. Since $\tilde{S}$ commutes with this action of G_S we deduce that there exists $g_x \in G_S$ such that

$$\tilde{S}^n \tilde{x} = g_x \cdot \tilde{x}, \quad \forall \tilde{x} \in \pi^{-1}(x).$$

Following [32, 33] we define the *twisted Lefschetz index* of S^n

$$\hat{L}(S^n) = \sum_{x \in \mathrm{Fix}_n} L(S^n, x) g_x \in \mathbb{Z}[G_S],$$

where $L(S^n, x)$ denotes the local Lefschetz index of the fixed point x of S^n. More precisely

$$L(S^n, x) = \mathrm{sign}(1 - S_0^n) = \mathrm{sign}(2 - \mathrm{Tr}\, S_0^n) =: \epsilon_n.$$

It is convenient to write

$$\hat{L}(S^n) = \sum_{g \in G_S} \hat{L}_g(S^n) g.$$

Observe that

$$\hat{L}_g(S^n) = \frac{1}{|G_S|} L(\tilde{S}^n - g).$$

The homeomorphisms $\tilde{S}^n$ and $\tilde{S}^n - g$ of Σ_S are homotopic and using the Lefschetz fixed point theorem we deduce

$$L(\tilde{S}^n - g) = L(\tilde{S}^n) = 2 - \mathrm{Tr}(\tilde{S}_0^n) = 2 - \mathrm{Tr}\, S_0^n.$$

Hence

$$\hat{L}(S^n) = L(\tilde{S}^n)\mathfrak{I}_S, \quad \mathfrak{I}_S = \frac{1}{|G_S|} \sum_{g \in G_S} g. \tag{2.27}$$

Define the twisted Lefschetz zeta function

$$\hat{\zeta}(S) = \exp\Big(\sum_{n>0} \hat{L}(S^n) \frac{t^n}{n} \Big) \in \mathbb{Q}[G_S][[t]]. \tag{2.28}$$

The results in [32, 33] show that the Reidemeister torsion of M_S is equal to the above twisted zeta function. To obtain a more explicit description we introduce a new variable

$$T := \mathfrak{I}_S t.$$

Since $\mathfrak{I}_S^2 = \mathfrak{I}_S$ in $\mathbb{Q}[G_S]$ we deduce that

$$T^n = \mathfrak{I}_S t^n.$$

We can now rewrite (2.28) as

$$\hat{\zeta}(S) = \exp\Big(\sum_{n>0} L(S^n)\frac{T^n}{n}\Big) \in \mathbb{Q}[G_S][[T]].$$

As shown in Appendix §A.2 the last expression simplifies to

$$\hat{\zeta}(S) = \frac{\det(1 - TS_0)}{(1-T)^2} = \frac{1 - (\mathrm{Tr}\, S_0)T + T^2}{(1-T)^2} = \frac{T^2 - 2T + 1 + (2 - \mathrm{Tr}\, S_0)T}{(1-T)^2}.$$

If we now recall that $|G_S| = |\det(1 - S_0)| = |2 - \mathrm{Tr}\, S_0| = |2 - \mathrm{Tr}\, S|$ and we set $\mathfrak{S}_S = \sum_{g \in G_S} g$ we can now rewrite the last equality as

$$\mathfrak{T}_{M_S} = \hat{\zeta}(S) = 1 + \mathrm{sign}(2 - \mathrm{Tr}\, S)\frac{\mathfrak{S}_S}{(1-T)^2}.$$

The last quantity belongs to the ring $\mathfrak{N}_2(H_1(M_S))$ as predicted by Theorem 2.37. □

Remark 2.59. The computational examples presented in this section conspicuously avoided plumbings defining rational homology spheres, i.e. 3-manifolds with finite H_1. These plumbings graphs are trees, and all the vertices have genus zero. In the next chapter we will deal with this issue in great detail and explain an algorithm for computing the torsion of any rational homology 3-sphere. □

Exercise 2.11. Compute the Alexander polynomial of the link in Figure 2.22, and then compute the Reidemeister torsion of the 3-manifold described by the surgery presentation indicated in this figure. Compare with the computation in the previous example. □

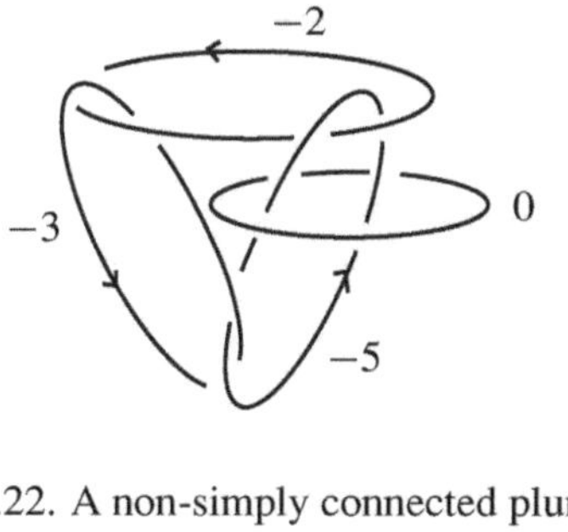

Figure 2.22. A non-simply connected plumbing.

§2.7 Applications

As mentioned in the introduction, the torsion captures rather subtle topological interactions. We will illustrate the strength of this invariant by presenting the classifications of the 3-dimensional lens spaces.

Recall that $L(p,q)$ is defined as the quotient

$$L(p,q) := S^3/\diamond_{1,q}$$

where for $(r,p) = (s,p) = 1$ we denote by $\diamond_{r,s}$ the action of cyclic group $\mathbb{Z}_p$ on

$$S^3 := \{(z_1, z_2) \in \mathbb{C}^2;\ |z_1|^2 + |z_2|^2 = 1\}$$

defined by the rule

$$\zeta \diamond_{r,s} (z_1, z_2) = (\zeta^r z_1, \zeta^s z_2), \quad \forall \zeta^p = 1.$$

Observe that

$$L(p,q) \cong S^3/\diamond_{k,qk}, \quad \forall (p,k) = 1.$$

This shows immediately that $\pi_1(L(p,q)) = \mathbb{Z}_p$ so that the integer p is a homotopy invariant of the lens space. The lens spaces

$$L(p,q), \quad 1 \le q < p,\ (p,q) = 1$$

have identical fundamental groups and homology so these classical invariants alone do not suffice to distinguish them.

Theorem 2.60 (Franz–Rueff–Whitehead, [30, 94, 120]). *Two lens spaces $L(p,q_0)$ and $L(p,q_1)$ are homotopically equivalent if and only if*

$$q_0 = \pm \ell^2 q_1 \bmod p \tag{2.29}$$

for some $\ell \in \mathbb{Z}$.

Proof. We denote the homotopy equivalences by $\simeq$.
Step 1.

$$q_1 = \pm \ell^2 q_0 \bmod p \Longrightarrow L(p,q_0) \simeq L(p,q_1)$$

For every integers k_1, k_2 such that $(p,k_i) = 1$ define map $f_{k_1,k_2} : S^3 \to S^3$ by

$$f_k(z_1,z_2) = \left(|z_1|^{(1-k_1)} z_1^{k_1}, |z_2|^{(1-k_2)} z_2^{k_2}\right).$$

Observe that f has degree k_1k_2, and it is equivariant, i.e.

$$f_{k_1,k_2}(\zeta \diamond_{r,s} (z_1,z_2)) = \zeta \diamond_{k_1 r, k_2 s} (z_1,z_2).$$

This shows that f_{k_1,k_2} induces a map

$$[f_{k_1,k_2}]\colon S^3/\diamond_{r,s} \to S^3/\diamond_{k_1 r,k_2 s}$$

of degree k_1k_2. Fix a point $p \in S^3$ and a small ball B centered at p such that $\zeta^k \diamond_{r,s} B \cap B = \emptyset, \forall 0 < k < p$. Set

$$U = \bigcup_{k=0}^{p-1} \zeta^k \diamond_{r,s} B.$$

We can equivariantly modify f_{k_1,k_2} in U to change its degree by any multiple of p. Suppose for simplicity that $q_1 = \ell^2 q_0 \mod p$. Denote by r_0 the inverse of q_0 mod p. The map $f_{\ell,\ell q_1 r_0}$ is $(\diamond_{1,q_0}, \diamond_{\ell,\ell q_1})$-equivariant so that it induces a map

$$[f_{\ell,\ell q_1 r_0}]\colon L(p,q_0) \to L(p,q_1).$$

We can arrange that $\deg f_{\ell,\ell q_1 r_0}$ is any number congruent to $\ell^2 q_1 r_0 = 1 \mod p$. In particular, we can arrange so that it has degree 1. Thus $f_{\ell,\ell q_1 r_0}$ induces an isomorphisms $H_*(S^3) \to H_*(S^3)$. Using the Hurewicz and Whitehead's theorems we deduce that f is a homotopy equivalence.

Clearly, $[f_{\ell,\ell q_1 r_0}]$ induces an isomorphism between fundamental groups. Since $\pi_k(L(p,q_i)) = \pi_k(S^3)$, $i = 0, 1, k \geq 2$, and the morphisms

$$[f_{\ell,\ell q_1 r_0}]_*\colon \pi_k(L(p,q_0) \to \pi_k\big(L(p,q_1)\big)$$

coincide with the morphisms $(f_{\ell,\ell q_1 r_0})_*\colon \pi_k(S^3) \to \pi_k(S^3)$ which are isomorphisms we deduce from Whitehead's theorem that f is a homotopy equivalence.

Step 2.

$$L(p,q_0) \simeq L(p,q_1) \Longrightarrow q_0 = \pm \ell^2 q_1 \bmod p.$$

To see this we will use the *linking form* of $L(p,q)$ (see the classical but very intuitive [57, Chap. V] or [99, §77] or the more formal [5, p.366] for details). This is a symmetric, bilinear map

$$\Lambda_{p,q}\colon H_1\big(L(p,q)\big) \times H_1\big(L(p,q)\big) \to \mathbb{Q}/\mathbb{Z}$$

defined as follows. Pick $c, d \in H_1(L(p,q))$ represented by smoothly embedded circles then pd bounds a 2-chain D which we can represent as an embedded surface with boundary pd. Denote by $c \cdot D$ the (signed) intersection number of c and D and set

$$\Lambda_{p,q}(c,d) := \frac{1}{p} c \cdot D \bmod \mathbb{Z}.$$

Up to a sign, the linking form is a homotopy invariant of the lens space. In fact, we have (see [99] or Example B.8)

$$\Lambda_{p,q}(ku,ku) = -k^2 q/p \bmod \mathbb{Z},$$

where u is a generator of $H_1(L(p,q))$. If $L(p,q_0) \simeq L(p,q_1)$ the linking forms must be isomorphic. Thus there exists generators u_i of $H_1(L(p,q_i)) \cong \mathbb{Z}_p$, $i=0,1$ such that $u_0 = \ell u_1$ in $\mathbb{Z}_p$ and

$$q_0/p = \Lambda_{p,q_0}(u_0,u_0) = \pm\Lambda_{p,q_1}(\ell u_1,\ell u_1) = \pm\ell^2 q_1/p \bmod \mathbb{Z}. \qquad \square$$

Since ± 2 is not a quadratic residue modulo 5 we deduce

$$L(5,1) \not\simeq L(5,2).$$

On the other hand, since 2 is a quadratic residue modulo 7 we deduce

$$L(7,1) \simeq L(7,2).$$

The reader can verify easily that two lens spaces $L(p,q_i)$, $i=0,1$ such that

$$q_0 = \pm q_1 \quad \text{or} \quad q_0 q_1 = \pm 1 \bmod p$$

are homeomorphic. We can thus parametrize the homeomorphism classes of lens spaces by pairs (p,q) such that

$$1 \le q < p/2, \quad (p,q) = 1. \tag{2.30}$$

In this list, some spaces are homotopically equivalent, e.g.$L(7,1) \simeq L(7,2)$. We will show that no two lens spaces in this list are homeomorphic. In fact, we have the following result.

Theorem 2.61 (Reidemeister–Franz [29, 90]). *If* (p_0,q_0) *and* (p_1,q_1) *satisfy* (2.30) *then*

$$\mathcal{T}_{L(p_0,q_0)} \sim \mathcal{T}_{L(p_1,q_1)} \iff (p_0,q_0) = (p_1,q_1).$$

Proof. The implication $\Longleftarrow$ is obvious. Conversely, if $\mathcal{T}_{L(p_0,q_0)} \sim \mathcal{T}_{L(p_1,q_1)}$ then clearly $p_0 = p_1 =: p$. We have to show that if there exist $r \in \mathbb{Z}$, $-p/2 \le r \le p/2$ and $\epsilon \in \{\pm 1\}$

$$(1-\zeta)(1-\zeta^{q_0}) = \epsilon\zeta^r(1-\zeta)(1-\zeta^{q_1}), \quad \forall \zeta^p = 1$$

then $q_0 = q_1$. The above identities are equivalent to

$$1 - \zeta^{q_0} = \epsilon\zeta^r(1-\zeta^{q_1})$$

for all $\zeta^p = 1$. Assume for simplicity that $r \ge 0$. This implies that the polynomial of degree $< p$

$$P(x) = \epsilon x^{r+q_1} - x^{q_0} - \epsilon x^r + 1.$$

has p distinct roots. This implies $r=0$, $\epsilon = 1$ and $q_0 = q_1$. $\square$

The last result implies that the Reidemeister torsion distinguishes the homeomorphism types of the homotopically equivalent spaces $L(7,1)$ and $L(7,2)$.

Remark 2.62. The *Hauptvermutung* (Main Conjecture) asks if two homeomorphic simplicial complexes are necessarily combinatorially equivalent. The answer is known to be positive for manifolds of dimension ≤ 3. In a very beautiful paper [69], J. Milnor has shown that the *Hauptvermutung* is false in dimensions ≥ 6. More precisely he considered the manifolds with boundary

$$X_1 := L(7,1) \times D^n, \quad X_2 = L(7,2) \times D^n, \quad n \geq 3$$

and then defined Y_i as the (simplicial) space obtained from X_i by adjoining the cone on ∂X_i. He showed that the simplicial spaces Y_i are homeomorphic but not combinatorially equivalent. The *relative Reidemeister* torsion captures the finer combinatorial structure. Surprisingly, the absolute torsion is a *topological* invariant (see [101, 13]) and thus it is blind to the combinatorial structure. A few years later, R. Kirby and L. Siebenmann (see [101]) have constructed *topological manifolds* violating the *Hauptvermutung*. □

Chapter 3
Turaev's refined torsion

In the previous chapter we have defined the torsion of a cell complex X as an element of $\mathbb{Q}(H_1(X))/\pm H_1(X)$.

In the beautiful paper [113], Vladimir Turaev has explained the $\bullet/H_1(X)$ ambiguity of the torsion in terms Euler structures. In the special case of 3-manifolds, these Euler structures are equivalent to spin^c-structures. In other words, the Reidemeister torsion of a 3-manifold is rather an invariant of a spin^c structure. In this chapter we will survey these results of Turaev. We assume the reader is familiar with the basic facts concerning spin and spin^c structures on smooth manifolds, as discussed for example in [37].

§3.1 Combinatorial Euler structures

Suppose X is a connected, finite simplicial complex. Denote by $|X|$ the associated topological space, and by X' the first barycentric subdivision of X. For each simplex σ of X we denote by $[\sigma]$ its barycenter. Form the 0-chain

$$e_X = \sum_{\sigma \in X} (-1)^{\dim \sigma}[\sigma] \in C_0(X').$$

If $|X|$ were a compact, oriented manifold without boundary then, according to [43], e_X would be the Poincaré dual of the Euler class of X. Observe that $\chi(X) = 0$ implies that e_X is a boundary.

Definition 3.1 (V. Turaev, [113]). Suppose $\chi(X) = 0$.

(a) An *Euler chain* on X is a singular 1-chain $c \in C_1(|X|)$ such that

$$\partial c = e_X.$$

(b) Two Euler chains c, c' are called *homologous* if the chain $c - c'$ is a boundary.

(c) A *combinatorial Euler structure* is a homology class of Euler chains. We denote by $\mathfrak{Eul}_c(X)$ the set of combinatorial Euler structures. □

A special case of Euler chain is a star-shaped 1-chain (suggestively called *spider* by Turaev), consisting of a center $O \in |X|$, and paths from O to $[\sigma]$ for $\dim \sigma$ even and paths from $[\sigma]$ to O, for $\dim \sigma$ odd (see Figure 3.1). One can prove easily that any Euler structure is homologous to a spider. If Y is a subcomplex of X such that

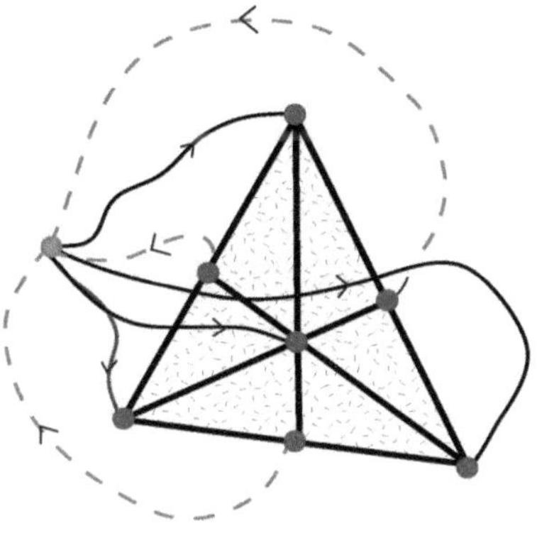

Figure 3.1. A spider.

$\chi(X, Y) = 0$ then a *relative Euler structure* is a singular 1-chain c in $|X|$ such that

$$\partial c = e_{X/Y} := \sum_{\sigma \in X \setminus Y} (-1)^{\dim \sigma} [\sigma].$$

We can define similarly a homology relation between relative Euler structure and obtain a space $\mathfrak{Eul}_c(X, Y)$.

The first homology group $H := H_1(X)$ acts on $\mathfrak{Eul}_c\,(X, Y)$ in a natural way

$$\ker \partial \times \partial^{-1}(e_X) \ni (z, c) \mapsto z \cdot c := z + c.$$

This action is clearly free and transitive so that $\mathfrak{Eul}_c\,(X, Y)$ is an H-torsor.

Denote by $|\tilde{X}|$ the universal Abelian cover of $|X|$. $|\tilde{X}|$ is equipped with a triangulation $\tilde{X}$. A family $\mathcal{F}$ of simplices of $\tilde{X}$ is called *fundamental* if it defines a $\mathbb{Z}[H]$ basis of the simplicial chain complex $C(\tilde{X})$ viewed in a natural way as a $\mathbb{Z}[H]$-module. Equivalently, this means that each simplex of X is covered by exactly one simplex in $\mathcal{F}$.

Fix $O \in |X|$, and $\tilde{O} \in |\tilde{X}|$ above O. If c is a spider with center at O, then any path γ of c admits an unique lift $\tilde{\gamma}$ in $|\tilde{X}|$ starting at $\tilde{O}$. The family of endpoints of the lifts $\tilde{\gamma}$ are the barycenters of a fundamental family. Conversely, if $\mathcal{F} \subset \tilde{X}$ is a fundamental family, then any collection of paths $\tilde{\gamma}$ starting at $\tilde{O}$, and ending at the barycenters of the simplices in $\mathcal{F}$ is the lift of a spider, whose homology class is independent of the choices of $\tilde{\gamma}$. We can thus identify[1] the space of combinatorial Euler structures on X with the set of fundamental families of $\tilde{X}$.

One can prove that if X_1 is a subdivision of X then there exists a natural H-equivariant isomorphism

$$\mathfrak{Eul}(X) \to \mathfrak{Eul}(X_1).$$

[1]The idea of using spiders to construct $\mathbb{Z}[H]$-bases goes back to Reidemeister [91]. See [102] for a particularly nice presentation.

Exercise 3.1. Prove the above claim. □

Since any two piecewise smooth triangulations of a smooth manifold have a common subdivision, the above considerations unambiguously define the space of combinatorial Euler structures on a smooth manifold.

In the next section we will present a different description of the notion of Euler structure on a smooth manifold where the combinatorial structure does not intervene.

§3.2 Smooth Euler structures

Suppose X is a compact, oriented, m-dimensional manifold, possibly with boundary. We assume that the space of components of Y is decomposed into two disjoint parts (possibly empty) and we write this $\partial X = \partial_+ X \cup \partial_- X$. It is convenient to think of X as an oriented cobordism between the two distinguished parts $\partial_\pm$ of its boundary (see Figure 3.2).

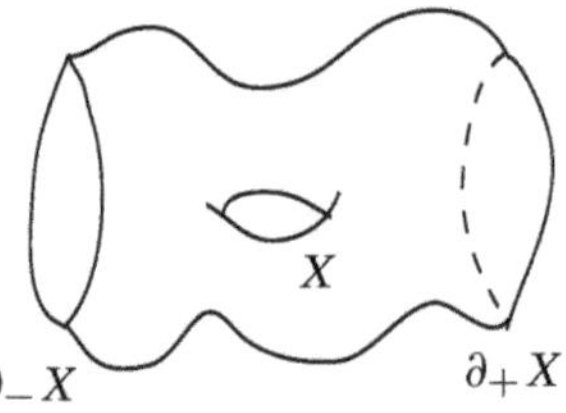

Figure 3.2. An oriented cobordism.

Definition 3.2 (V. Turaev, [113]). A *smooth Euler structure* on $(X, \partial_+ X, \partial_- X)$ is a nowhere vanishing vector field V on X pointing outwards on $\partial_+ X$ and inwards on $\partial_- X$. □

By the Poincaré–Hopf theorem we deduce that smooth Euler structures exist if and only if $\chi(X, \partial_\pm X) = 0$.

Two smooth Euler structures V, V' are called *homologous* if there exists a closed m-dimensional ball $D \subset \operatorname{Int} M$ such that the restrictions of V and V' to $M \setminus \operatorname{Int} D$ are homotopic as nowhere vanishing vector fields pointing outwards along $\partial_+ X$ and inwards along $\partial_- X$. (The homotopy can behave arbitrarily along ∂D.) One can verify easily that this is an equivalence relation. We denote by $\mathfrak{Eul}_s(X, \partial_+ X)$ the space of homology classes of smooth Euler structures on this oriented cobordism. When $\partial_- X = \emptyset$ we write simply $\mathfrak{Eul}_s(X, \partial X)$.

It is useful to compare the relation of being homologous with the stronger relation of being homotopic. Consider a cellular decomposition of $(X, \partial X)$ which has a single

m-dimensional cell. Given two nonsingular vector fields V, V', the first obstruction to them being homotopic is given by an element (see [103])

$$V/V' \in H^m\big((X, \partial_+X) \times (I, \partial I); \pi_{m-1}(S^{m-1})\big)$$
$$\cong H^{m-1}(X, \partial_+X) \overset{\boldsymbol{PD}}{\cong} H_1(X, \partial_-X),$$

where ***PD*** denotes the Poincaré duality. This obstruction vanishes if the vector fields are homologous. If this happens, there is still a secondary obstruction

$$\delta'(V, V') \in H^{m+1}\big((X, \partial_+X) \times (I, \partial I); \pi_m(S^{m-1})\big)$$
$$\cong H^m\big(X, \partial_+X; \pi_m(S^{m-1})\big),$$

where

$$\pi_m(S^{m-1}) = \begin{cases} 0 & m = 2 \\ \mathbb{Z} & m = 3 \\ \mathbb{Z}_2 & m \geq 4. \end{cases}$$

The above discussion shows that we have a well defined map

$$/ : \mathfrak{Eul}_s\,(X, \partial_+X) \times \mathfrak{Eul}_s\,(X, \partial_+X) \to H_1(X, \partial_-X), \quad (U, V) \mapsto U/V,$$

which describes the first obstruction to U being homotopic to V. This operation satisfies a few elementary properties. (Below we will think of $H_1(X, \partial_-X)$ as a *multiplicative* group.)

$$U/V = 1 \iff U = V. \tag{E_1}$$

$$(U/V) \cdot (V/W) = U/W. \tag{E_2}$$

$\forall h \in H_1(X, \partial X), \quad \forall V \in \mathfrak{Eul}_s\,(X, \partial_+X),$

$$\exists \text{ a unique } U \in \mathfrak{Eul}_s\,(X, \partial_+X) \quad \text{such that } h = U/V. \tag{E_3}$$

We will denote by $h \cdot V$ the *unique* element U postulated by (E_3).

We have thus obtained a free and transitive action of $H_1(X, \partial_-X)$ on $\mathfrak{Eul}_s\,(X, \partial_+X)$,

$$H_1(X, \partial_-X) \times \mathfrak{Eul}_s\,(X, \partial_+X) \ni (h, V) \mapsto h \cdot V \in \mathfrak{Eul}_s\,(X, \partial_+X).$$

In particular, if $\partial X = \emptyset$, the spaces of combinatorial and smooth structures on (X, ∂_+X) must be isomorphic. A little bit later we will prove that there exists a *canonical* isomorphism between these two spaces of Euler structures. We want to present an explicit description of the action of $H_1(X)$ on $\mathfrak{Eul}_s(X, \partial_+X)$. Assume $\partial_-X = \emptyset$, so that $\partial_+X = \partial_-X$.

Consider an element $h \in H_1(X, \mathbb{Z})$, and an Euler structure represented by a vector field U. Choose an oriented, simple closed curve ℓ representing h and denote by N a tubular neighborhood of $\ell \hookrightarrow \operatorname{Int} X$. Thus $N \cong D^{m-1} \times S^1$ where $\ell = \{0\} \times S^1$.

S^1 acts in an obvious way by rotations on N, and we denote by $\vec{R}$ the infinitesimal generator of this 1-parameter group of rotations on N. Also, we denote by $\vec{\nu}$ the obvious extension to N of the (origin pointing) radial vector field on D^{m-1} to a vector field on N. (Think of $\vec{\nu}$ as the gradient of the function $d_g(x) = \mathrm{dist}_g\,(x, \ell)$ where both the distance and the gradient are computed using a Riemann metric g. This shows that the choice of $\vec{\nu}$ is unique up to a homotopy.) Modulo a homotopy we can assume that

$$U = -\vec{R} \quad \text{on } N.$$

Assuming D^{m-1} is the disk of radius π define

$$V = \begin{cases} U & \text{on } X \setminus \mathrm{Int}\, N \\ \cos(r)\vec{R} + \sin(r)\vec{\nu} & \text{on } \mathrm{Int}\, N, \end{cases}$$

where $r\colon D^{m-1} \to \mathbb{R}$ denotes the radial distance. We call this operation the *Reeb surgery* along h. In Figure 3.3 we tried to illustrate the differences between the flow of U, on the left, and the flow of V, on the right. Then $V/U = h \in H^{m-1}(X, \partial X; \mathbb{Z})$. To see this notice that given any smooth $(m-1)$-cell σ of $(X, \partial X)$ we get a map

$$f\colon S^{m-1} \to S^{m-1}, \quad f|_{S^{m-1}_+} = V, \quad f|_{S^{m-1}_-} = U$$

such that $\deg(f) = \#(\ell \cap \sigma)$. The degree of the map f is precisely the obstruction to deforming $V|_\sigma$ to $U|_\sigma$ keeping $V|_{\partial\sigma}$ fixed.

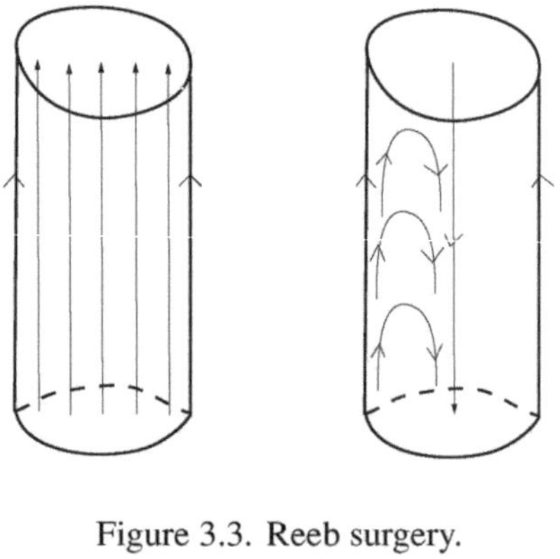

Figure 3.3. Reeb surgery.

Remark 3.3. In dimension $m = 3$ there are countably many possibilities of framing $N \cong D^2 \times S^1$, and any two differ by a sequence of Dehn twists. Thus the choice $\vec{R}$ may not be canonical. However, any two such choices will be homotopic as nowhere vanishing vector fields on $D^2 \times S^1$ because the only possible obstruction lives in $H^2(S^1 \times D^2; \mathbb{Z}) \cong 0$. Thus the Reeb surgery operation is a well defined operation on homology classes of Euler structures. □

Let us observe that the involution $U \mapsto -U$ on the space of vector fields induces a bijection

$$\mathfrak{Eul}_s(X, \partial_+ X) \to \mathfrak{Eul}_s(X, \partial_- X).$$

We will denote this bijection by $U \mapsto \bar{U}$, and we will call this map the *conjugation* of Euler structures When X is closed the above bijection defines an involution on $\mathfrak{Eul}(X)$,

$$\mathfrak{Eul}_s(X) \ni \mathfrak{e} \mapsto \bar{\mathfrak{e}} \in \mathfrak{Eul}_s(X).$$

For every $\mathfrak{e} \in \mathfrak{Eul}_s(X)$ we set

$$c(\mathfrak{e}) := \mathfrak{e}/\bar{\mathfrak{e}} \in H_1(X).$$

Proposition 3.4. *If $\mathfrak{e} \in \mathfrak{Eul}_s(X)$ is represented by the nonsingular vector field U on X then $c(\mathfrak{e})$ is the Poincaré dual of the Euler class $\boldsymbol{e}(U^\perp) \in H^{m-1}(X)$, where $m = \dim X$, and $U^\perp$ denotes the $(m-1)$-plane sub-bundle of TX orthogonal to U.*

Proof. Fix a CW-decomposition of X with a single m-cell. Denote by $S_1(U^\perp)$ the unit sphere bundle of $U^\perp$ and by $S_1(TX)$ the unit sphere bundle of TX. Set $U_\pm = \pm U$ and denote by V a section of $S_1(U^\perp)$ over the $(m-2)$-skeleton. The section V defines a natural homotopy (see the left-hand side of Figure 3.4)

$$\tilde{U} \colon [-1, 1] \times X^{(m-2)} \to S_1(TX), \quad (t, x) \mapsto \tilde{U}_t(x), \quad \tilde{U}_{\pm 1} = U_\pm,$$

connecting U_- to U_+ inside the plane spanned by $U_\pm$ and V.

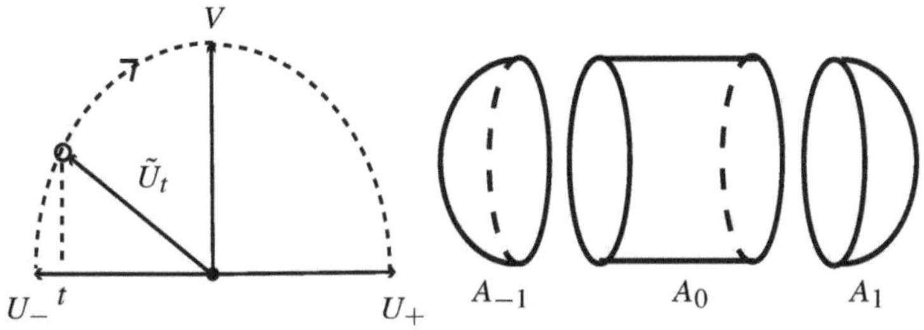

Figure 3.4. $\boldsymbol{PD}\, c(\mathfrak{e}) = \boldsymbol{e}(U^\perp)$.

Suppose σ is an $(m-1)$-cell of X with attaching map $f \colon \partial\sigma \to X^{(m-2)}$ and set $Y = X^{(m-2)} \cup_f \sigma$. Then both $U_\pm$ extend to Y. Set

$$\boldsymbol{S} = \{-1\} \times \sigma \cup [-1, 1] \times \partial\sigma \cup \{1\} \times \sigma =: A_-1 \cup A_0 \cup A_1.$$

Fix a trivialization of TX over σ so that we can view $U_\pm$ as constants maps $\sigma \to S^{m-1}$ and $V|_{\partial\sigma}$ as a map $V \colon \partial\sigma \to S^{m-2}$, where we identify S^{m-2} to the Equator on S^{m-1} perpendicular to U. Now define (see right-hand side of Figure 3.4)

$$H \colon \boldsymbol{S} \to S^{m-1} = \begin{cases} U_\pm & \text{on } A_{\pm 1} \\ \tilde{U} & \text{on } A_0. \end{cases}$$

Observe that deg H is precisely the obstruction to extending the homotopy $\tilde{U}$ over the σ, i.e.

$$\deg H = \langle \boldsymbol{PD}(\mathfrak{e}/\bar{\mathfrak{e}}), \sigma \rangle = \langle \boldsymbol{PD}(c(\mathfrak{e})), \sigma \rangle.$$

On the other hand, H is homotopic to the suspension ΣV of the map $V : \partial\sigma \to S^{m-2}$. By Freudenthal suspension theorem [5, 44] we deduce

$$\deg V = \deg \Sigma V = \deg H.$$

Now observe that deg V is precisely the obstruction to extending V over σ, i.e.

$$\deg V = \langle \boldsymbol{PD}(c(\mathfrak{e})), \sigma \rangle.$$

This concludes the proof of the proposition. □

Theorem 3.5 (Turaev, [113]). *Suppose X is a compact, oriented, smooth, m-dimensional manifold (possibly with boundary) equipped with a smooth triangulation*

$$(K, L) \longleftrightarrow (X, \partial X).$$

Assume $\chi(X, \partial X) = 0$. Then there exists a natural $H_1(X)$-equivariant isomorphism

$$\rho : \mathfrak{Eul}_c(K, L) \to \mathfrak{Eul}_s(X, \partial X).$$

This isomorphism is compatible in a natural way with the barycentric subdivisions.

Proof. We will describe only the construction of ρ. For simplicity, we will do this only in the case $\partial X = \emptyset$. First we need to introduce a bit of terminology.

Consider a line segment $[\alpha, \beta] \subset \mathbb{R}^m$ which we can assume to be of length 3. Denote by V the set of points in $\mathbb{R}$ situated at a distance ≤ 1 from this segment but at a distance ≥ 1 from its endpoints α and β; see Figure 3.5. We denote by D_α

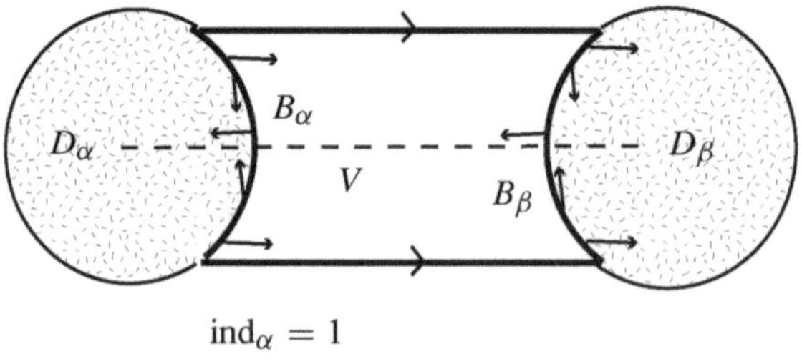

Figure 3.5. A special vector field near a 1-dimensional simplex.

(resp. D_β) the closed disk of radius 1 centered at α (resp. β). We set $B_\alpha := D_\alpha \cap V$, $B_\beta := D_\beta \cap V$ (see Figure 3.5).

A *special vector field* near this segment is by definition a nowhere vanishing vector field $\vec{u}$ on V such that $\vec{u} = \overrightarrow{\alpha\beta}$ on $\partial V \setminus (B_\alpha \cup B_\beta)$. A special vector field defines a map

$$g_\alpha : B_\alpha \to S^{m-1}, \quad x \mapsto \frac{1}{|\vec{u}(x)|}\vec{u}(x)$$

which is constant on ∂B_α. We can regard g_α as a continuous map $B_\alpha / \partial B_\alpha \to S^{m-1}$. As such it has a degree which we denote by $\operatorname{ind}_{\alpha,\beta} \vec{u}$. We can define $\operatorname{ind}_{\beta,\alpha} \vec{u}$ in a similar fashion. Observe that

$$\operatorname{ind}_{\alpha,\beta} + \operatorname{ind}_{\beta,\alpha} = 0.$$

Clearly, for every $n \in \mathbb{Z}$ we can find a special vector field $\vec{v}_n$ near $[\alpha, \beta]$ such that $\operatorname{ind}_{\alpha,\beta} \vec{v}_n = n$.

Denote by K' the first barycentric subdivision of the triangulation K. For each simplex σ of K we denote by $[\sigma]$ its barycenter. If $S = \langle [\sigma_0], [\sigma_1], \dots, [\sigma_p] \rangle$ is a simplex of K', ($\sigma_0 < \sigma_1 < \cdots < \sigma_p$ are simplices of K) then define a vector field V_1 on Int S by

$$V_1(x) = \sum_{0 \le i < j \le p} \lambda_i(x)\lambda_j(x)([\sigma_j] - x).$$

Above, $\lambda_0(x), \lambda_1(x), \dots, \lambda_p(x)$ denote the barycentric coordinates of $x \in \operatorname{Int} S$. These vector fields define a flow on K we will refer to as the *Stiefel flow*. The vertices of K' coincide with the stationary points of this flow (see Figure 3.6).

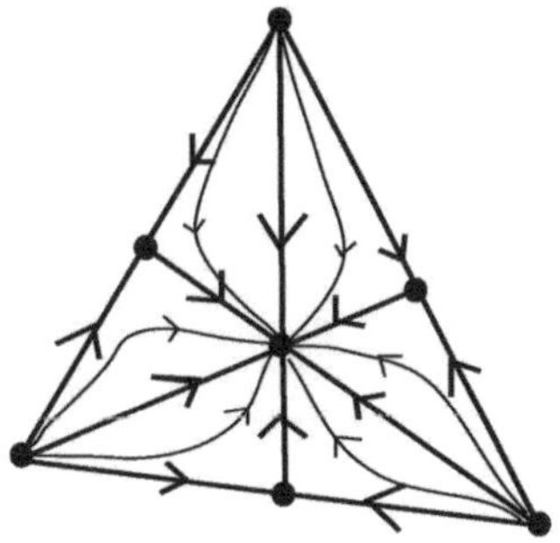

Figure 3.6. The Stiefel flow on a 2-simplex.

For any 1-dimensional simplex $\langle [\sigma_0], [\sigma_1] \rangle$ of K', the vector field V_1 is special near this segment. Moreover, we have

$$\operatorname{ind}_{[\sigma_0],[\sigma_1]\rangle} V_1 = 0.$$

Surround every vertex $[\sigma]$ of K' by a tiny open ball $D(\sigma)$. For every edge $\langle [\sigma_0], [\sigma_1] \rangle$ of K' we denote by $T(\sigma_0, \sigma_1)$ as

$$T(\sigma_0, \sigma_1) := \big(\text{tubular neighborhood of } \langle [\sigma_0], [\sigma_1] \rangle\big) \setminus \big(D(\sigma_0) \cup D(\sigma_1)\big).$$

Suppose now that

$$\xi = \sum_{\sigma_0<\sigma_1} \xi(\sigma_0, \sigma_1)\langle[\sigma_0], [\sigma_1]\rangle$$

is an Euler chain. We can find a nowhere vanishing vector field V_ξ on $K \setminus \bigcup_{\sigma<K} D(\sigma)$ such that the following hold.

- V_ξ is special near each segment $\langle[\sigma_0], [\sigma_1]\rangle$, $\sigma_0 < \sigma_1$, and

$$\mathrm{ind}_{[\sigma_0],[\sigma_1]} V_\xi = \xi(\sigma_0, \sigma_1).$$

- Outside the union of tubes $\bigcup_{\sigma_0<\sigma_1} T(\sigma_0, \sigma_1)$ we have

$$V_\xi \equiv V_1.$$

We will show that V_ξ extends to a nowhere vanishing vector field on K. This is equivalent to the fact that the induced map

$$V_\xi : \partial D(\sigma) \to S^{m-1}$$

is homotopically trivial, i.e.

$$d(\sigma) := \deg\bigl(V_\xi : \partial D(\sigma) \to S^{m-1}\bigr) = 0.$$

Set

$$d_0(\sigma) := \deg\bigl(V_1 : \partial D(\sigma) \to S^{m-1}\bigr).$$

In [43] it was shown that $d_0(\sigma) = (-1)^{\dim\sigma}$. (This identity is intuitively clear in Figure 3.6.) Observe that

$$\begin{aligned} d(\sigma) &= d_0(\sigma) + \sum_{\sigma<\eta}(\mathrm{ind}_{\sigma,\eta} V_\xi - \mathrm{ind}_{\sigma,\eta} V_1) + \sum_{\eta<\sigma}(\mathrm{ind}_{\sigma,\eta} V_\xi - \mathrm{ind}_{\sigma,\eta} V_1) \\ &= (-1)^{\dim\sigma} + \Bigl(\sum_{\sigma<\eta} \xi(\sigma, \eta) - \sum_{\eta<\sigma} \xi(\eta, \sigma)\Bigr). \end{aligned}$$

On the other hand

$$\sum_\sigma (-1)^{\dim\sigma}[\sigma] = \partial\xi = \sum_\sigma\Bigl(\sum_{\eta<\sigma} \xi(\eta, \sigma) - \sum_{\sigma<\eta} \xi(\sigma, \eta)\Bigr)[\sigma]$$

from which it follows that $d(\sigma) = 0$, $\forall\sigma < K$. Thus V_ξ extends to a nowhere vanishing vector field on X. The correspondence

$$\xi \mapsto V_\xi$$

establishes the isomorphism postulated in Theorem 3.5. □

In the sequel we will frequently switch between combinatorial and/or smooth Euler structures so that we will drop the subscripts c and s in $\mathfrak{Eul}_{c/s}$.

Remark 3.6. (a) One can define a notion of Euler structure which combines both the combinatorial and the differential combinatorial aspects. If X is a closed, compact, oriented smooth manifold such that $\chi(X) = 0$, then following [49], we can define an Euler structure as a pair (V, c) where V is a smooth vector field on X with nondegenerate zero set $v^{-1}(0)$ and c is a smooth 1-chain such that

$$\partial c = v^{-1}(0)$$

where the zeros of v are weighted by the Poincaré–Hopf signs. The notion of isomorphism is defined in an obvious way.

(b) One can give a combinatorial description of the conjugation of Euler structure. Suppose X is a smooth, closed oriented manifold such that $\chi(X) = 0$. Fix a smooth triangulation of X so we can identify X with a polyhedron. Suppose the Euler structure $\mathfrak{e}$ is represented by the Euler chain $c \in C_1(X)$. Then the Euler structure $\bar{\mathfrak{e}}$ is represented by the Euler chain

$$\bar{c} = \sum_{\sigma_0 < \sigma_1} (-1)^{\dim \sigma_0 + \dim \sigma_1} \langle \sigma_0, \sigma_1 \rangle + (-1)^{\dim X} c.$$

If we think of combinatorial Euler structures in terms of fundamental families of simplices in the universal Abelian cover, then we can give an even simpler description of this involution.

Suppose $\mathfrak{F}$ is a fundamental family representing the Euler structure $\mathfrak{e}$. Then the Euler structure $\bar{\mathfrak{e}}$ is represented by the fundamental family $\check{\mathfrak{F}}$, were $\check{\mathfrak{F}}$ consists of the cells dual to the cells in $\mathfrak{F}$. For proofs of these facts we refer to [113, Appendix B]. □

§3.3 U(2) and Spin^c(3)

V. Turaev observed in [114] that the space of smooth Euler structures on 3-manifolds can be identified with the space of isomorphisms of spin^c structures. This identification has its origin in some low dimensional representation theoretic "accidents" which we will be discussed in the present section. In the following section we will explain in detail the connection between spin^c structures and smooth Euler structures on 3-manifolds.

Recall that

$$\mathrm{Spin}^c(n) \cong \mathrm{Spin}(n) \times \mathrm{U}(1)/\mathbb{Z}_2, \quad \mathbb{Z}_2 \cong \{(1,1), (-1,-1)\} \subset \mathrm{Spin}(n) \times \mathrm{U}(1).$$

In dimension 3 we have an isomorphism $\mathrm{Spin}(3) \cong \mathrm{SU}(2)$, and the natural map

$$\mathrm{SU}(2) \times S^1 \ni (T, z) \mapsto zT \in \mathrm{U}(2)$$

descends to an isomorphism $\mathrm{Spin}^c(3) \cong \mathrm{U}(2)$.

We want to discuss several facets of this isomorphism. We denote by $\underline{u}(n)$ the Lie algebra of $\mathrm{U}(n)$, $\underline{su}(n)$ the Lie algebra of $\mathrm{SU}(n)$ etc. We begin by presenting a more explicit description of the morphism $\mathrm{U}(2) \to \mathrm{SO}(3)$.

Consider the adjoint representation

$$\mathrm{Ad}\colon\ \mathrm{U}(2) \to \mathrm{Aut}(\underline{u}(2)).$$

The diagonal $\mathrm{U}(1) \hookrightarrow \mathrm{U}(2)$ is the center of $\mathrm{U}(2)$ and thus $\underline{u}(2)$ splits into irreducible parts

$$\underline{u}(2) = \underline{u}(1) \oplus \big(\underline{u}(2)/\underline{u}(1)\big).$$

We denote by $\mathrm{Ad}_0\colon\ \mathrm{U}(2) \to \mathrm{SO}(3)$ the morphism induced by the above representation of $\mathrm{U}(2)$ on the real 3-dimensional space $\underline{u}(2)/\underline{u}(1)$.

More explicitly, the space $\underline{u}(2)/\underline{u}(1)$ can be identified with the orthogonal complement of $\delta_*\underline{u}(1)$ inside $\underline{u}(2)$. This complement is precisely $\underline{su}(2)$. A matrix $A \in \underline{su}(2)$ has the form

$$A = \begin{bmatrix} \mathbf{i}x & \bar{z} \\ -z & -\mathbf{i}x \end{bmatrix}, \qquad x \in \mathbb{R},\ z \in \mathbb{C}.$$

From this description we get a natural decomposition $\underline{su}(2) \cong \mathbb{R} \oplus \mathbb{C}$. The representation Ad_0 associates to each unitary frame $\vec{f} := (f_1, f_2)$ of $\mathbb{C}^2$ a matrix $\mathrm{Ad}_0(\vec{f}) \in \mathrm{Aut}(\underline{su}(2))$ as follows. If

$$f_i = \begin{bmatrix} u_i \\ v_i \end{bmatrix}, \qquad u_i, v_i \in \mathbb{C},\ |u_i|^2 + |v_i|^2 = 1,\ i = 1, 2,$$

$$u_1\bar{u}_2 + v_1\bar{v}_2 = 0.$$

so that

$$\vec{f} := \begin{bmatrix} u_1 & u_2 \\ v_1 & v_2 \end{bmatrix} \in \mathrm{U}(2),$$

then $\mathrm{Ad}_0(\vec{f})$ acts on $\underline{su}(2)$ by

$$\mathrm{Ad}_0(\vec{f}) \cdot \begin{bmatrix} \mathbf{i}x & \bar{z} \\ -z & -\mathbf{i}x \end{bmatrix} = \vec{f} \cdot \begin{bmatrix} \mathbf{i}x & \bar{z} \\ -z & -\mathbf{i}x \end{bmatrix} \cdot \vec{f}^*$$

$$= \begin{bmatrix} u_1 & u_2 \\ v_1 & v_2 \end{bmatrix} \cdot \begin{bmatrix} \mathbf{i}x & \bar{z} \\ -z & -\mathbf{i}x \end{bmatrix} \cdot \begin{bmatrix} \bar{u}_1 & \bar{v}_1 \\ \bar{u}_2 & \bar{v}_2 \end{bmatrix}$$

$$= \begin{bmatrix} \mathbf{i}\big((|u_1|^2 - |u_2|^2)x + 2\mathfrak{Im}(u_1\bar{u}_2 z)\big) & \mathbf{i}(u_1\bar{v}_1 - u_2\bar{v}_2)x + u_1\bar{v}_2\bar{z} - u_2\bar{v}_1 z \\ * & * \end{bmatrix}.$$

Observe that if

$$\vec{f} = \begin{bmatrix} u & -\bar{v} \\ v & \bar{u} \end{bmatrix} \in \mathrm{SU}(2)$$

then

$$\vec{f} \cdot \begin{bmatrix} \mathbf{i}x & z \\ -\bar{z} & -\mathbf{i}x \end{bmatrix} \cdot \vec{f}^* = \begin{bmatrix} \mathbf{i}(|u|^2 - |v|^2)x - 2\mathbf{i}\mathfrak{Im}(uvz) & 2\mathbf{i}u\bar{v}x + u^2\bar{z} - \bar{v}^2 z \\ * & * \end{bmatrix}.$$

In particular,

$$\mathrm{Ad}_0(\vec{f}) \begin{bmatrix} \mathbf{i} & 0 \\ 0 & -\mathbf{i} \end{bmatrix} = 2\mathbf{i} \begin{bmatrix} \frac{1}{2}(|u|^2 - |v|^2) & u\bar{v} \\ \bar{u}v & -\frac{1}{2}(|u|^2 - |v|^2) \end{bmatrix}.$$

Let us point out that the matrix which appears on the right hand side of the above equality is precisely the quadratic term which enters into the formulation of the Seiberg–Witten equations.

The above description of the morphism $U(2) \mapsto SO(3)$ is obviously coordinate dependent. We will now present a coordinate free description of this isomorphism. Suppose now that V is a real, 3-dimensional, oriented, Euclidean vector space. Observe first that, as an $SO(V)$-module, V is isomorphic to the adjoint representation of $SO(V)$ on its Lie algebra. (This is a purely 3-dimensional phenomenon.) This is given by the correspondence

$$V \ni v \mapsto X_v := (v \times \bullet) \in \underline{so}(V). \tag{3.1}$$

Above, "$\times$" denotes the cross product. It can be alternatively defined by

$$u \times v := *(u \wedge v),$$

where $*$ is the Hodge operator. By fixing a nonzero vector $\tau \in V$ we determine several things.

- A subgroup $G_\tau \subset SO(V)$, the stabilizer of τ with respect to the tautological action of $SO(V)$ on V. G_τ is a maximal torus in $SO(V)$ so that $G_\tau \cong S^1$. The Lie algebra of G_τ is generated by the infinitesimal rotation X_τ in (3.1).
- An action of $G_\tau \cong S^1$ on the orthogonal complement V_τ of τ in V. Thus, V_τ is equipped with a complex structure J, and a Hermitian metric

$$\langle u, v \rangle := (u, v) - \mathbf{i}(Ju, v),$$

where $(\bullet, \bullet)$ denotes the Euclidean (real) inner product.

Now form the space

$$\hat{V}_\tau := \mathbb{R} \oplus V \cong \big(\mathbb{R} \oplus \mathbb{R} \cdot \tau\big) \oplus V_\tau \cong \mathbb{C} \oplus V_\tau.$$

$\hat{V}_\tau$ is equipped with a complex structure and Hermitian metric which depend on τ. We will construct a representation

$$\varphi_\tau : U(\hat{V}_\tau) \to SO\,(\mathbb{R}\tau \oplus V_\tau) = SO\,(V)$$

as follows. Define a vector space isometry

$$I_\tau : V \to \underline{su}(\hat{V}_\tau), \qquad \mathbb{R}\tau \oplus V_\tau \ni (t, \phi) \mapsto \begin{bmatrix} \mathbf{i}t \cdot \diamond & \langle \bullet, \phi \rangle \\ - \diamond \cdot \phi & -\mathbf{i}t \cdot \bullet \end{bmatrix}.$$

If $T \in \mathrm{U}(\hat{V}_\tau)$ then $\varphi_\tau(T) \in \mathrm{SO}(V)$ is defined by the commutative diagram

$$\begin{array}{ccc} \underline{su}(\hat{V}_\tau) & \xrightarrow{\mathrm{Ad}_0(T)} & \underline{su}(\hat{V}_\tau) \\ \uparrow I_\tau & & \uparrow I_\tau \\ V & \xrightarrow{\varphi_\tau(T)} & V. \end{array}$$

More explicitly, if

$$T := \begin{bmatrix} z_1 & \langle \bullet, \phi_2 \rangle \\ \phi_1 & z_2 \end{bmatrix}, \quad z_i \in \mathbb{C}, \ \phi_i \in V_\tau,$$

then

$$\varphi_\tau(T) \begin{bmatrix} t \\ \phi \end{bmatrix} = \begin{bmatrix} (|z_1|^2 - |\phi_2|^2)t + 2\mathfrak{Im}\langle z_1\phi, \phi_2 \rangle \\ -\mathbf{i}t(\bar{z}_1\phi_1 - z_2\phi_2) + \bar{z}_1 z_2 \phi - \langle \phi_1, \phi \rangle \phi_2 \end{bmatrix}.$$

Let us point out a confusing fact. We have produced *two* $\mathrm{U}(2)$-representations on $\mathbb{R} \oplus V$. The first one is the tautological representation

$$\theta : \mathrm{U}(\mathbb{C} \oplus V_\tau) \to \mathrm{Aut}(\mathbb{C} \oplus V_\tau),$$

and it is a *complex* representation. The second one is

$$\mathbf{1} \oplus \varphi_\tau : \mathrm{U}(2) \to \mathrm{Aut}(\mathbb{R} \oplus V),$$

and it is *real*. The first representation is the complex spinor representation of $\mathrm{Spin}^c(3)$ and has (infinitesimal) weights $\{\theta_1, \theta_2\}$. The second representation is precisely the adjoint representation and its complexification has weights $\{0, 0, \pm(\theta_2 - \theta_1)\}$.

We have thus shown that a choice of an unit vector $\tau \in V$ canonically defines a complex structure J_τ on $\mathbb{R} \oplus V$ and a morphism

$$\mathrm{U}(\mathbb{R} \oplus V, J_\tau) \to \mathrm{SO}(V).$$

Let us point out another low dimensional accident.

Denote by $\boldsymbol{Herm}(V)$ the space of hermitian structures on $\hat{V} := \mathbb{R} \oplus V$ compatible with the natural orientation. More precisely,

$$\boldsymbol{Herm}(V) = \{J \in \mathrm{SO}(\mathbb{R} \oplus V);\ J^2 = -1\}.$$

Denote by e_0 the vector $1 \oplus 0 \in \mathbb{R} \oplus V$. One can check that the map

$$\boldsymbol{Herm}(V) \ni J \mapsto Je_0 \in S_1(V) = \text{unit sphere in } V,$$

is a bijection. We denote its inverse by

$$S_1(V) \ni \tau \mapsto J_\tau \in \boldsymbol{Herm}(V).$$

The complex space $\hat{V}_\tau$ is precisely $(\hat{V}, J_\tau)$. Moreover, φ_τ is a morphism

$$\varphi_\tau\colon \mathrm{U}(\hat{V}, J_\tau) \to \mathrm{SO}(V).$$

Finally, we want to explain why the map between classifying spaces

$$\pi\colon B\mathrm{U}(2) \to B\mathrm{SO}(3)$$

induced by the morphism $\mathrm{Ad}_0\colon \mathrm{U}(2) \to \mathrm{SO}(3)$ is a homotopic fibration with homotopic fiber BS^1. We will prove a more general result.

Lemma 3.7. *Suppose*

$$1 \hookrightarrow H \hookrightarrow \hat{G} \stackrel{\phi}{\to} G \twoheadrightarrow 1$$

is an extension of compact Lie groups. Then the induced map between the corresponding classifying spaces

$$B\phi\colon B\hat{G} \to BG$$

is (homotopically) a fibration with homotopic fiber BH.

Proof.[2] Denote by $EG \to BG$ (resp. $E\hat{G} \to B\hat{G}$) the universal (classifying) G-bundle (resp. $\hat{G}$-bundle). The natural projection

$$EG \times E\hat{G} \to EG$$

is naturally $\hat{G}$-equivariant, where $\hat{G}$ acts on EG via ϕ and diagonally on $EG \times E\hat{G}$. We thus have a map

$$B\hat{G} \cong (EG \times E\hat{G})/\hat{G} \to EG/\phi(\hat{G}) \cong BG.$$

One can check easily this is a fibration with fiber $E\hat{G}/H \cong BH$. □

[2] I am indebted to Stephan Stolz for this simple argument.

§3.4 Euler structures on 3-manifolds

Suppose X is a smooth, compact, oriented 3-manifold. Assume for the purpose of this preliminary discussion that $\partial X = \emptyset$. Then $\mathfrak{Eul}\,(X)$ is an $H_1(X)$-torsor. On the other hand, the space $\mathrm{Spin}^c(X)$ of isomorphism classes of spin^c structures on X is an $H^2(X)$-torsor. By Poincaré duality we have

$$H_1(X) \cong H^2(X)$$

which shows there exist bijections $\mathfrak{Eul}\,(X) \to \mathrm{Spin}^c(X)$. In this subsection we will construct one such canonical bijection. This will require a fresh look at spin^c structures.

As we have explained in the previous section, $\mathrm{Spin}^c(3) \cong \mathrm{U}(2)$ and

$$\mathrm{SO}(3) \cong \mathrm{U}(2)/\,\mathrm{U}(1),$$

where $\mathrm{U}(1)$ lies inside $\mathrm{U}(2)$ as the diagonal subgroup. We denote by $\mathrm{Ad}_0\colon\ \mathrm{U}(2) \to \mathrm{SO}(3)$ the ensuing morphism.

Definition 3.8. Suppose X is a finite cell complex and $P \to X$ is a principal $\mathrm{SO}(3)$-bundle. We define a spin^c structure on (X, P) as a pair (F, α) where $F \to X$ is a principal $\mathrm{U}(2)$-bundle over X, and α is a surjective, $\mathrm{U}(2)$-equivariant map

$$\alpha\colon F \to P.$$

where $\mathrm{U}(2)$ acts on P via the morphism $\mathrm{Ad}_0\colon\ \mathrm{U}(2) \to \mathrm{SO}(3)$. □

The notion of isomorphism of spin^c-structures is obvious. We will denote by $\mathrm{Spin}^c(X, P)$ set of isomorphism classes of spin^c structures on X.

To obtain a homotopic theoretic description of $\mathrm{Spin}^c(X, P)$ we need to use the classifying spaces $B\,\mathrm{SO}(3)$ and $B\,\mathrm{U}(2)$. The morphism $\mathrm{U}(2) \to \mathrm{SO}(3)$ induces a map

$$\pi\colon B\mathrm{U}(2) \to B\,\mathrm{SO}(3)$$

which is a homotopic fibration with fiber $BS^1 \cong \mathbb{CP}^\infty \cong K(\mathbb{Z}, 2)$. Since the fiber is 2-connected there is only one obstruction to the lifting problem below.

$$\begin{array}{ccc} & & B\mathrm{U}(2) \\ & \overset{?}{\nearrow} & \downarrow \pi \\ X & \xrightarrow[f]{} & B\mathrm{SO}(3). \end{array}$$

It is given by a three dimensional integral class, the third Stiefel–Whitney class $W_3(f) \in H^3(X, \mathbb{Z})$. Two such lifts will be homotopic once they are homotopic over the two skeleton. The obstruction to homotopy is given by a single primary obstruction in $H^2(X, \mathbb{Z})$. We obtain the following result, very similar in spirit to J. Milnor's characterization of spin structure in [71].

Proposition 3.9 (Gompf, [40]). *Suppose X is a compact* CW*-complex, and $P \to X$ is a principal* $\mathrm{SO}(3)$*-bundle. Denote by $X^{(k)}$ the k-skeleton of X. A spin^c-structure on P is a* $\mathrm{U}(2)$ *structure on $P|_{X^{(2)}}$ which extends to $P|_{X^{(3)}}$.*

Two spin^c-structures (F_i, α_i), $i = 1, 2$ on P are isomorphic if and only if the restrictions of F_i to the 2-skeleton of X are isomorphic as $\mathrm{U}(2)$*-bundles. Moreover, the group $H^2(X, \mathbb{Z})$ acts freely and transitively on* $\mathrm{Spin}^c(X, P)$*, i.e.* $\mathrm{Spin}^c(X, P)$ *is naturally an $H^2(X, \mathbb{Z})$-torsor.*

Suppose $Y \hookrightarrow X$ is a subcomplex, $P \to X$ is a principal $\mathrm{SO}(3)$-bundle on X, and s is a homotopy class of sections of $P|_Y$. Then P defines an $\mathrm{SO}(3)$-bundle $[P] \to X/Y$. We define a spin^c structure on P relative to (Y, s) to be a spin^c structure on P induced by a spin^c structure on $[P]$ via the natural projection $X \to X/Y$. The notion of isomorphism is the obvious one.

We can provide a more geometric description of the notion of spin^c structure. Denote by V the rank 3 real vector bundle $P \times_\rho \mathbb{R}^3$ where $\rho\colon \mathrm{SO}(3) \to \mathrm{Aut}(\mathbb{R}^3)$ is the tautological representation. Any nowhere vanishing section τ of V defines a spin^c-structure on σ_τ on P as follows.

• Form the rank 2-real vector sub-bundle $V_\tau \hookrightarrow V$ spanned by the vectors orthogonal to τ. We orient V_τ using the convention

$$\mathbf{or}(V) = \langle\tau\rangle \wedge \mathbf{or}(V_\tau).$$

We have thus equipped V_τ with a $\mathrm{U}(1)$-structure.

• Form the oriented, rank 4 real vector bundle

$$\hat{V}_\tau = \underline{\mathbb{R}} \oplus V \cong \underline{\mathbb{R}} \oplus \langle\tau\rangle \oplus V_\tau.$$

The above decomposition equips $\hat{V}_\tau$ with a complex structure defining a principal $\mathrm{U}(2)$-bundle $F_V = \mathrm{Fr}(\hat{V}_\tau) \to X$. As explained in §3.3, the vector field τ defines a lift

$$\varphi_\tau\colon \mathrm{Fr}(\hat{V}_\tau) \to \mathrm{Fr}(V) = P$$

which is the spin^c structure associated to the vector field τ. We denote it by σ_τ. The associated bundle of complex spinors is the complex bundle $\hat{V}_\tau$ and it has determinant line bundle V_τ. We denote it by $\det(\sigma) = \det(\sigma_\tau)$. Observe that if $h \in H^2(X)$ and $\sigma \in \mathrm{Spin}^c(X, P)$ then

$$\det(h \cdot \sigma) = \det(\sigma) + 2h, \tag{3.2}$$

where

$$H^2(X) \times \mathrm{Spin}^c(X, P) \ni (h, \sigma) \mapsto h \cdot \sigma \in \mathrm{Spin}^c(X, P)$$

denotes the action of $H^1(X)$ on $\mathrm{Spin}^c(X, P)$, and we have used the identification between complex line bundles on X and $H^2(X)$ given by the (integral) first Chern class.

The discussion at the end of §3.3 shows that a choice of spin^c structure is equivalent to a choice of a nowhere vanishing section of TX. Two choices of such sections lead to isomorphic spin^c-structures if and only if they are homotopic over the two skeleton of X.

Exercise 3.2. Formulate and prove the counterparts of the above statements for relative spin^c structures. □

In the special case when X is a closed, oriented three manifold such that $\chi(X) = 0$, and $V \cong TX$ we deduce that the correspondence

$$\text{nowhere vanishing vector field } \tau \text{ on } X \to \sigma_\tau \in \mathrm{Spin}^c(X)$$

induces a bijection

$$\mathfrak{Eul}(X) \to \mathrm{Spin}^c(X), \quad \mathfrak{e} \mapsto \sigma_{\mathfrak{e}}.$$

Proposition 3.4 implies that

$$\boldsymbol{PD}\, c(\mathfrak{e}) = \det(\sigma_{\mathfrak{e}}).$$

The above discussion also shows that the above map is $H^2(X, \mathbb{Z})$-equivariant.

Finally, we can equivalently describe a spin^c-structure as an element $u \in H^2(\mathrm{Fr}_X)$, $\mathrm{Fr}_X :=$ the principal bundle of oriented orthonormal frames of TX, whose restriction to each fiber is a generator of $H^2(\mathrm{SO}(3)) \cong \mathbb{Z}_2$. The correspondence between these two descriptions is clear.

A nowhere vanishing section τ of TX defines a trivial complex line sub-bundle of the rank two complex vector bundle $\hat{V}_\tau := \langle\tau\rangle \oplus TX$. We thus obtain a trivial $\mathrm{U}(1)$ sub-bundle of $F = \mathrm{Fr}_{\hat{V}_\tau}$. We then construct the line bundle

$$\mathcal{L}_\tau := F \to F/\mathrm{U}(1) \overset{\tau}{\cong} \mathrm{Fr}_X\,.$$

Then the class $u := c_1(\mathcal{L}_\tau) \in H^2(\mathrm{Fr}_X)$ restricts to the generator on each fiber.

In terms of the second interpretation, the action of $H^2(X, \mathbb{Z})$ on $\mathrm{Spin}^c(X)$ has the description

$$x \cdot u = \pi_X^* x + u, \quad u \in H^2(\mathrm{Fr}_X),\ x \in H^2(X),$$

where $\pi_X \colon \mathrm{Fr}_X \to X$ is the natural projection. (The above action of $H^2(X)$ on $H^2(\mathrm{Fr}_X)$ obviously preserves $\mathrm{Spin}^c(X) \subset H^2(\mathrm{Fr}_X)$.)

The group of orientation preserving diffeomorphisms of X induces a natural action on $\mathfrak{Eul}(X)$, and thus an action on $\mathrm{Spin}^c(X)$.

The case of 3-manifolds with boundary deserves special consideration. Suppose X is a compact, oriented 3-manifold with boundary ∂X a union of 2-tori. Let us now point out the remarkable fact that the tangent bundle of a torus S has a canonical framing induced by an arbitrary diffeomorphism

$$\phi \colon S \to S^1 \times S^1.$$

Exercise 3.3. Prove that the homotopy class of the framing of TS described above is independent of the diffeomorphism $\phi \colon S \to S^1 \times S^1$. □

The restriction of the SO(3) bundle TX to the boundary ∂X has a canonical trivialization induced by the outer normal section

$$\vec{\nu}\colon \partial X \to TX|_{\partial X},$$

and the canonical framing of $T\partial X$. We can thus define spin^c structures on X relative to ∂X and the above canonical framing of $TX|_{\partial X}$. The notion of relative spin^c structure can be given a more geometric description.

Fix for convenience a Riemann metric g on X and denote by $\vec{\boldsymbol{n}}$ the unit outer normal. Recall that $\underline{\mathbb{R}}$ denotes the trivial real line bundle over a (generic) space. Observe that the rank 4- vector bundle

$$V_0 := \big(\underline{\mathbb{R}} \oplus TX\big)|_{\partial X} \cong \big(\underline{\mathbb{R}} \oplus \langle \vec{\boldsymbol{n}} \rangle\big) \oplus T\partial X,$$

is canonically a *trivialized* U(2)-bundle. We denote by J_0 the complex structure on V_0. A *relative spin^c structure* is then a U(2)-structure J on $V := \underline{\mathbb{R}} \oplus TX$ together with an isomorphism $\phi\colon V|_{\partial X} \to V_0$. Two relative spin^c structures $\sigma_i := (J_i, \phi_i)$, $i = 0, 1$ are called *isomorphic* if there exists an isomorphism $\Phi\colon (V, J_0) \to (V, J_1)$ which makes the following diagram commutative.

$$\begin{array}{ccc} (V, J_0)|_{\partial X} & \xrightarrow{\quad\Phi\quad} & (V, J_1)|_{\partial X} \\ & {\scriptstyle\phi_0}\searrow \quad \swarrow{\scriptstyle\phi_1} & \\ & (V_0, J_0). & \end{array}$$

The space $\mathrm{Spin}^c(X, \partial X)$ of isomorphism classes of relative spin^c structures is naturally an $H^2(X, \partial X)$-torsor. We have an obvious $H^2(X, \partial X)$ map

$$\mathfrak{Eul}(X, \partial X) \to \mathrm{Spin}^c(X, \partial X), \quad \mathfrak{e} \mapsto \sigma(\mathfrak{e}),$$

which must be an isomorphism. The group of orientation preserving diffeomorphisms of $(X, \partial X)$ acts naturally $\mathfrak{Eul}(X, \partial X)$, and thus on the space of relative spin^c-structures as well.

For any $\sigma = (V, J) \in \mathrm{Spin}^c(X, \partial X)$, the determinant line bundle $\det(\sigma)|_{\partial X}$ is equipped with a canonical nowhere vanishing section so that we have a well defined class

$$c_1(\det(\sigma)) \in H^2(X, \partial X).$$

The identity (3.2) shows that the map

$$c\colon \mathrm{Spin}^c(X, \partial X) \to H^2(X, \partial X), \quad \sigma \mapsto c(\sigma) := c_1(\det \sigma)$$

is one-to-one if $H^2(X, \partial X)$ has no 2-torsion.

Example 3.10. Suppose X is the solid torus $S^1 \times D^2$. The relative spin^c-structures σ on X are uniquely determined by

$$c(\sigma) := c_1(\det \sigma) \in H^2(X, \partial X) \cong H_1(X) \cong \mathbb{Z}.$$

Denote by (r, θ) the polar coordinates on D^2 and by φ the angular coordinate on S^1. Consider the nowhere vanishing vector field V on X defined by

$$V(r, \theta, \varphi) := \cos(\pi r/2)\partial_\varphi + \sin(\pi r/2)\partial_r.$$

The vector field $\sin(\pi r/2)\partial_r - \cos(\pi r/2)\partial_\varphi$ is a section of $V^\perp$ which vanishes transversally exactly along the core of the solid torus. Thus, if σ_V denotes the relative spin^c structure determined by V then

$$c_{\sigma_V} = \pm[C] \in H_1(X)$$

where $[C]$ denotes the oriented core, i.e. the cycle $S^1 \times \{0\} \in X$. We deduce that for any spin^c structure $\sigma \in \mathrm{Spin}^c(X, \partial X)$ we have

$$c(\sigma) = (2n+1)[c], \quad n \in \mathbb{Z}.$$

The canonical spin^c structure on the solid torus is the spin^c structure σ_{can} uniquely determined by the condition

$$c(\sigma_{\mathrm{can}}) = -[C].$$

□

Example 3.11. Suppose

$$X = I \times S^1 \times S^1, \quad I = [-1, 1].$$

We can regard it as a trivial cobordism between $\partial_- X := \{-1\} \times T^2$ and $\partial_+ X := \{1\} \times T^2$. The longitudinal vector field ∂_t induces a canonical spin^c structure $\sigma_0 \in \mathrm{Spin}^c(X, \partial_+ X)$.

On the other hand, we can regard X as a 3-manifold with (disconnected) boundary ∂X. The space $\mathrm{Spin}^c(X, \partial X)$ is an $H_1(X) = \mathbb{Z}^2$-torsor. Since $H_1(X) \cong H^2(X, \partial X)$ has no 2-torsion we deduce that the map

$$\mathrm{Spin}^c(X, \partial X) \ni \sigma \mapsto c_1(\det(\sigma)) \in H^2(X, \partial X) \cong \mathbb{Z}$$

is one-to-one. Hence, in this case a relative spin^c-structure is uniquely specified by the associated determinant line bundle.

Observe first that the image of $\mathrm{Spin}^c(X, \partial X)$ in $\mathbb{Z}^2$ via the above map is

$$\varepsilon + 2\mathbb{Z}, \quad \vec{\varepsilon} \in \mathbb{Z}^2.$$

We claim that $\vec{\varepsilon} = 0$. To see this, frame TX using the moving frame

$$(e_1, e_2, e_3) := (\partial_t, \partial_{\theta^1}, \partial_{\theta^2}).$$

Now define $\Gamma_0 \in \mathrm{Vect}\,(X)$ by

$$\Gamma_0(t, \theta^1, \theta^2) = \sin(\pi t/2)e_0 + \cos(\pi t/2)e_1.$$

Clearly Γ_0 points outwards on ∂X. Moreover, the vector e_2 defines a nowhere vanishing section of $\langle \Gamma_0 \rangle^{\perp}$. This shows that the relative spin^c structure induced by Γ_0 has trivial determinant and thus $\vec{\varepsilon} = 0$.

We will denote by σ_0 the spin^c structure induced by Γ_0 and we will refer to it as the *trivial spin^c structure on the trivial cobordism.* □

The vector field Γ_0 constructed in the above example has the following obvious universality property.

Lemma 3.12. *Suppose X is an oriented 3-manifold with boundary an union of tori. Fix a tubular neighborhood U of $\partial X \hookrightarrow X$ of the form $[-1, 1] \times \partial X$ oriented such that ∂_t is the outward pointing longitudinal vector field. Then any nowhere vanishing vector field V pointing outward on ∂X is homologous to a vector which coincides with Γ_0 along U.*

Fix a tubular neighborhood U of ∂X as in the above lemma. Suppose V is a nowhere vanishing vector field on X which is equal to ∂_t along U. Define the vector field $\bar{V}$ by

$$\bar{V} = \begin{cases} -V & \text{in } X \setminus U \\ \Gamma_0 & \text{in } U. \end{cases}$$

This operation induces an involution

$$\mathfrak{Eul}(X, \partial X) \to \mathfrak{Eul}(X, \partial X), \quad \mathfrak{e} \mapsto \bar{\mathfrak{e}}.$$

Proposition 3.13. *Let $\mathfrak{e} \in \mathfrak{Eul}(X, \partial X)$. Then*

$$\mathfrak{e} = c(\sigma(\mathfrak{e})) \cdot \bar{\mathfrak{e}}$$

i.e. $c(\sigma(\mathfrak{e})) = c(\mathfrak{e}/\bar{\mathfrak{e}})$.

Exercise 3.4. Prove the above result. □

The conjugation operation on Euler structure translates to an involution

$$\mathrm{Spin}^c(X, \partial X) \to \mathrm{Spin}^c(X, \partial X), \quad \sigma \mapsto \bar{\sigma}.$$

Suppose now that a closed, oriented 3-manifold X is decomposed into two, manifolds with boundary by an embedded torus (or union of tori) Σ,

$$X = X_0 \cup X_1, \quad \partial X_i \cong (-1)^i \Sigma, \quad i = 0, 1.$$

Fix a tubular neighborhood U of $\Sigma \hookrightarrow X$ which is orientedly diffeomorphic to $I \times \Sigma$. Enlarge

$$X_i \to \hat{X}_i := X_i \cup I \times \Sigma, \quad i = 0, 1.$$

Any nowhere vanishing vector field V_i on $\hat{X}_i$ which points outwards on $\partial \hat{X}_i$ can be assumed to coincide with Γ_0 on $U = I \times \Sigma \subset \hat{X}_i$, $i = 0, 1$. Thus

$$V_0 = V_1 \quad \text{on } I \times \Sigma = \hat{X}_0 \cap \hat{X}_1,$$

so that we can form the glued vector field on X

$$V := V_0 \# V_1.$$

It is easy to see that this induces a map

$$\#\colon \mathfrak{Eul}(X_0, \partial X_0) \times \mathfrak{Eul}(X_1, \partial X_1) \to \mathfrak{Eul}(X).$$

This pairing is not necessarily injective and/or injective. We will refer to this pairing as the *gluing operation*.

§3.5 The Reidemeister–Turaev torsion of Euler structures

Suppose X is a connected, finite simplicial complex such that $\chi(X) = 0$. Fix $p_0 \in X$. Set $H := H_1(X)$ and denote by

$$\pi : \hat{X} \to X \cong \hat{X}/H$$

the universal Abelian cover. Fix $\hat{p}_0 \in \hat{X}$ covering p_0.

Pick now an Euler structure $\mathfrak{e} \in \mathfrak{Eul}(X)$ which we can represent by a spider $\mathfrak{s}$ centered at p_0. $\mathfrak{s}$ admits a unique lift to a spider $\hat{\mathfrak{s}}$ on $\hat{X}$ centered at $\hat{p}_0$. The 0-chain $\partial\hat{\mathfrak{s}}$ depends *only on the homology class of the spider* $\mathfrak{s}$, i.e. *only on the Euler structure* $\mathfrak{e}$! Every point $q \in \partial\hat{\mathfrak{s}}$ is the barycenter of a simplex $\hat{c}_q$ of the triangulation of $\hat{X}$ induced by the triangulation of X. It is clear that if $q_1 \neq q_2$ then $\hat{c}_{q_1}$ and $\hat{c}_{q_2}$ do not cover the same simplex of X. This means that the collection

$$\underline{\mathfrak{c}}_{\mathfrak{e}} = \{\hat{c}_q;\ q \in \partial\hat{q}\}$$

is a *geometric basis* of the $\mathbb{Z}[H]$-module $\underline{C}(\hat{X})$, and we can now define

$$\mathcal{T}_{X,\mathfrak{e},\hat{p}_0} := \mathcal{T}(\underline{C}(\hat{X}), \underline{\mathfrak{c}}_{\mathfrak{e}}, \hat{p}_0) \in \mathbb{Q}(H)/\pm 1.$$

The ± 1 is due to the multiple choices of orderings/orientations of $\underline{\mathfrak{c}}$. Moreover, since $\chi(X) = 0$ we can see that $\mathcal{T}_{X,\mathfrak{e},\hat{p}_0}$ is independent of $\hat{p}_0$. Thus, we can use the notation

$\mathcal{T}_{X,\mathfrak{e}}$. We will call it the *Reidemeister–Turaev torsion* of the Euler structure $\mathfrak{e}$. Observe that

$$\mathcal{T}_{X,h\cdot\mathfrak{e}} \approx h \cdot \mathcal{T}_{X,\mathfrak{e}}, \quad \forall h \in H, \ \mathfrak{e} \in \mathfrak{Eul}_c(X).$$

Above, "$\approx$" denotes the equality in $\mathbb{Q}(H)/\pm 1$.

This refined torsion was defined in terms of a simplicial structure on X. One can prove, much like in the un-refined situation, that this torsion is invariant in an obvious sense under subdivisions and simple homotopy equivalences. For details we refer to [113, §3,4].

One can get rid of the ± 1 ambiguity by ordering and orienting the simplices of X. A choice of ordering and orientations on $\underline{c}$ clearly induces an ordering and orientation on any lift $\underline{c}_{\mathfrak{e}}$. An equivalence class of orderings and orientations of the simplices of X is completely determined by an orientation of the homology space $H_*(X, \mathbb{R})$. This can be seen using the Euler isomorphism

$$\boldsymbol{Eul}\colon \operatorname{Det}(C_*(S(X)) \otimes \mathbb{R} \to \operatorname{Det}(H_*(X, \mathbb{R})).$$

We define a *homology orientation* on a simplicial complex X to be a trivialization of the determinant line $\operatorname{Det}_s H_*(X, \mathbb{R})$. Fix a homology orientation $\mathfrak{o}$. For any geometric basis $\underline{c}$ of $C_*(X)$ we define as in Remark 2.40.

$$\epsilon(\underline{c}, \mathfrak{o}) = \operatorname{sign} \boldsymbol{Eul}(\det(\underline{c})) \in \operatorname{Det}(H_*(X, \mathbb{R})) \cong \mathbb{R}.$$

If $\underline{c}$ is a geometric basis of $C_*(X)$ and $\mathfrak{s}$ is a spider representing a fixed combinatorial Euler structure $\mathfrak{e}$, then we get a geometric basis $\hat{\underline{c}}_{\mathfrak{s}}$ of the $\mathbb{Z}[H_1(X)]$-module $C_*(\hat{X})$ which covers $\underline{c}$. We use this basis to compute the torsion, and we define the *sign-refined* torsion of $(X, \mathfrak{e})$ to be

$$\mathcal{T}_{X,\mathfrak{e},\mathfrak{o}} := \epsilon(\underline{c}, \mathfrak{o}) \mathcal{T}(C_*(\hat{X}), \hat{\underline{c}}_{\mathfrak{s}}).$$

This quantity is independent of the geometric basis $\underline{c}$. The relative Reidemeister–Turaev torsion is defined in a similar way (see [113, 117]).

§3.6 Arithmetic properties of the Reidemeister–Turaev torsion of 3-manifolds

This section is a refinement of §2.4 where we proved several arithmetic properties of the torsion of 3-manifolds. We take-up this subject again, emphasizing the new aspects due to the sign, and spin^c-refinements introduced in the previous section. For more information, and details we refer to [114, 116] which served as our main sources of information.

Denote by $\mathfrak{X}^+$ the collection consisting of triplets $(M, \sigma, \mathfrak{o})$ satisfying the following conditions.

• M is a compact, smooth, oriented 3-manifold, possibly with boundary consisting of an union of tori.
• $\sigma \in \mathrm{Spin}^c(M, \partial M)$.
• $\mathfrak{o}$ is *an enhanced homology orientation.* This means that if $b_1(M) \neq 1$, then $\mathfrak{o}$ is an usual homology orientation, and if $b_1(M) = 1$, then $\mathfrak{o}$ is an orientation of the one-dimensional real vector space $H_1(M, \mathbb{R})$.

We denote by $\mathfrak{X}_1^+$ the subfamily of $\mathfrak{X}^+$ consisting of manifolds with positive b_1.

Remark 3.14. A closed oriented 3-manifold admits a natural homology orientation defined by the Poincaré duality. Similarly, the complement of an oriented link in a rational homology sphere admits a natural homology orientation (see [116, §3]) for details. In the sequel if an admissible manifold is either closed or it is the complement of a link in a rational homology sphere we will tacitly assume it is equipped with the natural orientation, unless indicated otherwise. □

For $(M, \sigma, \mathfrak{o}) \in \mathfrak{X}^+$ we denote by $\mathfrak{T}_{M,\sigma,\mathfrak{o}} \in \mathbb{Q}(H_1(M))$ the sign refined Reidemeister–Turaev torsion of the pair $(M, \partial M)$ and the Euler structure σ. It satisfies the following properties.

$$\mathfrak{T}_{M,h\cdot\sigma,\mathfrak{o}} \approx h\mathfrak{T}_{M,\sigma,\mathfrak{o}}, \quad h \in H_1(M), \ \sigma \in \mathrm{Spin}^c(M, \partial M), \tag{3.3}$$

where we recall that $\approx$ denotes equality up to a sign. In particular

$$\overline{\mathfrak{T}_{M,\sigma,\mathfrak{o}}} \approx \mathfrak{T}_{M,\bar{\sigma},\mathfrak{o}} = c(\bar{\sigma}/\sigma)\mathfrak{T}_{M,\sigma,\mathfrak{o}} = c(\sigma)^{-1}\mathfrak{T}_{M,\sigma,\mathfrak{o}}. \tag{3.4}$$

We can be much more precise about the signs in the above formula. More precisely, we have (see [113, Appendix B], [116, Appendix 3])

$$\overline{\mathfrak{T}_{M,\sigma,\mathfrak{o}}} = (-1)^{b_0(\partial M)}\mathfrak{T}_{M,\bar{\sigma},\mathfrak{o}} = (-1)^{b_0(\partial M)}c(\sigma)^{-1}\mathfrak{T}_{M,\sigma}. \tag{3.5}$$

Example 3.15. We have defined the canonical spin^c structure σ_{can} on the homologically oriented solid torus $Z = D^2 \times S^1$ with axis $K = \{0\} \times S^1$ by the equality (see Example 3.10)

$$c(\sigma_{\mathrm{can}}) = K^{-1} \in H_1(Z).$$

The torsion of the canonical spin^c structure σ_{can} is then

$$\mathfrak{T}_Z \approx (1 - K)^{-1}.$$

□

Remark 3.16. Suppose M is closed (and equipped with the canonical homology orientation). We denote by $\mathrm{Spin}(M)$ the space of isomorphism classes of spin structures on M. It is naturally an $H^1(M, \mathbb{Z}_2)$-torsor. There exists a natural map

$$\mathrm{Spin}(M) \to \mathrm{Spin}^c(M), \quad \mathrm{Spin}(M) \ni \epsilon \mapsto \sigma(\epsilon) \in \mathrm{Spin}^c(M).$$

Its image consists exactly of the fixed points of the involution $\sigma \mapsto \bar{\sigma}$ on $\mathrm{Spin}^c(M)$. For every spin structure ϵ on M we set

$$\mathfrak{T}_{M,\epsilon} := \mathfrak{T}_{M,\sigma(\epsilon)} \in \mathbb{Q}(H).$$

The symmetry properties (3.4) and (3.5) imply

$$\mathfrak{T}_{M,\epsilon} = \bar{\mathfrak{T}}_{M,\epsilon}, \quad \forall \epsilon \in \mathrm{Spin}(M).$$

When M is a rational homology sphere the map $\epsilon \mapsto \sigma(\epsilon)$ is an injection. □

To list the other properties of the sign-refined Reidemeister–Turaev torsion we need to discuss separately several cases. Let $M, \sigma, \mathfrak{o}) \in \mathfrak{X}^+$, and set $H := H_1(M)$.

A. $b_1(M) \geq 2$. We already know that $\mathfrak{T}_{M,\sigma,\mathfrak{o}} \in \mathfrak{N}_2[H]$ (see §2.4).

B. $b_1(M) = 1$. Here we distinguish two subcases.

B.1. $\partial M = \emptyset$. We already know that $\mathfrak{T}_{M,\sigma} \in \mathfrak{N}_2(H)$ (see §2.4). We can be much more precise. The orientation $\mathfrak{o}$ on $H_1(M, \mathbb{R})$ defines a bijection $H/\operatorname{Tors} H \to \mathbb{Z}$, and thus a surjection

$$\deg_{\mathfrak{o}} : H \to \mathbb{Z}.$$

Fix an element $T \in H$ such that $\deg_{\mathfrak{o}} T = 1$. As in §1.5 we set

$$\mathfrak{S}_H := \sum_{h \in \mathrm{Tors}(H)} h \in \mathbb{Z}[H].$$

Then (see [114, §4.2])

$$\mathfrak{T}_{M,\sigma} + \frac{\deg_{\mathfrak{o}}(c(\sigma)) + 2}{2}(1 - T)^{-1}\mathfrak{S}_H - (1 - T)^{-2}\mathfrak{S}_H \in \mathbb{Z}[H]$$

Suppose that $\sigma = \sigma(\epsilon)$, $\epsilon \in \mathrm{Spin}(M)$. Then $\deg_{\mathfrak{o}}(c(\epsilon)) = 0$, and the above equality takes the form

$$\mathfrak{T}_{M,\epsilon,\mathfrak{o}} - \frac{T}{(T-1)^2}\mathfrak{S}_H \in \mathbb{Z}[H].$$

Set $\deg_{\mathfrak{o}}^+ := \max(\deg, 0)$, and define

$$W_M := \sum_{h \in H} \deg_{\mathfrak{o}}^+(h^{-1})h = \frac{T}{(T-1)^2}\mathfrak{S}_H \in \mathbb{Q}(H), \quad \mathfrak{T}^0_{M,\epsilon} = \mathfrak{T}_{M,\epsilon} - W_M.$$

Observe that

$$W_M = \bar{W}_M$$

which implies that $\mathfrak{T}^0_{M,\epsilon}$ is an element of $\mathbb{Z}[H]$ symmetric with respect to the conjugation in $\mathbb{Z}[H]$. We will refer to $\mathfrak{T}^0_{M,\epsilon}$ as the *modified Reidemeister–Turaev* torsion of M. It is independent of the orientation of $H_1(M, \mathbb{R})$ For reasons which will become apparent in §4.1, we will refer to W_M as the *wall crossing term* defined by the orientation of $H_1(M, \mathbb{R})$. Thus

$$\mathfrak{T}_{M,\epsilon,\mathfrak{o}} = \mathfrak{T}^0_{M,\epsilon,\mathfrak{o}} + W_M. \tag{3.6}$$

We set

$$\Delta_{M,\epsilon,\mathfrak{o}} := (1-T)^2\, \mathfrak{T}_{M,\epsilon,\mathfrak{o}} \in \mathbb{Z}[H].$$

As our choice of notation suggests, one should think of $\Delta_{M,\epsilon,\mathfrak{o}}$ as a refined version of the Alexander polynomial of M. This intuition agrees with the identities in Theorem 2.37. We deduce from the equality (3.6) that

$$\Delta_{M,\epsilon,\mathfrak{o}} = (1-T)^2\, \mathfrak{T}^0_{M,\epsilon} + T\mathfrak{S}.$$

If we take the Fourier transform of the above equality we deduce

$$\hat{\Delta}_{M,\epsilon}(1) = |\operatorname{Tors} H|$$

which is precisely the Alexander formula.

B.2. $\partial M = S^1 \times S^1$. In this case M can be viewed as the complement of a knot in a rational homology sphere. An orientation of the knot induces a natural homology orientation. In this case $H_k(M) = 0$ for $k > 1$ and the homology orientation defines as above a surjection $\deg_{\mathfrak{o}} H \to \mathbb{Z}$. Choose an element T such that $\deg_{\mathfrak{o}} T = 1$. Then (see [114, §4.2])

$$\mathfrak{T}_{M,\sigma,\mathfrak{o}} - (1-T)^{-1}\mathfrak{S}_H \in \mathbb{Z}[H]. \tag{3.7}$$

In particular, this implies $\mathfrak{T}_{M,\sigma} \in \mathfrak{N}_1(H)$ as established in Theorem 2.35.

C. $b_1(M) = 0$. Thus $\partial M = \emptyset$ and we know that $\mathfrak{T}_{M,\sigma} \in \mathfrak{N}(H)$. In terms of Fourier transform this means that

$$\hat{\mathfrak{T}}_{M,\sigma}(1) = 0.$$

This time $\mathfrak{T}_{M,\sigma} \notin \mathbb{Z}[H]$ but the torsion still has some extra arithmetical properties. More precisely, if $\mathfrak{o}_0$ is the canonical homology orientation, then (see [114])

$$\mathfrak{T}_{M,\sigma,\mathfrak{o}_0}(g-1)(h-1) = -\boldsymbol{lk}_M(g,h) \bmod \mathbb{Z}, \quad \forall g, h \in H_1(M), \tag{3.8}$$

where $\boldsymbol{lk}_M : H_1(M) \times H_1(M) \to \mathbb{Q}/\mathbb{Z}$ is the linking form of the rational homology 3-sphere M.

Observe that if $(M, \sigma, \mathfrak{o}) \in \mathfrak{X}^+$, then any orientation preserving diffeomorphism f of M induces a new enhanced homology orientation $f^*\mathfrak{o}$ and a new spin^c structure. Define $\epsilon(f) \in \pm 1$ by the equality $f^*\mathfrak{o} = \epsilon(f)\mathfrak{o}$. Then

$$\mathfrak{T}_{M,f^*\sigma,f^*\mathfrak{o}} = \epsilon(f)c(f^*\sigma/\sigma)\, f_*\big(\mathfrak{T}_{M,\sigma,\mathfrak{o}}\big). \tag{3.9}$$

§3.7 Axiomatic description of the Reidemeister–Turaev torsion of 3-manifolds

Denote by $\mathfrak{X}$ the collection consisting of pairs (M, σ) satisfying the following conditions.

• M is a compact, smooth, oriented 3-manifold, possibly with boundary consisting of an union of tori. We will refer to such a manifold as *admissible*.
• $\sigma \in \mathrm{Spin}^c(M, \partial M)$.

We denote by $\mathfrak{X}_1$ the subfamily of $\mathfrak{X}$ consisting of manifolds with positive b_1. We summarize the results established so far. The Reidemeister–Turaev torsion is an invariant

$$\mathfrak{X} \ni (M, \sigma) \mapsto \mathcal{T}_{M,\sigma} \in \mathbb{Q}(H_1(M))/\pm 1$$

satisfying the following properties.

Axiom 0. *Integrality.*

$$\mathcal{T}_{M,\sigma} \in \begin{cases} \mathfrak{N}_2[H] & \text{if } \partial M = \emptyset \\ \mathfrak{N}_1[H] & \text{if } \partial M \neq \emptyset, \end{cases} \qquad \forall \sigma \in \mathrm{Spin}^c(M, \partial M).$$

Axiom 1. *Topological invariance.* The map

$$\mathcal{T}_{M,\bullet} \colon \mathrm{Spin}^c(M, \partial M) \to \mathfrak{N}[H]/\pm 1, \quad \sigma \mapsto \mathcal{T}_{M,\sigma},$$

is *$H_1(M)$-equivariant*, and moreover, if $f \colon M \to M'$ is an orientation preserving diffeomorphism then

$$\mathcal{T}_{M,f^*\sigma'} \approx c(f^*\sigma/\sigma) f_*(\mathcal{T}_{M',\sigma'}).$$

Axiom 2. *Excision.* Suppose M is an admissible 3-manifold, and $L = L_1 \cup \cdots \cup L_n$ is an oriented link in M such that $[L_1], \ldots, [L_n]$ have infinite orders in $H_1(M)$. (In other words, M is the result of a nondegenerate Dehn surgery.) Denote by Z_i a small, open tubular neighborhood of $L_i \hookrightarrow M$ and set

$$E := M \setminus \Big(\bigcup_{i=1}^{n} Z_i\Big).$$

Denote by σ_i the canonical spin^c structure on the solid torus. Then, the elements $(1 - [L_i])$ are invertible in $\mathfrak{N}_1(H_1(M))$ (cf. Lemma 2.42 in §2.5), and for every $\sigma \in \mathrm{Spin}^c(E, \partial E)$ we have

$$\mathcal{T}_{M,\sigma\#\sigma_1\#\ldots\#\sigma_n} \approx \frac{i_\#(\mathcal{T}_{E,\sigma})}{\prod_{i=1}^{n}(1 - [L_i])},$$

where $\boldsymbol{i}: H_1(E) \to H_1(M)$ denotes the inclusion induced morphism.

Axiom 3. *Normalization.* Suppose $L \hookrightarrow S^3$ is a link with at least two components. Denote by E the exterior of E. Then for every $\sigma \in \operatorname{Spin}^c(E, \partial E)$ the torsion $\mathcal{T}_{E,\sigma}$ is a representative of the Alexander polynomial of the link E, i.e.

$$\mathcal{T}_{E,\sigma} \sim \Delta_L,$$

where we recall (see page 45) that $\sim$ denotes the equality in $\mathfrak{N}_2(H)/\pm H$, while $\approx$ denotes the equality in $\mathfrak{N}_2(H)/\pm 1$.

Suppose H is a finitely generated Abelian group of rank ≥ 1. In §1.5 we have identified $\mathfrak{N}_2(H)$ with a ring of functions

$$f: H \to \mathbb{Z}$$

with semi-infinite support, with multiplication given by the convolution product. For every $f \in \mathbb{Z}[[H]]$ we can define

$$\operatorname{supp}(f) := \{h \in H;\ f(h) \neq 0\}.$$

Observe that if $(M, \sigma_0) \in X$ we define

$$\begin{aligned}\mathbf{supp}(M) &:= \left\{\sigma \in \operatorname{Spin}^c(M, \partial M);\ \sigma := h \cdot \sigma_0,\ h \in \operatorname{supp} \mathcal{T}_{M,\sigma_0}\right\} \\ &= \left\{\sigma \in \operatorname{Spin}^c(M, \partial M);\ 0 \in \operatorname{supp}(\mathcal{T}_{M,\sigma})\right\}.\end{aligned}$$

Clearly, $\mathbf{supp}(M)$ is independent of the initial choice $\sigma_0 \in \operatorname{Spin}^c(M, \partial M)$. The group Γ_M of isotopy classes of orientation preserving diffeomorphisms of M preserves $\operatorname{Spin}^c(M, \partial M)$ and Axiom 1 implies that $\mathbf{supp}(M)$ is Γ_M invariant. This is a very powerful restriction when $b_1(M) \geq 2$ because in this case

$$\mathcal{T}_{M,\sigma} \in \mathbb{Z}[H]$$

so that $\mathbf{supp}(M)$ is a *finite* Γ_M-invariant subset of $\operatorname{Spin}^c(M, \partial M)$.

Theorem 3.17 (Uniqueness Theorem; Turaev [115]). *If Ξ_1, Ξ_2 are two invariants on $\mathfrak{X}_1$ satisfying the above axioms then*

$$\Xi_1 \approx \Xi_2.$$

Proof. Let us first observe that if H is a finitely generated Abelian group of rank ≥ 1 then an element $U \in \mathfrak{N}(H)$ is trivial if and only if there exists a non-torsion class h such that

$$(1 - h)U = 0.$$

Define as above $\mathbf{supp}_{\Xi_i}(M) \subset \operatorname{Spin}^c(M, \partial M)$ for any admissible M and set

$$\mathfrak{Y} := \left\{(M, \sigma) \in \mathfrak{X}_1;\ \Xi_1(M, \sigma) \approx \Xi_2(M, \sigma)\right\},$$

$$\mathfrak{Y}' := \{(M,\sigma) \in \mathfrak{X}_1;\ \Xi_1(M,\sigma) \sim \Xi_2(M,\sigma)\}.$$

Clearly $\mathfrak{Y} \subset \mathfrak{Y}'$.

The normalization axiom implies that $\mathfrak{Y}'$ contains the complements of links in S^3 with at least two components. Moreover, the $H_1(M)$-equivariance implies

$$(M,\sigma) \in \mathfrak{Y} \text{ (respectively } \mathfrak{Y}') \iff (M,\sigma') \in \mathfrak{Y} \text{ (respectively } \mathfrak{Y}'),$$

$\forall \sigma' \in \mathrm{Spin}^c(M, \partial M)$. For this reason, we will say that $M \in \mathfrak{Y}$ (or $\mathfrak{Y}'$) if $(M,\sigma) \in \mathfrak{Y}$ (or $\mathfrak{Y}'$) for some σ.

Since any admissible 3-manifold can be obtained by a nondegenerate Dehn surgery we deduce from the excision axiom that $\mathfrak{Y}'$ contains all the admissible manifolds. At this point however, it is not even clear that $\mathfrak{Y} \neq \emptyset$. The uniqueness theorem is equivalent to the equality $\mathfrak{Y}' = \mathfrak{Y}$ whose proof will be carried out in several steps.

Step 1. $I \times S^1 \times S^1 \in \mathfrak{Y}$.

Step 2. $D^2 \times S^1 \in \mathfrak{Y}$.

Step 3. If $K_1, \dots, K_m$, $m \geq 2$ are disjoint unknots in S^3 such that

$$\boldsymbol{Lk}(K_1, K_m) \neq 0, \quad \forall i = 2, \dots, m$$

then the exterior of the link $L = \cup_i K_i$ belongs to $\mathfrak{Y}$.

Step 4. The exterior of any *weakly trivial* link in S^3 in $\mathfrak{Y}$. (A link is called weakly trivial if its components are unknots.)

Step 5. The exterior of any link in S^3 is in $\mathfrak{Y}$.

Step 6. $\mathfrak{Y}' = \mathfrak{Y}$.

The proof of Step 1 is based on the observation that $I \times S^1 \times S^1$ is the exterior E of the Hopf link in S^3. Moreover, using Axiom 3 we deduce $\mathbf{supp}_{\Xi_i}(E)$ consists of a single $\sigma_i \in \mathrm{Spin}^c(E, \partial E)$ which must be Γ_E-invariant. There is only one such Euler structure, namely the trivial one constructed in Example 3.11. Using Step 1 and the excision axiom we deduce that if K denotes the core of a solid torus $X = D^2 \times S^1$ then

$$\big(1 - [K]\big)\Xi_1(X) \approx \Xi_1(X \setminus K) \approx \Xi_2(X \setminus K) \approx \big(1 - [K]\big)\Xi_2(X),$$

so that

$$\big(1 - [K]\big)\big(\Xi_1(X) \pm \Xi_2(X)\big) = 0, \quad \Xi_i(X) \in \mathfrak{N}_2(H_1(X)).$$

Thus $\Xi_1(X) \approx \Xi_2(X)$, and this completes Step 2.

Step 3 follows by induction on the number m of components. The case $m = 1$ is covered by Step 2. We assume the claim is true for $k < m$ and we prove it for links with m components. Denote by E the exterior of a link $L = \bigcup_{i=1}^m L_i$ with m components such that $\boldsymbol{Lk}(L_1, L_j) \neq 0$, $\forall j \neq 1$. Let us first show that $\Xi_1(L) \neq 0$. Denote by M the complement of L_1 in S^3. Then we can regard E as the complement of $L_2 \cup \cdots \cup L_m$ in M. Since $\boldsymbol{Lk}(L_1, L_i) \neq 0$ we deduce that L_i determines a nontrivial homology class in $H_1(M)$. If $\Xi_1(E) = 0$ then the excision axiom would imply $\Xi_1(M) \neq 0$ which we know is not the case.

Fix $\sigma_0 \in \mathrm{Spin}^c(E, \partial E)$. Since $E \in \mathfrak{Y}'$, there exists $g \in H_1(E)$ such that

$$\Xi_1(E, \sigma_0) = \varepsilon g \Xi_2(E, \sigma_0), \quad \varepsilon = \pm 1.$$

From the $H_1(E)$-equivariance we deduce

$$\Xi_1(E, \sigma) = \varepsilon g \Xi_2(E, \sigma), \quad \forall \sigma \in \mathrm{Spin}^c(E, \partial E).$$

g is uniquely determined by the above equality since $\Xi_1(E) \neq 0$. Denote by μ_i an oriented meridian of L_i. The cycles μ_i form a basis of $H_1(E)$ and thus we can write

$$g = \prod_{i=1}^{m} \mu_i^{k_i}, \quad k_i \in \mathbb{Z}.$$

We can now conclude by gluing back to E the tubular neighborhood of L_i, $i > 1$, we have removed and then using the excision axiom. We get a link with fewer components to which we apply the induction hypothesis to conclude

$$k_j = 0, \quad \forall j \neq i.$$

Step 4 follows from the excision axiom and Step 3. Step 5 follows from Step 4 using the excision axiom, and the fact that given any link $L \hookrightarrow S^3$ there exists a disjoint link $K \hookrightarrow S^3$ such that the exterior of $K \cup L$ is diffeomorphic to the exterior of a weakly trivial link; see Lemma 3.18 below. Finally, Step 6 follows from Step 5 using the excision axiom and the fact that any admissible 3-manifold can be obtained by a nondegenerate Dehn surgery. □

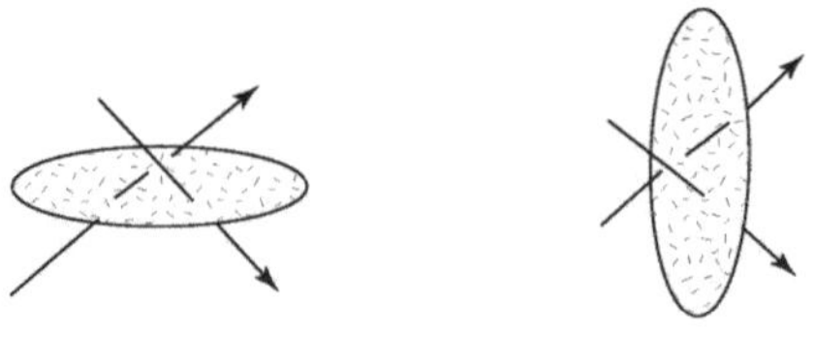

Figure 3.7. Two spanning disks with different piercing properties.

Lemma 3.18. *For any link $L \hookrightarrow S^3$ there exists a disjoint link $K \hookrightarrow S^3$ such that the exterior of $K \cup L$ is diffeomorphic to the exterior of a weak link.*

Proof. Present L by a link diagram. We can transform L into a weak link L' by switching certain over/under- crossings into under/over-crossings. In fact, we only need to do this at self-crossings of the components of L. At each such crossing q consider a small unknotted circle C_q bounding a small disk D_q pierced twice by L. We choose C_q so that D_q is pierced in the same direction; see Figure 3.7.

The circle C_q represents a nontrivial element in $S^3 \setminus L$. We denote by K the link formed by all these unknotted circles. Clearly $K \cup L'$ is trivial. Also it is clear that the exterior of $K \cup L$ is diffeomorphic to the complement of $K \cup L'$ because the change of an over/under-crossing to an under/over-crossing can be performed by a Dehn twist of the complement of C_q localized on the fattened spanning disk. □

Remark 3.19. In concrete problems, the most difficult to deal with is the normalization axiom because the Alexander polynomial of a link is computationally very involved. Fortunately V. Turaev has indicated in [112, §4] an elegant way to bypass this difficulty.

Denote by $\mathfrak{L}$ the family of links in S^3 and for each L denote by $A(L)$ its Alexander polynomial. To prove that

$$\Xi(S^3 \setminus L) = A(L), \quad \forall L \in \mathfrak{L}$$

it suffices to know that $\Xi(S^3 \setminus \text{unknot}) = A(\text{unknot})$ and that $\Xi(S^3 \setminus L)$ changes exactly as the Alexander polynomial when the link L is subjected to some elementary universal transformations which can be described by certain universal Dehn surgeries on $S^3 \setminus L$. Thus the difficulty in proving that an invariant coincides with the refined Reidemeister–Turaev torsion boils down to computing that invariant in the for the complement of the unknot in S^3 and to proving a few surgery formulæ. □

By design, the above approach cannot deal with rational homology spheres due mainly to the excision axiom. In §4.1 we will outline an uniqueness statement of a totally different nature, which involves *additive* gluing formulae, but only *closed* manifolds satisfying $b_1 \leq 1$. For now we are content to do the next best thing, that is to explain how to compute the torsion of a rational homology 3-spheres relying on surgery presentations.

§3.8 The torsion of rational homology 3-spheres. Part 1.

Suppose N is a rational homology 3-sphere described by the Dehn surgery on the oriented link $\mathcal{K} = \mathcal{K}_1 \cup \cdots \cup \mathcal{K}_n \subset S^3$ with rational surgery coefficients

$$\vec{r} = (p_1/q_1, \dots, p_n/q_n) \in \mathbb{Q}^n, \quad q_i > 0,\ (p_i, q_i) = 1,\ \forall i = 1, \dots, n.$$

We denote by E the complement of this link, and we set $G = H_1(E)$, $H = H_1(N)$. We denote by $\mu_i \in G$ the meridian of K_i, oriented by the condition

$$\boldsymbol{Lk}_{S^3}(\mathcal{K}_i, \mu_i) = 1.$$

The manifold N is obtained from E by attaching n solid tori which we denote by $Z_1, \dots, Z_n$. The meridians μ_i determine homology classes $\mu_i \in G$, and $[\mu_i]$ in H. The collection $\{[\mu_i]\}_{1\leq i\leq n}$ generates H, while the collection $\{\mu_i\}_{1\leq i\leq n}$ is an integral basis of G. We thus have a natural isomorphism $G \cong \mathbb{Z}^n$. In particular $G \otimes \mathbb{R} \cong \mathbb{R}^n$ is equipped with a natural Euclidean inner product which we denote by $(\bullet, \bullet)$.

The linking matrix of the link is the $n \times n$ symmetric matrix L with entries

$$\ell_{ij} = \begin{cases} \boldsymbol{Lk}_{S^3}(\mathcal{K}_i, \mathcal{K}_j) & \text{if } i \neq j \\ 0 & \text{if } i = j. \end{cases}$$

For any vector $\vec{x} \in \mathbb{Q}^n$ we denote by $D_{\vec{x}}$ the diagonal matrix

$$D_{\vec{x}} = \operatorname{diag}(x_1, \dots, x_n).$$

We form the symmetric $n \times n$ matrix

$$P := LD_{\vec{q}} + D_{\vec{p}}. \tag{3.10}$$

More explicitly, its entries are

$$p_{ij} = \begin{cases} q_j \ell_{ij} & \text{if } i \neq j \\ p_i & \text{if } i = j. \end{cases}$$

Define

$$P_0 := PD_{\vec{q}}^{-1} = L + D_{\vec{r}}.$$

Note that P_0 is a *symmetric* matrix with rational coefficients. As explained in Appendix §B.2, $\det P \neq 0$, and in fact $|H| = |\det P|$. Moreover the linking form of N is completely determined by the inverse of P_0, in the sense that

$$\boldsymbol{lk}_N([\mu_i], [\mu_j]) = -(P_0^{-1}\mu_i, \mu_j) \bmod \mathbb{Z}.$$

Choose vectors $\vec{\alpha}, \vec{\beta} \in \mathbb{Z}^n$ such that

$$p_i\beta_i - q_i\alpha_i = 1.$$

Moreover, if $q_i = 1$ we choose $(\alpha_i, \beta_i) = (-1, 0)$. Form the matrix

$$K := L \cdot D_\beta + D_\alpha. \tag{3.11}$$

The matrix K has the following interpretation. Denote by $\pi : G \to H$ the natural projection. The cores of the attached solid tori Z_i determine homology classes $\boldsymbol{k}_i \in H$, $i = 1, \dots, n$. The columns of K define elements $K_1, \dots, K_n \in G$. Then (see Appendix §B.2)

$$\boldsymbol{k}_i = \pi K_j, \quad i = 1, \dots, n.$$

For every $S \subset \overline{1,n} := \{1, \dots n\}$ we denote by E_S the manifold obtained by performing the surgery only along the knots K_i, $i \in \bar{S} := \overline{1,n} \setminus S$. Equivalently, E_S is the exterior of the link in N determined by the cores of the attaching solid tori $\{Z_i;\ i \in S\}$.

We set $G_S = H_1(E_S)$. Thus $N = E_\emptyset$ and $H = G_\emptyset$. Observe that to every inclusion $S_1 \hookrightarrow S_2$ there corresponds a projection $G_{S_2} \twoheadrightarrow G_{S_1}$, and an inclusion $\hat{G}_{S_1} \hookrightarrow \hat{G}_{S_2}$. In particular we get a projection $\pi_S\colon G \twoheadrightarrow G_S$, and an injection $\boldsymbol{i}_S\colon \hat{H} \hookrightarrow \hat{G}_S, \forall S$. From the identity

$$D_{\vec{p}} D_{\vec{\beta}} - D_{\vec{q}} D_{\vec{\alpha}} = 1$$

we deduce that

$$P D_{\vec{\alpha}} - K D_{\vec{q}} = 1.$$

In particular, this implies that

$$q_i \pi_S K_i = -\pi_S \mu_i \in G_S, \quad \forall i \in \bar{S}. \tag{3.12}$$

When $S = \{j\}$ we set $E_S := E_j$, $G_S := G_j$, we denote the projection $G \to G_j$ by π_j, and the injection $\hat{H} \hookrightarrow \hat{G}_j$ by $\boldsymbol{i}_j$.

Definition 3.20. A surgery presentation of a rational homology sphere is called *nondegenerate* if for any $i \neq j$ the homology class $\pi_j K_i$, has infinite order in G_j. □

Here is an algebraic criterion for recognizing nondegenerate surgeries.

Lemma 3.21. *The following statements are equivalent.*

(i) *The surgery is nondegenerate.*

(ii) *The matrix P is nondegenerate, i.e. every off-diagonal element of P^{-1} is nontrivial.*

Proof. (ii) $\Longrightarrow$ (i). We argue by contradiction. Suppose there exist $i_0 \neq j_0$ such that the class $\pi_{j_0} K_{i_0}$ has finite order in G_{j_0}. The equality (3.12) implies that $\pi_{j_0}\mu_{i_0}$ has finite order in G_j. Then there exists $n \in \mathbb{Z}^*$ and $\vec{v} \in \mathbb{Z}^n$ such that

$$v_{j_0} = 0, \quad n\mu_{i_0} = P \cdot \vec{v} \iff \vec{v} = nP^{-1}\mu_{i_0}. \tag{3.13}$$

Thus, the coordinates of $\vec{u} := \frac{1}{n}\vec{v}$ are given by the i_0-th column of P^{-1}. Since P is nondegenerate we deduce that $v_k \neq 0, \forall k \neq i_0$. This contradicts the condition $v_{j_0} = 0$ proving that (ii) $\Longrightarrow$ (i). The implication (i) $\Longrightarrow$ (ii) is proved in a similar fashion. □

Exercise 3.5. Prove that any rational homology 3-sphere can be described by a nondegenerate Dehn surgery. □

Definition 3.22. Suppose G is finitely generated Abelian group. Two meromorphic function f_1, f_2 on $\tilde{G} := \mathrm{Hom}(G, \mathbb{C}^*)$ are called *t-equivalent*, and we write this $f_1 \sim f_2$ if there exists $g_0 \in G$ and $\epsilon = \pm 1$ such that

$$f_1(\chi) = \epsilon\chi(g_0) f_2(\chi), \quad \forall \chi \in \tilde{G} \setminus \big(f_1^{-1}(\infty) \cup f_2^{-1}(\infty)\big). \qquad \square$$

Suppose now that the surgery presentation (3.10) is nondegenerate. Then $\tilde{G}$ is a complex n-dimensional torus, and the complex Fourier transform of the torsion of E is a holomorphic function $\hat{\mathfrak{T}}_E(\chi)$ on $\tilde{G}$. Observe that for every $g \in G$ the complex Fourier transform of g viewed as element in $\mathbb{Z}[G]$ is the holomorphic function $\tilde{G} \ni \chi \mapsto \delta_g(\chi) = \chi(g)^{-1} \in \mathbb{C}^*$. The complex Fourier transforms of $1 - K_i \in \mathbb{Z}[G]$, $1 \leq i \leq n$, are the holomorphic functions on $\tilde{G}$,

$$\chi \mapsto 1 - \delta_{K_i}(\chi) = 1 - \chi(K_i)^{-1}.$$

Since rank $G_S = |S|$, we deduce that the space of representations $\tilde{G}_S := \mathrm{Hom}(G_S, \mathbb{C}^*)$ is an union of complex tori of dimension $|S|$ and, according to Corollary 2.38, the complex Fourier transform of $\mathfrak{T}_{E_S}$ is a holomorphic function $\hat{\mathfrak{T}}_{E_S}(\chi)$ on $\tilde{G}_S \setminus \{1\}$. Since the elements $\pi_S K_i, i \in \bar{S}$ have infinite orders in G_S we deduce from the surgery formula Theorem 2.41 and Lemma 2.42 that $\hat{\mathfrak{T}}_{E_S}$ is t-equivalent to *the unique holomorphic extension of the meromorphic function*

$$\tilde{G}_S \setminus \{1\} \ni \chi \mapsto \frac{\hat{\mathfrak{T}}_E(\chi)}{\prod_{i \in \bar{S}}(1 - \chi^{-1}(\tilde{K}_i))}.$$

Let $\chi \in \hat{H} \setminus \{1\}$. Since the collection $\big\{[K_1], \dots, [K_n]\big\}$ generates H there exists j such that $\chi([K_j]) \neq 1$. We deduce from the surgery formula Theorem 2.41 that

$$\hat{\mathfrak{T}}_N(\chi) \sim \frac{1}{1 - \chi([K_j])^{-1}} \hat{\mathfrak{T}}_{E_j}(\boldsymbol{i}_j \chi), \quad \forall \chi \in \hat{H} \setminus \{1\}.$$

The above observations show that for nondegenerate surgeries, the computation of the torsion of E_j simplifies considerably. We have thus proved the following surgery formula.

Theorem 3.23. *Suppose that the rational homology 3-sphere N is described by a nondegenerate Dehn surgery on an oriented link $\mathcal{K} = \mathcal{K}_1 \cup \cdots \cup \mathcal{K}_n$. Set $E := S^3 \setminus K$, $G := H_1(E)$, $H := H_1(N)$, $\tilde{G} := \mathrm{Hom}(G, \mathbb{C}^*)$, and define K_i by (3.11). Denote by π the natural surjection $G \to H$. Its dual $\hat{\pi}$ is an embedding $\hat{H} \hookrightarrow \tilde{G}$. Then the complex Fourier transform of $\mathfrak{T}_E$ is a holomorphic function on $\tilde{G} \setminus \{1\}$, and the meromorphic function*

$$F_P(\chi) := \frac{\hat{\mathfrak{T}}_E(\chi)}{\prod_{i=1}^n (1 - \chi^{-1}(K_i))}$$

is regular at the points $\chi \in \hat{H} \setminus \{1\} \hookrightarrow \tilde{G}$. Moreover if we set $F_P(1) = 0$, then the restriction of F_P to $\hat{H}$ is t-equivalent to the Fourier transform of the torsion of N.

Let us now explain how to use the above theoretical result in concrete computations. For each $\chi \in \hat{H} \setminus \{1\}$ we set

$$S_\chi := \{i;\ \chi(\pi K_i) \neq 1\}.$$

Pick $i \in S_\chi$. Then χ belongs to $\tilde{G}_i \setminus \{1\}$. The group $\tilde{G}_i$ is an union of complex tori, and we denote by $\mathbb{T}_{\chi,i}$ the connected component containing χ. More precisely there exists $\vec{w}_i \in \operatorname{Hom}(G, \mathbb{Z})$ such that

$$\mathbb{T}_{\chi,i} = \{t *_{w_i} \chi;\ t \in \mathbb{C}^*\},$$

where[3]

$$(t *_{\vec{w}_i} \chi)(\vec{v}) := t^{-\langle \vec{w}_i, \vec{v}\rangle} \chi(\vec{v}), \quad \forall t \in \mathbb{C}^*,\ \vec{v} \in G.$$

We think of $\vec{w}_i$ as a *row* vector, and of μ_i as *column* vectors. Observe the following. If we set $z_j := \chi(\mu_j)$, and $n_j := \langle \vec{w}_i, \mu_j\rangle$, and $g = \sum_j \gamma_j \mu_j \in G$ then

$$\delta_g(\chi) = \prod_j z_j^{-\gamma_j}, \quad \delta_g(t *_{\vec{w}_i} \chi) = \prod_j t^{n_j \gamma_j} z_j^{\gamma_j} = t^{\langle \vec{w}_i, g\rangle} \delta_g(\chi). \tag{3.14}$$

The weight $\vec{w}_i$ is determined as follows. G_i is an Abelian group of rank 1. Then its dual $\check{G}_i := \operatorname{Hom}(G_i, \mathbb{Z})$ is a *free* Abelian group of rank 1 which injects in $\operatorname{Hom}(G, \mathbb{Z})$. Then $\vec{w}_i$ is nontrivial integral multiple of one of the two generators of $\operatorname{Hom}(G_i, \mathbb{Z}) \hookrightarrow \operatorname{Hom}(G, \mathbb{Z})$. More explicitly this means that

$$\langle \vec{w}_i, P_i\rangle \neq 0, \quad \langle \vec{w}_i, P_j\rangle = 0, \quad \forall j \neq i,$$

where P_j denotes the j-th column of the presentation matrix P. If we consider the basis $\boldsymbol{e}^j \in \operatorname{Hom}(G, \mathbb{Z})$,

$$\langle \boldsymbol{e}^j, \mu_i\rangle = \delta_i^j,$$

we deduce that there exists $k \in \mathbb{Z}^*$ such that

$$\vec{w}_i = k \cdot n_i \boldsymbol{e}^i \cdot P^{-1}, \quad \forall i,$$

where n_i is the least common multiple of the denominators of the entries on the i-th row of P^{-1}. In other words $\vec{w}_i \in \mathbb{Z}^n \setminus \{0\}$ must be an integral multiple of the i-th row of P^{-1}. We see that $t \mapsto t *_{\vec{w}_i} \chi$ is a complex curve inside $\tilde{G}_i$ which passes through χ at $t = 1$. Hence

$$F_P(\chi) = \frac{1}{\left(1 - \chi(K_i)^{-1}\right)} \cdot \lim_{t \to 1} \frac{\hat{\mathfrak{T}}_E(t *_{w_i} \chi)}{\prod_{j \neq i}\left(1 - t^{\langle \vec{w}_i, K_j\rangle} \chi(K_j)^{-1}\right)}.$$

We can write the above equality in a more symmetric form. Namely

$$i \in S_\chi \Longrightarrow F_P(\chi) = \lim_{t \to 1} \frac{\hat{\mathfrak{T}}_E(t *_{\vec{w}_i} \chi)}{\prod_j\left(1 - t^{\langle \vec{w}_i, K_j\rangle} \chi(K_j)^{-1}\right)}, \quad \forall m \in \mathbb{Z}^*. \tag{3.15}$$

[3] Pay attention to the negative sign in the definition of $t *_{\vec{w}_i}$.

We can improve this formula, to take into account all $i \in S_\chi$. More precisely, the above argument shows that for every weight $\vec{w} \in \mathrm{Hom}(G, \mathbb{Z}) \setminus \{0\}$ such that

$$\langle \vec{w}, P_j \rangle = 0, \quad \langle \vec{w}, K_j \rangle \neq 0, \quad \forall j \in \bar{S}_\chi, \tag{3.16}$$

we have

$$F_P(\chi) = \frac{1}{\prod_{i \in S_\chi} \left(1 - \chi(K_i)^{-1}\right)} \cdot \lim_{t \to 1} \frac{\hat{\mathfrak{T}}_E(t *_w \chi)}{\prod_{j \in \bar{S}_\chi} \left(1 - t^{\langle \vec{w}, K_j \rangle}\right)}. \tag{3.17}$$

Definition 3.24. (a) A weight $\vec{w} \in \mathrm{Hom}(G, \mathbb{Z}) \setminus \{0\}$ satisfying (3.16) is called *admissible* for χ.

(b) We will refer to the process of computing $F_P(\chi)$ described in (3.17) as *regularization along an admissible weight.* □

Remark 3.25. (a) The function F_P determines the Reidemeister–Turaev torsion of N only up to a sign and spin^c structure ambiguity. Later on in §3.9 and §3.10 we will explain how to remove these ambiguities.

(b) The nondegeneracy condition is a technical assumption which we use only because it simplifies considerably the final appearance of the surgery formula. In the recent preprint [116], V. Turaev has established very general surgery formulæ which do not require the nondegeneracy condition. As one can expect, for degenerate surgeries they involve many more terms then our (3.17) and are computationally more complex. □

Before we present several concrete computations based on the above formula, we want to describe an important class of nondegenerate surgery presentations which arises in singularity theory.

Proposition 3.26. *Suppose P is an $n \times n$ matrix with rational entries satisfying the following conditions.*

(i) *P is symmetric and negative definite.*

(ii) *Every off-diagonal element is non negative.*

(iii) *For any collection $S \subsetneq \overline{1, n}$, there exist $i \in S$ and $j \in \bar{S}$ such that $p_{ij} > 0$.*

Then the matrix P is nondegenerate.

Proof. Denote by $\boldsymbol{e}_1, \dots, \boldsymbol{e}_n$ the canonical basis of $\mathbb{Q}^n$. We define an order relation on $\mathbb{Q}^n$

$$\vec{u} = \sum_i u_i \boldsymbol{e}_i \geq \vec{v} = \sum_i v_i \boldsymbol{e}_i \iff u_i \geq v_i, \quad \forall i = 1, \dots, n.$$

We will first prove that if $\vec{u}$ is a vector such that $P\vec{u} \geq 0$ then $\vec{u} \leq 0$. We write

$$\vec{u} = \vec{a} - \vec{b}, \quad \vec{a}, \vec{b} \in \mathbb{Q}_+^n,\ a_i \cdot b_i = 0,\ \forall i.$$

Then

$$0 \leq (P\vec{u}, \vec{a}) = (P(\vec{a} - \vec{b}), \vec{a}) = (P\vec{a}, \vec{a}) - (P\vec{a}, \vec{a}) \leq (P\vec{a}, \vec{a}).$$

On the other hand

$$(P\vec{a}, \vec{b}) = \sum_{i \neq j} p_{ij} a_i b_j \geq 0.$$

Hence we conclude that $\langle P\vec{a}, \vec{a}\rangle = 0$, and since P is negative *definite* we deduce $\vec{a} = 0$.

We can now prove that form every $i \neq j$ we have $(P^{-1}\boldsymbol{e}_i, \boldsymbol{e}_j) \neq 0$. Set $\boldsymbol{f} := P^{-1}\boldsymbol{e}_i$. Then

$$P\boldsymbol{f} = \boldsymbol{e}_i \geq 0$$

so that $\boldsymbol{f} \leq 0$. We will prove that

$$f_j := (\boldsymbol{f}, e_j) = (P^{-1}\boldsymbol{e}_i, \boldsymbol{e}_j) < 0, \quad \forall j.$$

We argue by contradiction. Define

$$S := \{j \in \overline{1, n};\ f_j < 0\} \neq \emptyset.$$

If $S \neq \overline{1, n}$ then we can find $i_0 \in \bar{S}$ such that $p_{i_0 j_0} > 0$ for some $j_0 \in S$. Then $f_{i_0} := (\boldsymbol{f}, \boldsymbol{e}_{i_0}) = 0$.

$$0 \leq (P\boldsymbol{f}, \boldsymbol{e}_{i_0}) = \sum_{j \neq i_0} p_{j i_0} f_j \leq p_{i_0 j_0} f_{j_0} < 0.$$

This completes the proof of the proposition. □

We will now focus exclusively on a very special class of surgery presentations, namely *rational plumbings along trees*. Consider a connected tree $(G, \mathcal{V}, E)$ whose vertices weighted by rational numbers $r_v = p_v/q_v$, $v \in \mathcal{V}$. As in §2.6, we associate to such a weighed graph the 3-manifold $M(G, \vec{r})$ described by a surgery on an oriented link $\mathcal{K} = \mathcal{K}_G \subset S^3$ defined as follows.

- There is a bijection between the vertices of G and the components of $\mathcal{K}$, $v \mapsto \mathcal{K}_v$.
- All the components are unknots.
- If $u, v \in \mathcal{V}$ are connected by an edge then the sublink $\{\mathcal{K}_u, \mathcal{K}_v\}$ is the Hopf link. Otherwise these two components are unlinked.
- The surgery coefficient corresponding to $\mathcal{K}_v$ is r_v.

When all the surgery coefficients are integral we re-obtain the surgery description of the usual plumbings. The plumbing matrix $P_0(G) = P_0(G, \vec{r})$ is defined by

$$p_{uv} = \begin{cases} \boldsymbol{Lk}(\mathcal{K}_u, \mathcal{K}_v) & \text{if } u \neq v \\ p_u/q_u & \text{if } u = v. \end{cases}$$

Set $P(G, \vec{r}) = D_{\vec{q}} \cdot P_0(G, \vec{r})$. Observe that the matrix $P_0(G)$ coincides with what we called $P_0 = L + D_{\vec{r}}$, and $P(G, \vec{r})$ coincides with the presentation matrix associated to the surgery. We say that the weighted graph $(G, \vec{r})$ is nondegenerate if the associate surgery it describes is nondegenerate. Note the following consequence of Proposition 3.26.

Corollary 3.27. *Suppose that the plumbing matrix $P_0(G, \vec{r})$ is negative definite. Then the presentation matrix $P(G, \vec{r}) = D_{\vec{q}}L + D_{\vec{p}}$ is nondegenerate.*

Plumbings defined by negative definite matrices arise naturally in the resolution of isolated singularities of complex surfaces.

We can use the *slam-dunk* operation in [37, §5.3] to transform one weighted graph to an equivalent one. This operation is described in Figure 3.8, where n is an *integral* surgery coefficient. We have the following elementary fact whose proof is left to the reader.

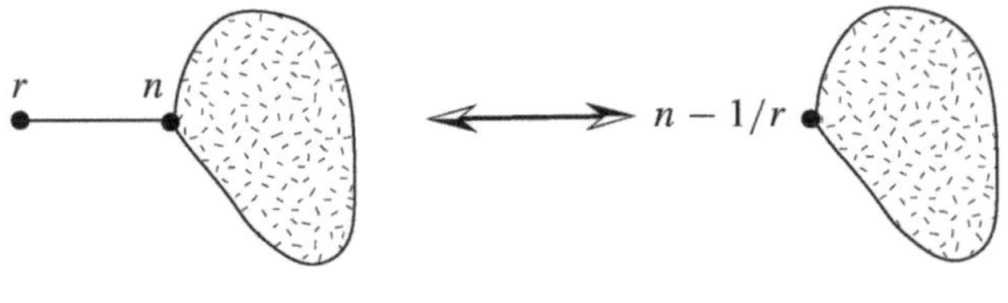

Figure 3.8. Slam-dunk.

Proposition 3.28. *If the weighted tree $(G', \vec{r}\,')$ is obtained from $(G, \vec{r})$ by a slam-dunk, and $P_0(G, \vec{r})$ is nondegenerate or negative definite, then so is $P_0(G', \vec{r}\,')$.*

Exercise 3.6. Prove Proposition 3.28.

We illustrate the above theoretical facts on a concrete example.

Example 3.29 (A plumbed rational homology 3-sphere). Considered the 3-manifold M described by the plumbing in Figure 3.9.

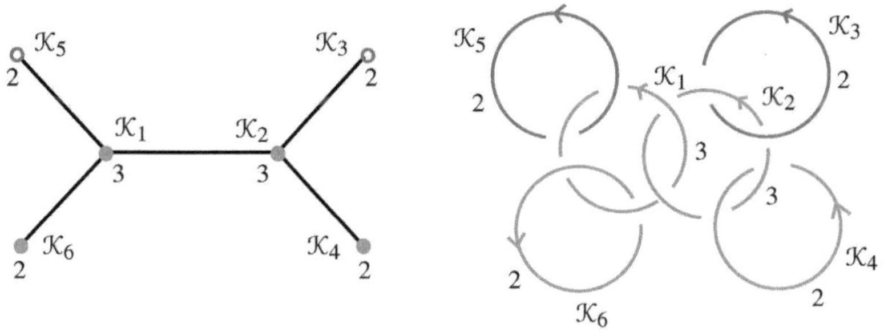

Figure 3.9. A plumbed rational homology 3-sphere.

In this case the matrix P is

$$P := \begin{bmatrix} 3 & 1 & 0 & 0 & 1 & 1 \\ 1 & 3 & 1 & 1 & 0 & 0 \\ 0 & 1 & 2 & 0 & 0 & 0 \\ 0 & 1 & 0 & 2 & 0 & 0 \\ 1 & 0 & 0 & 0 & 2 & 0 \\ 1 & 0 & 0 & 0 & 0 & 2 \end{bmatrix}$$

with determinant 48 so that M is a rational homology 3-sphere. Its inverse is

$$P^{-1} := \begin{bmatrix} 2/3 & -1/3 & 1/6 & 1/6 & -1/3 & -1/3 \\ -1/3 & 2/3 & -1/3 & -1/3 & 1/6 & 1/6 \\ 1/6 & -1/3 & 2/3 & 1/6 & -1/12 & -1/12 \\ 1/6 & -1/3 & 1/6 & 2/3 & -1/12 & -1/12 \\ -1/3 & 1/6 & -1/12 & -1/12 & 2/3 & 1/6 \\ -1/3 & 1/6 & -1/12 & -1/12 & 1/6 & 2/3 \end{bmatrix}, \tag{3.18}$$

so we see that the surgery is nondegenerate. Invoking the `MAPLE` procedure ismith we deduce that $H := H_1(M) := \mathbb{Z}_4 \oplus \mathbb{Z}_{12}$,

$$P := UDV,$$

where

$$D := \begin{bmatrix} 1 & 0 & 0 & 0 & 0 & 0 \\ 0 & 1 & 0 & 0 & 0 & 0 \\ 0 & 0 & 1 & 0 & 0 & 0 \\ 0 & 0 & 0 & 1 & 0 & 0 \\ 0 & 0 & 0 & 0 & 4 & 0 \\ 0 & 0 & 0 & 0 & 0 & 12 \end{bmatrix}, \quad U := \begin{bmatrix} 0 & 0 & 0 & 0 & 0 & 1 \\ 1 & 0 & 0 & 0 & 0 & -3 \\ 0 & -1 & 1 & 2 & 1 & 0 \\ -1 & 0 & 0 & 1 & 0 & 3 \\ -2 & 0 & 0 & 2 & 1 & 5 \\ 14 & 2 & -1 & -19 & -1 & -43 \end{bmatrix}.$$

The above decomposition of P defines a new integral basis $\boldsymbol{e}_1, \ldots, \boldsymbol{e}_6$, and the coordinates of μ_j in this new basis are given by the entries in the j-th column U_j of U.

$\boldsymbol{e}_5$ defines a generator of the $\mathbb{Z}_4$ summand, while $\boldsymbol{e}_6$ defines a generator of the $\mathbb{Z}_{12}$-summand.

The manifold M can also be described as a Dehn surgery on the link depicted in Figure 3.9. Using the formula (2.22) on page 90 we deduce that the torsion of the complement E of this link is

$$\mathcal{T}_E \sim (\mu_1 - 1)^2(\mu_2 - 1)^2,$$

where μ_i denotes the meridian of the component $\mathcal{K}_i$ of this link. Denote by E the complement of this link, and by G its first homology group. In this case, the matrix K is -1. For every $\chi \in \hat{G} = \mathrm{Hom}(G, \mathbb{C}^*)$ we set

$$\zeta_i := \chi(\boldsymbol{e}_i), \quad z_i := \chi(\mu_i) \in \mathbb{C}^*, \quad i = 1, \dots, 6.$$

The Fourier transform of the Alexander polynomial defines the holomorphic function

$$\hat{\mathcal{T}}_E(z) = (z_1^{-1} - 1)^2(z_2^{-1} - 1)^2.$$

Note that if $\chi \in \hat{H}$ then $\zeta_i = 1$ for $1 \leq i \leq 4$. We set $(u, v) := (\zeta_5, \zeta_6) \in \mathbb{C}^* \times \mathbb{C}^*$. Observe now that $u^4 = v^{12} = 1$. Then $z_i = \zeta^{U_i}$ and

$$\begin{aligned} \zeta^{U_1} &= u^{-2}v^{14} = u^2v^2, & \zeta^{U_2} &= v^2, & \zeta^{U_3} &= v^{-1}, \\ \zeta^{U_4} &= u^2v^{-19} = u^2v^5, & \zeta^{U_5} &= uv^{-1}, & \zeta^{U_6} &= uv^5. \end{aligned}$$

We conclude that for $\chi \in \hat{G}$.

$$\hat{\mathcal{T}}_M(\chi) \sim F_P(\chi) = \frac{(z_1^{-1} - 1)^2(z_2^{-1} - 1)^2}{\prod_{j=1}^6(1 - z_j)} = z_1^{-2}z_2^{-2}\frac{(z_1 - 1)(z_2 - 1)}{\prod_{j=3}^6(z_j - 1)}.$$

Recall that this means that for every spin^c structure σ on M, there exist $\vec{k}_\sigma \in \mathbb{Z}^6$ and $\varepsilon = \pm 1$ (independent of σ) such that

$$\hat{\mathcal{T}}_{M,\sigma}(\chi) = \varepsilon\chi^{-1}(\mu^{\vec{k}_\sigma})F_P(\chi), \quad \forall \chi \in \hat{H} \setminus \{1\}.$$

The value of F_P at χ is obtained by regularization along a χ-admissible weight. We consider a special case.

Suppose for example that the character χ is such that $v^{-1} = 1 = v$, but $u^2 \neq 1$. Then $u^2 = -1$ and $z_1 \neq 1$, i.e. $1 \in S_\chi$. Then

$$\begin{aligned} \zeta^{U_1} &= -1, & \zeta^{U_2} &= v^2 = 1, & \zeta^{U_3} &= 1, \\ \zeta^{U_4} &= -1, & \zeta^{U_5} &= u, & \zeta^{U_6} &= u, \quad u = \pm\mathbf{i}. \end{aligned} \tag{3.19}$$

We use the surgery formula (3.15) with the admissible weight $\vec{w}$ described by an integral multiple of the first row of P^{-1}. We take

$$\vec{w} := (4, -3, 1, 1, -2, -2) =: (w_1, \dots, w_6), \quad w_i := \langle \vec{w}, \mu_i \rangle.$$

$$
\begin{aligned}
F_P(t *_{\vec{w}} \chi) &= t^{2w_1+2w_2} z_1^{-2} z_2^{-2} \frac{(t^{-w_1}z_1 - 1)(t^{-w_2}z_2 - 1)}{\prod_{j=3}^{6}(t^{-w_j}z_j - 1)} \\
&= t^2 \frac{(-t^{-4}-1)(t^3-1)}{(t^{-1}-1)(-t^{-1}-1)(ut^2-1)(ut^2-1)} \\
&= -t^3 \frac{t^{-3}(t^4+1)(t^3-1)}{(t-1)(t+1)(ut^{-2}-1)(ut^{-2}-1)}.
\end{aligned}
$$

We conclude

$$
\lim_{t\to 1} F_P(t *_{\vec{w}} \chi) = -\frac{3}{(u-1)^2}.
$$

In Example 3.49 on page 162 we explain how to determine the vector $\vec{k}_\sigma$ for a particular spin^c structure. □

There is one important lesson to be learned from the above example. It is possible that different characters of H may not have a common admissible weight, and thus the computation of F_P for these characters may require regularizations along different weights. If however the plumbing graph has a rich symmetry the computations simplify considerably. We will describe below one such class of surgery presentations which arises in the study of isolated quasihomogeneous singularities. In this case a miracle happens. We can find a weight which is admissible for *all* characters!

Example 3.30 (Seifert fibered rational homology 3-spheres). Consider the surgery presentation in Figure 3.10. It describes a Seifert fibered rational homology 3-sphere. We assume

$$
\ell := -\sum_{i=1}^{\nu} \frac{\beta_i}{\alpha_i} < 0
$$

since this condition arises naturally in the study of the singularities and guarantees the nondegeneracy of the surgery presentation.

Using the surgery trick (2.19) on page 90 we deduce that the Reidemeister–Turaev torsion of the exterior of this link is

$$
\Delta_{\mathcal{K}} \sim (\mu_0 - 1)^{\nu-1}.
$$

We denote by N the 3-manifold obtained by the surgery in Figure 3.10, by E the complement of the link, and we set $G := H_1(E)$, $H := H_1(N)$. Then H admits the presentation

$$
\langle \mu_j,\ j = 0, 1, \dots, \nu;\ \mu_1 \dots \mu_\nu = 1,\ \mu_i^{\alpha_i}\mu_0^{\beta_i} = 1,\ i = 1, \dots, \nu \rangle. \tag{3.20}
$$

We now pick integers (p_i, q_i), $i = 1, \dots, \nu$ such that

$$
\alpha_i q_i - \beta_i p_i = 1,
$$

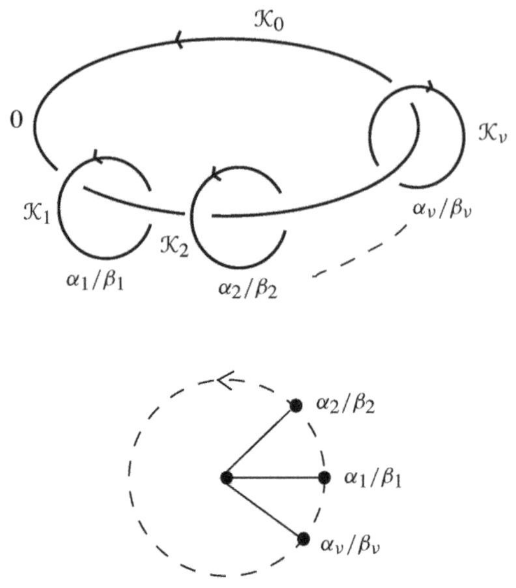

Figure 3.10. Surgery presentation of a Seifert fibered rational homology sphere.

and set as usual $K_i := \mu_i^{p_i}\mu_0^{q_i}$. The central surgery coefficient is integral and we have $K_0 = \mu_0^{-1}$. Later on we will be more specific about the choices p_i, q_i. For each $\chi \in \mathrm{Hom}(G, \mathbb{C}^*)$ we set

$$z_i = \chi^{-1}(\mu_i), \quad \zeta_i = \chi^{-1}(K_i) = z_0^{q_i} z_i^{p_i}, \quad i \in I_\nu := \{1\dots,\nu\}.$$

If $\chi \in \hat{H}$ then (3.20) implies

$$z_0 = \zeta_i^{\alpha_i}, \quad z_i = \zeta_i^{\beta_i} \quad \forall i \in I_\nu. \tag{3.21}$$

For every $\chi \in \hat{H}$ we define its *support* to be

$$S_\chi := \{i \in I_\nu;\ \zeta_i \neq 1\}.$$

Lemma 3.31 (Support lemma). *Let* $\chi \in \hat{H}\setminus\{1\}$ *such that* $\chi(\mu_0) = 1$. *Then* $|S_\chi| \geq 2$.

Proof. Since the classes K_i generate H we deduce $S_\chi \neq \emptyset$. Suppose $|S_\chi| = 1$, say $S_\chi = \{1\}$. Thus $\zeta_i = 1, \forall i \neq 1$. The relations (3.21) imply that

$$1 = \zeta_i^{\alpha_i}, \quad \forall i \in I_\nu$$

The equalities $z_i = \zeta_i^{\beta_i}$ together with the relation $z_1 \dots z_n = 1$ imply $\zeta_1^{\beta_1} = 1$. Hence

$$\zeta_1^{\alpha_1} = \zeta_1^{\beta_1} = 1.$$

Since $\gcd(\alpha_1, \beta_1) = 1$ we must conclude that $\zeta_1 = 1$ as well. This contradicts the condition $S_\chi \neq \emptyset$ and concludes the proof of the lemma. □

Suppose now that $\chi \in \hat{H} \setminus 1$. We distinguish two cases.

1. $z_0 \neq 1$. Using (3.21) we conclude that $\zeta_i \neq 1, \forall i = 0, 1, \dots, \nu$. We deduce

$$F_P(\chi) = \frac{(z_0 - 1)^{\nu-2}}{\prod_{i=1}^{\nu}(1 - \zeta_i)}.$$

2. $z_0 = 1$. Using the Support Lemma we deduce $|S_\chi| \geq 2$. Assume $\{1, 2\} \subset S_\chi$, i.e. $\zeta_1 \neq 1$, $\zeta_2 \neq 1$. We want to choose a weight $\vec{w} \in \operatorname{Hom}(G, \mathbb{Z}) \setminus \{0\}$ satisfying the conditions (3.16). In this case these conditions take the form

$$w_0\beta_i + w_i\alpha_i = 0, \quad \forall i > 2.$$

where $w_i := \vec{w}(\mu_i)$. In other words $w_i = -\frac{\beta_i}{\alpha_i} w_0$. Observe that for every $i > 2$ we have

$$n_i := \langle \vec{w}, K_i \rangle = w_0 q_i + p_i w_i = w_0 (q_i - \frac{p_i \beta_i}{\alpha_i}) = \frac{w_0}{\alpha_i}.$$

Let $\alpha = \operatorname{lcm}(\alpha_1, \dots, \alpha_\nu)$. We set $w_0 = -\alpha$ so that $n_i = -\frac{\alpha}{\alpha_i}$. Using (3.17) with $\langle \vec{w}, K_1 \rangle = \frac{\alpha}{\alpha_1}$, $\langle \vec{w}, K_2 \rangle = \frac{\alpha}{\alpha_2}$, we deduce

$$F_P(\chi) = \lim_{t \to 1} \frac{z_0 t^\alpha - 1}{\prod_{i=1}^{\nu}(1 - \zeta_i t^{\alpha/\alpha_i})}. \tag{3.22}$$

□

Remark 3.32. (a) Observe that the surgery formula (3.22) holds in both cases $z_0 = 1$ or $z_0 \neq 1$. In other words, the weight we have constructed is *admissible for all the characters!* This formula, first appeared in [75], where it plays a central role in the proof of some conjectures arising in singularity theory. For a complete and explicit description of the limit in the right hand side of (3.22) we refer to [76].

(b) The rational function in the right-hand-side of (3.22) also appears in [77] under an algebraic-geometric guise. More precisely, in that paper it is proved that this rational function (of t) coincides with the Poincaré series of associated to the graded ring of regular functions on the singular, quasihomogeneous, affine surface associated to the plumbing in Figure 3.10. This similarity played an important role in [76]. It would be interesting to know if there is a deeper connection between these two apparently unrelated incarnations of this rational function. □

§3.9 Quadratic functions, spin^c structures and charges

The surgery formulæ we have developed so far have one drawback. They produce the torsion up to a sign and a spin^c structure ambiguity. In this section we will describe several methods of keeping track of the spin^c structures when working with surgery presentations.

The first method of keeping track of spin^c structures is algebraic in nature. To present it we need an algebraic digression.

Definition 3.33. Suppose H is a finite Abelian group. A *quadratic function* on H is a function $q\colon H \to \mathbb{Q}/\mathbb{Z}$ such that the map $b = b_q\colon H \times H \to \mathbb{Q}/\mathbb{Z}$ defined by

$$H \times H \ni (x, y) \mapsto b(x, y) := q(xy) - q(x) - q(y) \in \mathbb{Q}/\mathbb{Z}$$

is a bilinear form $b\colon H \times H \to \mathbb{Q}/\mathbb{Z}$. We say that q is a *refinement* of b.

A *quadratic form* is a quadratic function q satisfying $q(nx) = n^2 q(x)$ for all $x \in H, n \in \mathbb{Z}$. In this case we say that q is a *quadratic refinement* of b_q.

Given a bilinear form b on H we denote by $\mathfrak{Q}^c(b)$ the set of refinements of b and by $\mathfrak{Q}(b)$ the set of quadratic refinements. Clearly $\mathfrak{Q}(b) \subset \mathfrak{Q}^c(b)$. □

Note that if $\chi \in \hat{H}$ and $q \in \mathfrak{Q}^c(b)$ then $\chi + q \in \mathfrak{Q}^c(b)$. Conversely, if $q_1, q_2 \in \mathfrak{Q}^c(b)$ then $q_1 - q_2 \in \hat{H}$. This shows that we have a free and transitive action

$$\hat{H} \times \mathfrak{Q}^c(b) \to \mathfrak{Q}^c(b), \quad \hat{H} \times \mathfrak{Q}^c(b) \ni (\chi, q) \mapsto \chi + q.$$

In other words, $\mathfrak{Q}^c(b)$ is a $\hat{H}$-torsor.

Suppose now that M is a rational homology sphere, $H := H_1(M)$. We set $\mathfrak{Q}^c(M) := \mathfrak{Q}^c(\boldsymbol{lk}_M)$. The identity (3.8) on page 130 implies that the sign refined Reidemeister–Turaev torsion defines a map

$$\mathfrak{q}_{\text{tors}}\colon \operatorname{Spin}^c(M) \to \mathfrak{Q}^c(M), \quad \sigma \mapsto \mathfrak{q}_{\text{tors}}(\sigma)$$

by setting

$$\mathfrak{q}_{\text{tors}}(\sigma)(h) = \mathfrak{T}_{M,\sigma,\mathfrak{o}_0}(1) - \mathfrak{T}_{M,\sigma,\mathfrak{o}_0}(h) \bmod \mathbb{Z},$$

where $\mathfrak{o}_0$ is the canonical homology orientation.

The linking form $\boldsymbol{lk}_M$ produces an isomorphism $H \to \hat{H}$, and thus we can regard $\operatorname{Spin}^c(M)$ as an $\hat{H}$-torsor via this isomorphism. The map $\mathfrak{q}_{\text{tors}}$ is then a $\hat{H}$-equivariant bijection. This fact suggests an algebraic encoding for spin^c structures. A spin^c structure is completely determined by the refinement $\mathfrak{q}_{\text{tors}}(\sigma)$ of $\boldsymbol{lk}_M$.

In [7], Brumfiel and Morgan have constructed another $\hat{H}$-equivariant bijection $\mathfrak{q}_{\text{top}}\colon \operatorname{Spin}^c(M) \to \mathfrak{Q}^c(M)$ which we now proceed to describe. Fix a spin^c structure on M. Then (see e.g. [37, §5.7]) there exists at least one simply connected spin^c 4-manifold $(\hat{M}, \hat{\sigma})$ such that $M \cong \partial\hat{M}$ (as oriented manifolds) and $\sigma \cong \hat{\sigma}|_{\partial\hat{M}}$. Set $c(\hat{\sigma}) := c_1(\det\sigma) \in H^2(M, \mathbb{Z})$ so that $c_1(\det\sigma) = c(\hat{\sigma})|_{\partial\hat{M}}$.

Let $L := H^2(\hat{M}, \partial\hat{M}) \overset{PD}{\cong} H_2(\hat{M})$, and denote by Q the intersection pairing

$$Q\colon L \times L \to \mathbb{Z}, \quad Q(x, y) = \langle x \cup y, [\hat{M}, \partial\hat{M}]\rangle.$$

Since M is a rational homology sphere the intersection pairing Q is nonsingular, i.e. the natural map $I_Q\colon L \to \check{L} := \mathrm{Hom}(L, \mathbb{Z})$ induced by Q is an injection. Observe that $\check{L} \cong H^2(\hat{M}) \overset{PD}{\cong} H_2(\hat{M}, \partial\hat{M}$. We have a short exact sequence

$$L \overset{I_Q}{\hookrightarrow} \check{L} \twoheadrightarrow H^2(M) \cong H \to 0.$$

For every $\check{x} \in \check{L}$ we denote by $[\check{x}]$ its image in H. Set $\nu := |H|$. Observe that for every $\check{x} \in \check{L}$ we have $\nu\check{x} \in I_Q(L)$ so that $I_Q^{-1}(\nu\check{x}) \in L$. The intersection form Q defines a nonsingular pairing

$$\check{Q}\colon \check{L} \times \check{L} \to \mathbb{Q}, \quad \check{Q}(\check{x}, \check{y}) = \frac{1}{\nu^2} Q\big(I_Q^{-1}(\nu\check{x}), I_Q^{-1}(\nu\check{y})\big)$$

We say that $\check{Q}$ is the *dual* of Q. Then

$$\boldsymbol{lk}_M([\check{x}], [\check{y}]) = -\check{Q}(\check{x}, \check{y}) \bmod \mathbb{Z}, \quad \forall \check{x}, \check{y} \in L.$$

A vector $\kappa \in \check{L}$ is called *characteristic* if

$$\langle \kappa, x\rangle = Q(x, x) \bmod 2, \quad \forall x \in L.$$

A characteristic vector κ defines a quadratic function

$$q_\kappa\colon H \to \mathbb{Q}/\mathbb{Z}, \quad q_\kappa([\check{x}]) = -\frac{1}{2}\big(\check{Q}(\kappa, \check{x}) + \check{Q}(\check{x}, \check{x})\big) \bmod \mathbb{Z}$$

Exercise 3.7. Prove that q_κ is well defined, i.e.

$$[\check{x}_1] = [\check{x}_2] \Longrightarrow \frac{1}{2}\big(\check{Q}(\kappa, \check{x}_1) + \check{Q}(\check{x}_1, \check{x}_1)\big) = \frac{1}{2}\big(\check{Q}(\kappa, \check{x}_2) + \check{Q}(\check{x}_2, \check{x}_2)\big) \bmod \mathbb{Z}. \quad \square$$

The quadratic function q_κ is a refinement of $\boldsymbol{lk}_M$. The element $c(\hat{\sigma})$ is characteristic, and we set

$$\mathfrak{q}_{\mathrm{top}}(\sigma) = q_{c(\hat{\sigma})}.$$

Lemma 3.34. *The refinement $\mathfrak{q}_{\mathrm{top}}(\sigma)$ is independent of the choice $(\hat{M}, \hat{\sigma})$.*

Proof. Suppose $(\hat{M}_i, \hat{\sigma}_i), i = 0, 1$ are two simply connected, oriented, spin^c-manifolds such that $\partial(\hat{M}, \hat{\sigma}_i) = (M, \sigma)$, $i = 0, 1$. Form as above L_i, Q_i, and $\kappa_i := c(\hat{\sigma}_i)$, $i = 0, 1$. Denote by $-\hat{M}_1$ the manifold M_1 equipped with the opposite orientation, and by Q_1^- the intersection form on $L_1 = H^2(-\hat{M}_1, -\partial\hat{M}_1; \mathbb{Z})$ induced by the opposite orientation. Then $Q_1^- = -Q_1$. Form the closed spin^c-manifold $\hat{M} := \hat{M}_0 \cup_M -\hat{M}_1$. The spin^c structures $\hat{\sigma}_i$ on $(-1)^i\hat{M}_i$ glue up to a spin^c structure $\hat{\sigma}$ on M. Set $\hat{L} := H^2(M, \mathbb{Z})$ and denote by $\hat{Q}$ the intersection form on $\hat{M}$. We have the identifications

$$\hat{L} \cong \{(\check{x}_0, \check{x}_1) \in \check{L}_0 \oplus \check{L}_1;\ [\check{x}_0] = [\check{x}_1]\},$$

and

$$\hat{Q} = (\check{Q}_0 \oplus -\check{Q}_1)|_{\hat{L}}.$$

The cohomology class $\kappa = c_1(\det\hat{\sigma}) \in \hat{L}$ is a characteristic element of $\hat{Q}$, and $\kappa = (\kappa_0, \kappa_1) \in \check{L}_0 \oplus \check{L}_1$.

Fix an element $h \in H$ and elements $\check{x}_i \in \check{L}_i$ such that $h = [\check{x}_i]$, $i = 0, 1$. These elements define $x = (\check{x}_0, \check{x}_1) \in \hat{L}$. Then

$$q_{\kappa_0}(h) - q_{\kappa_1}(h) = \frac{1}{2}\big(\hat{Q}(\hat{\kappa}, x) + \hat{Q}(x, x)\big) = 0 \bmod \mathbb{Z},$$

since κ is a characteristic element of $\hat{Q}$. □

Proposition 3.35. *The map* $\mathfrak{q}_{\mathrm{top}}\colon \mathrm{Spin}^c(M) \to \mathfrak{Q}^c(M)$ *is* $\hat{H}$*-equivariant.*

Proof. Let $\sigma \in \mathrm{Spin}^c(M)$ and $(\hat{M}, \hat{\sigma})$ as in the definition of $\mathfrak{q}_{\mathrm{top}}$. Set $\kappa := c_1(\det\hat{\sigma})$. Recall that $\mathrm{Spin}^c(M)$ is a $\hat{H}$-torsor via the isomorphism $H \cong \hat{H}$, $h \mapsto \chi_h := \boldsymbol{lk}_M(h, \bullet)$. Let $\chi \in \hat{H}$ and $h \in H$ such that $\chi = \chi_h$. We can find $\check{x} \in \check{L}$ such that $h = [\check{x}]$. Then

$$(M, h \cdot \sigma) = \partial(\hat{M}, \check{x} \cdot \hat{\sigma}), \quad c_1(\check{x} \cdot \hat{\sigma}) = 2\check{x} + \kappa.$$

Clearly $\mathfrak{q}_{\mathrm{top}}(h \cdot \sigma) = \chi_h + \mathfrak{q}_{\mathrm{top}}(\sigma)$. □

In §4.1 we will prove that $\mathfrak{q}_{\mathrm{top}} = \mathfrak{q}_{\mathrm{tors}}$. Now we want to provide a different, *intrinsic* description of $\mathfrak{q}_{\mathrm{top}}(\epsilon)$, $\epsilon \in \mathrm{Spin}(M)$.

Let us first recall that according to Milnor [71], any spin structure on an oriented vector bundle E of rank $r \geq 3$ over a compact CW-complex X is described by an extension to the 2-skeleton of X of the framing (trivialization) over the 1-skeleton defined by the orientation. Two such extensions define isomorphic spin structures if and only if they are homotopic. In particular, if X is a compact oriented 3-manifold, and $E = TX$, we deduce that a spin structure on X is defined by a framing[4] of TX. Two framings define isomorphic spin structures if and only if they are *homologous*, i.e. they are homotopic outside a three-dimensional ball. We will thus label spin structures by homology classes of framings of the tangent bundle.

---

[4]Here we have implicitly used the condition $\pi_2(\mathrm{SO}(n)) = 0$, $\forall n \geq 2$ which implies that any framing of TX outside a 3-ball extends to a framing of X everywhere.

To proceed further we need to recall another simple fact. Suppose E is an oriented, real vector bundle of rank $r \geq 3$ over S^1. There are two homotopy classes of framings of E. Of these two there is a *trivial* one described as follows. Extend E to an oriented vector bundle $\hat{E}$ over the 2-dimensional disk D. A framing of E is *trivial* if it *can* be extended to a framing of $\hat{E}$. We associate two each framing $\vec{F}$ of $E \to S^1$ its *parity*, denoted by $[\vec{F}] \in \mathbb{Z}_2$, by declaring the parity of the trivial framing to be 0. Note that two framings $\vec{F}_i$, $i = 1, 2$, on $E_i \to S_1$ induce a framing $\vec{F}_1 \oplus \vec{F}_2$ on $E_1 \oplus E_2$ and

$$[\vec{F}_1 \oplus \vec{F}_2] = [\vec{F}_1] + [\vec{F}_2].$$

We can extend the notion of parity of framings of to bundles over S^1 of rank ≤ 2 using the stabilization trick. More precisely if $\vec{F}$ is a framing of a real, oriented 2-plane bundle $E \to S^1$, then we define its parity as the parity of the induced framing on $\underline{\mathbb{R}}^s \oplus E$ where $\underline{\mathbb{R}}$=the trivial real line bundle, $s \geq 2$. In this case the condition $[\vec{F}] = 0$ implies that there exists a gauge transformation $g\colon S^1 \to \mathrm{SO}(2) \cong S^1$ of *even* degree such that the induced framing $\vec{F}' = \vec{F} \cdot g$ of E extends over the disk bounding S^1. Because of this fact, when speaking of even framings of a 2-plane bundle over S^1, we will always understand framings which extend over the disk.

A spin structure on S^1 is equivalent to a homotopy class of framings of its stable tangent bundle, $T_s S^1 \cong \underline{\mathbb{R}}^s \oplus TS^1$, s is an arbitrary integer ≥ 2. The canonical spin structure corresponds to the canonical framing of the stable tangent bundle.

Suppose now that M is a rational homology 3-sphere, $H := H_1(M)$, and $\epsilon \in \mathrm{Spin}(M)$. Fix a framing $\vec{F}^{\epsilon}_M$ of TM which induces this spin structure. Fix $h \in H\backslash\{1\}$. We represent h by an oriented knot $K \hookrightarrow M$. The chosen framing of TM defines a distinguished class of framings of the normal bundle ν_K of $K \hookrightarrow M$. We will refer to these framings of ν_K as *even* with respect to the spin structure ϵ. They are defined as follows. Equip the stable tangent bundle $T_s K = \underline{\mathbb{R}}^s \oplus TK$ of K with the trivial framing $\vec{F}_K$. Any framing $\vec{F}_\nu$ of ν_K defines two framings $\vec{F}_K \oplus \vec{F}_\nu$ and $\vec{F}^{\nu}_M$ of $\underline{\mathbb{R}}^s \oplus TM|_K$ via the isomorphisms

$$\big(\underline{\mathbb{R}}^s \oplus TK\big) \oplus \nu_K \cong \underline{\mathbb{R}}^{s-1} \oplus \big(\underline{\mathbb{R}} \oplus TM|_K\big).$$

Then the framing $\vec{F}_\nu$ of ν_K if called *even with respect to the spin structure* ϵ if the stable framings $\vec{F}^{\nu}_M$ and $\vec{F}^{\epsilon}_M$ are homotopic, that is

$$[\vec{F}_\nu] = [\vec{F}_M].$$

Suppose now that Z_K is a tubular neighborhood of $K \hookrightarrow M$. We can identify Z_K with the unit disk bundle of ν_K. If we represent a framing $\vec{F}_\nu$ of ν_K as an orthonormal pair of sections $(\boldsymbol{f}_1, \boldsymbol{f}_2)$, then the first section traces an oriented simply closed curve on ∂Z_K. We will denote it by $\vec{F}_\nu(K)$.

Lemma 3.36. *The* $\mathbb{Q}$-valued *linking number* $\boldsymbol{Lk}_M\big(K, \vec{F}_\nu(K)\big)$ *of* K *and* $\vec{F}_\nu(K)$ *is independent* mod $2\mathbb{Z}$ *of the even framing* $\vec{F}_\nu$*, and the homology class of* H.

Exercise 3.8. Prove the above result. □

We have thus obtained a map

$$\mathrm{Spin}(M) \times H \ni (\epsilon, H) \mapsto \mathfrak{q}'_{\mathrm{top}}(\epsilon)(h) = \frac{1}{2} \boldsymbol{Lk}_M\big(K, \vec{\boldsymbol{F}}_\nu(K)\big) \bmod \mathbb{Z},$$

$\vec{\boldsymbol{F}}_\nu$ = some framing of ν_K which is even with respect to ϵ.

Proposition 3.37. *For every* $\epsilon \in \mathrm{Spin}(M)$ *the map*

$$H \ni h \mapsto \mathfrak{q}'_{\mathrm{top}}(\epsilon)(h) \in \mathbb{Q}/\mathbb{Z}$$

is a quadratic refinement of $\boldsymbol{lk}_M$. *Moreover*

$$\mathfrak{q}_{\mathrm{top}}(\epsilon) = \mathfrak{q}'_{\mathrm{top}}(\epsilon).$$

Proof. First of all, let us recall that (see [37, Thm. 5.7.14]) that there exists a spin 4-manifold $(\hat{M}, \hat{\epsilon})$ with the following properties.

- $\partial(\hat{M}, \hat{\epsilon}) = (M, \epsilon)$.
- M is obtained by attaching 2-handlebodies $\{H_1, \dots, H_n\}$to the four-ball D^4.

Denote by Δ_i the co-cores of the handlebodies. These disks define a generating family $[\Delta_i]$ of $H_2(\hat{M}, \partial\hat{M})$, and their boundaries trace oriented knots $K_i \hookrightarrow M$, which define a generating family of $H_1(M)$. Denote by Q the intersection form on $H^2(\hat{M})$ and by $\check{Q}$ its dual on $\mathrm{Hom}\big(H_2(M), \mathbb{Z}\big) \cong H_2(\hat{M}, \partial\hat{M})$.

Since $\hat{M}$ has the homotopy type of a 2-dimensional CW-complex we deduce that spin structure $\hat{\epsilon}$ on M defines a framing $\vec{\boldsymbol{F}}_{\hat{M}}$ of $T\hat{M}$. We denote by $\vec{\boldsymbol{n}}$ the unit outer normal along $\partial\hat{M}$, and by $\vec{\boldsymbol{F}}_M$ a framing of TM defining the spin structure ϵ. The condition $\partial\hat{\epsilon} = \epsilon$ signifies that, for each surface Σ in $\hat{M}$ with boundary on $\partial\hat{M}$, the framing $\vec{\boldsymbol{n}} \oplus \vec{\boldsymbol{F}}_M$ of $T\hat{M}|_{\partial\hat{M}}$ extends over Σ to a framing homotopic to the $\hat{\epsilon}$-framing of $T\hat{M}|_\Sigma$.

Denote by $\hat{\nu}_i$ the normal bundle of $\Delta_i \hookrightarrow \hat{M}$, and by ν_i the normal bundle of $K_i \hookrightarrow M$. Observe that $T_s K_i \cong T_s \Delta_i|_{K_i}$. Denote by $\vec{\boldsymbol{F}}_{K_i}$ the trivial framing of $T_s\Delta_i|_{K_i}$. The framing of ν_i which is even with respect to the spin structure ϵ, is the framing $\vec{\boldsymbol{F}}_i$ such that the framing $\vec{\boldsymbol{F}}_{\hat{M}}|_{K_i}$, and the framing $\big(T\Delta_i|_{K_i}, \vec{\boldsymbol{F}}_{K_i}\big) \oplus (\nu_i, \vec{\boldsymbol{F}}_i)$ on $T_s\hat{M}|_{K_i}$ are homotopic. Since the framing $\vec{\boldsymbol{F}}_{\hat{M}}|_{K_i}$ extends over Δ_i we deduce that $[\vec{\boldsymbol{F}}_i] = 0$, i.e. $\vec{\boldsymbol{F}}_i$ extends (modulo an even degree gauge transformation) to a framing of $\hat{\nu}_i$.

Fix a pair of orthonormal sections $(\boldsymbol{f}_i, \boldsymbol{g}_i)$ defining the framing $\vec{\boldsymbol{F}}_i$. Extend $\boldsymbol{f}_i$ to a nowhere vanishing section $\hat{\boldsymbol{f}}_i$ of $\hat{\nu}_i$. Denote by $\boldsymbol{f}_i(K_i)$ the pushforward inside M of K_i along the normal vector $\boldsymbol{f}_i$. Define $\hat{\boldsymbol{f}}_(\Delta_i)$ in a similar fashion. Clearly $\Delta_i \cap \hat{\boldsymbol{f}}_i(\Delta_i) = \emptyset$. This last condition implies (see e.g. the proof of [20, Proposition A31]) that

$$\boldsymbol{Lk}_M\big(K_i, \boldsymbol{f}_i(K_i)\big) = -\check{Q}(\Delta_i, \Delta_i).$$

This shows

$$\begin{aligned}\mathfrak{q}'_{\mathrm{top}}(\epsilon)(K_i) &= \frac{1}{2}\boldsymbol{Lk}_M(K_i, \boldsymbol{f}_i(K_i)) \bmod \mathbb{Z} = -\frac{1}{2}\check{Q}\big(\Delta_i, \hat{\boldsymbol{f}}_i(\Delta_i)\big) \bmod \mathbb{Z}\\ &= -\frac{1}{2}\check{Q}\big(\Delta_i, \hat{\boldsymbol{f}}_i(\Delta_i)\big) \bmod \mathbb{Z} = \mathfrak{q}_{\mathrm{top}}(\epsilon)(K_i), \quad \forall i. \qquad \square\end{aligned}$$

The above result suggests a simple way of cataloging the spin structures on a rational homology 3-sphere M. It goes as follows. Fix a generating set $S \subset H$, and represent each $s \in S$ by a knot $K_s \subset H$. Then we can describe a spin structure ϵ on H by indicating the framings of the normal bundle of each K_s which ar even with respect to the spin structure ϵ. In the remainder of this section we will explain how to use this simple strategy to produce surgery descriptions of the spin structures on a rational homology spheres. We begin with a topological digression.

Suppose now that M is an oriented 3-manifold with boundary $\Sigma = \partial M$. Given a spin structure ϵ on Σ we can define a relative Stiefel–Whitney class

$$w_2(M, \epsilon) \in H^2(M, \partial M; \mathbb{Z}_2).$$

This class is the obstruction to the existence of a spin structure $\hat{\epsilon}$ on M such that $\partial\hat{\epsilon} = \epsilon$. We will identify $w_2(M, \epsilon)$ via the Poincaré–Lefschetz duality with an element in $H_1(M, \mathbb{Z}_2)$. Observe that the spin structures on a surface Σ can be equivalently described by homotopy classes of framings of the stable tangent bundle $T_s\Sigma$.

Example 3.38. Suppose $M = I \times \Sigma$, $I = [0, 1]$, and ϵ_i are spin structures of $TM|_{\{i\}\times\Sigma}$ defined by framings $\vec{\boldsymbol{F}}_i$ of $TM|_{\{i\}\times\Sigma} \cong T_s\Sigma$. We denote the relative class

$$w_2(M; \epsilon_0, \epsilon_1) \in H_1(M, \mathbb{Z}_2) \cong H_1(\Sigma, \mathbb{Z}_2)$$

by $\delta(\epsilon_1, \epsilon_0)$, or $\delta(\vec{\boldsymbol{F}}_1, \vec{\boldsymbol{F}}_0)$. It can be alternatively described as follows. These two framings define a map $g\colon M \to \mathrm{SO}(3)$ with the property

$$\vec{\boldsymbol{F}}_1 = \vec{\boldsymbol{F}}_0 \cdot g.$$

We obtain an element

$$g_*\colon \mathrm{Hom}\big(H_1(M), H_1(\mathrm{SO}(3))\big) = \mathrm{Hom}\big(H_1(M), \mathbb{Z}_2\big) \cong H^1(\Sigma, \mathbb{Z}_2).$$

Then g_* is the Poincaré dual of $\delta(\vec{\boldsymbol{F}}_1, \vec{\boldsymbol{F}}_0)$, that is

$$\langle g_*, c\rangle = \delta(\vec{\boldsymbol{F}}_1, \vec{\boldsymbol{F}}_0) \cdot c \bmod 2\mathbb{Z}, \quad \forall c \in H_1(\Sigma). \qquad \square$$

We have the following immediate result

Proposition 3.39. *Suppose M is a 3-manifold with boundary Σ. Denote by*

$$j_2\colon H_1(\Sigma, \mathbb{Z}_2) \to H_1(M, \mathbb{Z}_2)$$

the inclusion induced morphism. If ϵ_i, $i = 0, 1$, are two spin structures on Σ then

$$w_2(M, \epsilon_1) - w_2(M, \epsilon_0) = \boldsymbol{j}_2\big(\delta(\epsilon_1, \epsilon_0)\big).$$

On a torus S there are four spin structures

$$\big\{\epsilon_c;\ c \in H_1(S, \mathbb{Z}_2)\big\},$$

where ϵ_0 is the unique spin structure on with the property

$$(S, \epsilon_0) \neq 0 \in \Omega_2^{\mathrm{spin}}.$$

Equivalently, ϵ_0 is the spin structure induced by the canonical framing of TS defined at page 122. In particular, the tangent bundle of an union Σ of 2-tori admits a canonical framing which we will denote by ϵ_0. For every $c \in H^1(\Sigma, \mathbb{Z}_2)$ the spin structure ϵ_c is the unique spin structure on Σ such that

$$\delta(\epsilon_c, \epsilon_0) = c. \tag{3.23}$$

Example 3.40 (Spin structures on the solid torus). Consider the solid torus $Z = S^1 \times D^2$. Set $\lambda = [S^1 \times \{1\}] \in H_1(\partial Z)$, and $\mu = [\{1\} \times \partial D^2] \in H_1(\partial Z)$. It has a natural spin structure $\hat{\epsilon}_e$ induced by the obvious embedding of Z into the Euclidean space $\mathbb{R}$. Equivalently, $\hat{\epsilon}_e$ is the spin structure on the solid torus induced by the unique spin structure on the handlebody $D^2 \times D^2$. This defines a spin structure ϵ_e on the torus ∂Z. We will refer to it as the *Euclidean spin structure*. We want to compute $\delta(\epsilon_e, \epsilon_0) = \delta(\epsilon_0, \epsilon_e) \in H_1(\partial Z, \mathbb{Z}_2)$.

We denote by $\vec{\boldsymbol{F}}_0$ the framing of $\mathbb{R} \oplus T\partial Z$ which induces the spin structure ϵ_0 on ∂Z. We define $\vec{\boldsymbol{F}}_e$ in a similar fashion. Choose $g\colon \partial Z \to \mathrm{SO}(3)$ such that

$$\vec{\boldsymbol{F}}_0 = \vec{\boldsymbol{F}}_e \cdot g.$$

Define $g_* \in H^1(\partial Z, \mathbb{Z}_2)$ as in Example 3.38. Then a little soul searching shows that

$$\langle g_*, \mu\rangle = [\vec{\boldsymbol{F}}_0|_\mu] + [\vec{\boldsymbol{F}}_e|_\mu] = 1,$$

and

$$\langle g_*, \lambda\rangle = [\vec{\boldsymbol{F}}_e|_\lambda] + [\vec{\boldsymbol{F}}_0|_\lambda] = 1.$$

This shows that $\delta(\vec{\boldsymbol{F}}_e, \vec{\boldsymbol{F}}_0) = \mu + \lambda \bmod 2\mathbb{Z}$ so that $\epsilon_e = \epsilon_{\lambda+\mu}$.

Using Proposition 3.39 we deduce that for every $\kappa \in H_1(\partial Z, \mathbb{Z}_2)$ we have

$$w_2(Z, \epsilon_k) = \boldsymbol{j}_2(\kappa - \lambda - \mu) = \boldsymbol{j}_2\kappa + \lambda.$$

Thus ϵ_k extends to a spin structure on Z if and only if $k = \lambda + n\mu \bmod 2$, $n \in \mathbb{Z}$. □

Example 3.41. Consider the solid torus $Z = S^1 \times D^2$. Set $\lambda = [S^1 \times \{1\}] \in H_1(\partial Z)$, and $\mu = [\{1\} \times \partial D^2] \in H_1(\partial Z)$. If $\boldsymbol{f}$ is a framing of the normal bundle of the axis of solid torus, then the curve on ∂Z traced by the first vector in $\boldsymbol{f}$ carries the homology class $n\mu + \lambda$, $n \in \mathbb{Z}$. The integer n is called the *degree of the framing*, and completely characterizes its homotopy class.

A spin structure $\hat{\epsilon}$ on Z induces a spin structure ϵ on ∂Z. In particular, it has the form ϵ_κ, $\kappa \in H_1(\partial Z, \mathbb{Z}_2)$. Since ϵ extends over Z we deduce that κ has the form

$$\kappa = n_\kappa \mu + \lambda \bmod 2.$$

We claim that the framings of the normal bundle of axis K of the solid torus which are even with respect to the spin structure $\hat{\epsilon}$ are exactly the framings of degree $n = n_\kappa \bmod 2$. To prove this fact we need to verify this statement for a single spin structure. We will do this for the Euclidean spin structure $\hat{\epsilon}_e$ defined in Example 3.40. Denote by $\vec{\boldsymbol{F}}_e$ a framing of TZ which induces the Euclidean spin structure $\hat{\epsilon}_e$. In this case

$$\delta(\epsilon_e, \epsilon_0) = \lambda + \mu \bmod 2.$$

We have to show that the framings ν_K which are $\hat{\epsilon}_e$-even must have odd degrees.

Observe that framing $\vec{\boldsymbol{F}}_\nu$ of the normal bundle ν_K of $K \hookrightarrow Z$ which is even with respect to $\hat{\epsilon}_e$ is determined by the condition $[\vec{\boldsymbol{F}}] = [\vec{\boldsymbol{F}}_e] = 0$. The canonical framing $[\vec{\boldsymbol{F}}_{\mathrm{can}}]$ of ν_K given by the direct product description $S^1 \times D^2$ has degree 0, and parity $[\vec{\boldsymbol{F}}_{\mathrm{can}}] = 1$. Thus $\vec{\boldsymbol{F}}_\nu$ must have odd degree. □

Suppose now that M is a 3-manifold whose boundary $\Sigma = \partial M$ is an union of tori. Fix a relative spin^c structure $\sigma \in \mathrm{Spin}^c(M, \partial M)$. As explained on page 122, the canonical framing of $T\Sigma$ defines a section s_σ of $\det \sigma|_\Sigma$, and we have a relative Chern class

$$c(\sigma) = c_1(\det \sigma, s_\sigma) \in H^2(M, \partial M) \cong H_1(M).$$

The following result follows immediately from the description of a relative spin^c structure in terms of the canonical framing of the tangent bundle of a torus.

Proposition 3.42.

$$c(\sigma) = w_2(M, \epsilon_0) \bmod 2.$$

Exercise 3.9. Prove the above result. □

Suppose $\mathcal{K} = \mathcal{K}_1 \cup \cdots \cup \mathcal{K}_n$ is an oriented link in S^3. Denote by Z_i an open tubular neighborhood of $\mathcal{K}_i$. The exterior of the link is

$$E := S^3 \setminus \bigcup_{i=1}^n Z_i,$$

and set $G := H_1(E)$. ∂E has n boundary components which we denote by $\partial_i E$, $i = 1, \dots, n$. Denote by μ_i and λ_i the meridian and longitude of $\mathcal{K}_i$ oriented such that

$$\lambda_i \cdot \mu_i = 1,$$

where the intersection product is defined in terms of the orientation of ∂Z_i as boundary component of E. As in §3.8 we define the linking matrix of $\mathcal{K}$ to be the $n \times n$ symmetric matrix L with entries

$$\ell_{ij} = \begin{cases} \boldsymbol{Lk}(\mathcal{K}_i, \mathcal{K}_j) & \text{if } i \neq j \\ 0 & \text{if } i = j. \end{cases}$$

We denote the inclusion induced morphism $H_1(\partial E, \mathbb{Z}) \to H_1(E, \mathbb{Z})$ by $\boldsymbol{j}$. $\boldsymbol{j}_2$ will denote the similar morphism for $\mathbb{Z}_2$-coefficients. The images $\boldsymbol{j}\mu_i$ of the meridians define a basis of G, and thus an isomorphism $G \cong \mathbb{Z}^n$. Since there is no danger of confusion we will write μ_i instead of $\boldsymbol{j}\mu_i$. Denote by $\vec{\boldsymbol{e}} \in G \cong \mathbb{Z}^n$ the vector $(1, \dots, 1) = \sum_i \mu_i$. Then

$$\boldsymbol{j}\lambda_i = \sum_{k \neq i} \ell_{ki}\mu_k = L\vec{\boldsymbol{e}}.$$

We have the following result.

Proposition 3.43.

$$w_2(E, \epsilon_0) = \sum_{i=1}^{n} \boldsymbol{j}_2(\mu_i + \lambda_i) = \sum_{i=1}^{n} x_i\mu_i \in H_1(E, \mathbb{Z}_2), \quad x_i = 1 + \sum_{k \neq i} \ell_{ki} \bmod 2.$$

Proof. E is equipped with a natural spin structure $\hat{\epsilon}_e$ induced by the embedding $E \hookrightarrow S^3$. It induces on each boundary component exactly the Euclidean spin structure described in Example 3.40. Using the computations in that example we deduce

$$\partial\hat{\epsilon} = \epsilon_{\kappa_0},$$

where

$$\kappa_0 = \sum_{i=1}^{n} (\mu_i + \lambda_i) \bmod 2 \in H^1(\partial E, \mathbb{Z}_2).$$

Using Proposition 3.39 we deduce

$$w_2(E, \epsilon_0) = w_2(E, \epsilon_0) - w_2(E, \partial\hat{\epsilon}_e) = \boldsymbol{j}_2\kappa_0 = \vec{\boldsymbol{e}} + L\vec{\boldsymbol{e}} \bmod 2. \tag{3.24}$$

□

Remark 3.44. For a different proof of this result we refer to [116, Lemma 1.3]. □

Consider the spin structure $\epsilon = \epsilon_\kappa$ on ∂E labelled by an element $\kappa \in H_1(\partial E, \mathbb{Z}_2)$. The above identity coupled with Proposition 3.39 implies

$$w_2(E, \epsilon_\kappa) = 0 \iff \kappa - \kappa_0 \in \ker \boldsymbol{j}_2 \iff \boldsymbol{j}_2\kappa = \vec{e} + L\vec{e} \bmod 2. \tag{3.25}$$

Suppose we perform a Dehn surgery on this link with rational surgery coefficients $\vec{r} \in \mathbb{Q}^n$, $r_i = p_i/q_i$ where $q_i > 0$, $(p_i, q_i) = 1$, $\forall i = 1, \dots n$. We denote by Z_i the attaching solid tori, by $M = M_{\vec{r}}$ the 3-manifold obtained by this surgery, and by H its first homology group, $H := H_1(M)$. Set $\vec{p} = (p_1, \dots, p_n), \vec{q} = (q_1, \dots, q_n)$.

For every $\vec{r} \in \mathbb{Q}^n$ we denote by $D_{\vec{r}}$ the diagonal matrix

$$D_{\vec{r}} = \operatorname{diag}(r_1, \dots, r_n).$$

Then H admits the presentation

$$G \xrightarrow{P} G \overset{\pi}{\twoheadrightarrow} H \to 0,$$

where P is the $n \times n$ matrix

$$P := L \cdot D_{\vec{q}} + D_{\vec{p}}. \tag{3.26}$$

The axes of the attaching solid tori Z_i define homology classes in M which we denote by $\boldsymbol{k}_i$. Here is how one can determine them.

For each i we choose a pair of integers (α_i, β_i) such that

$$p_i\beta_i - \alpha_i q_i = 1, \quad \alpha_i \equiv q_i + \beta_i \equiv 1 \bmod 2 \tag{3.27}$$

If $q_i = 1$, i.e. the surgery coefficient r_i is integral, then we set $\alpha_i = -1, \beta_i = 0$. The 2×2 matrices

$$\Gamma_i := \begin{bmatrix} p_i & \alpha_i \\ q_i & \beta_i \end{bmatrix}, \quad i = 1, \dots, n,$$

define the attaching maps of the Dehn surgery. Note that the entries in the i-th column of P are the coordinates of $\boldsymbol{j}\Gamma_i\mu_i$ with respect to the natural basis in G. We set

$$K_i := \boldsymbol{j}(\Gamma_i\lambda_i) = \alpha_i\mu_i + \beta_i \sum_{j \neq i} \ell_{ji}\mu_j \in G.$$

Then $\boldsymbol{k}_i = \pi(K_i), \forall i$. We denote by K the matrix

$$K := L \cdot D_{\vec{\beta}} + D_{\vec{\alpha}}. \tag{3.28}$$

Note that the entries in the i-th column of K are the coordinates of K_i. Then

$$K\vec{e} = \sum_i K_i.$$

A spin structure $\hat{\epsilon}$ on M induces spin structures on $\hat{\epsilon}_{\mathrm{ext}}$ on E and $\hat{\epsilon}_i$ on Z_i with the property that

$$\Gamma_i \hat{\epsilon}_i|_{\partial Z_i} = \hat{\epsilon}_{\mathrm{ext}}|_{\partial_i E_K}.$$

We write $\hat{\epsilon}_i|_{\partial Z_i} = \epsilon_{c_i}$, $\hat{\epsilon}_{\mathrm{ext}}|_{\partial_i E} = \epsilon_{\kappa_i}$, where $c_i, \kappa_i \in H_1(\partial Z_i, \mathbb{Z}_2)$. Set

$$\kappa = \bigoplus_i \kappa_i \in H_1(\partial E, \mathbb{Z}_2),$$

so that $\hat{\epsilon}_{\mathrm{ext}}|_{\partial E} = \epsilon_\kappa$. The homology classes c_i, κ_i satisfy the compatibility conditions

$$\kappa_i = \Gamma_i c_i.$$

Since ϵ_{c_i} extends over Z_i we deduce

$$c_i = \lambda_i + u_i \mu_i \in H^1(\partial Z_i, \mathbb{Z}_2), \quad u_i \in \mathbb{Z}_2.$$

Since ϵ_{κ_i} extends over E we deduce that

$$\boldsymbol{j}_2 \kappa_i = \boldsymbol{j}_2(\mu_i + \lambda_i), \quad \forall i.$$

We conclude that

$$\boldsymbol{j}_2 \Gamma_i(\lambda_i + u_i \mu_i) = \boldsymbol{j}_2(\mu_i + \lambda_i) \iff K_i + u_i \boldsymbol{j}(p_i \mu_i + q_i \lambda_i) = \boldsymbol{j}_2(\mu_i + \lambda_i) \in G \otimes \mathbb{Z}_2,$$

for all i. If we set $\vec{u} = \sum_i u_i \mu_i \in G$ we can rewrite this in the more compact form

$$K\vec{e} + P\vec{u} = \vec{e} + L\vec{e} \bmod 2 \overset{(3.25)}{=} \boldsymbol{j}_2 \kappa_0 = \boldsymbol{j}_2 \kappa \bmod 2. \tag{3.29}$$

The next proposition summarizes the above observations.

Proposition 3.45. *Let M, P, K as above. Every spin structure $\hat{\epsilon}$ on M can be described by a vector $\vec{u} \in \mathbb{Z}$ satisfying*

$$P\vec{u} + K\vec{e} = \vec{e} + L\vec{e} \bmod 2. \tag{3.30}$$

Conversely, every vector $u \in \mathbb{Z}^n$ satisfying the above condition determines spin structure on $\hat{\epsilon}(\vec{u})$ on M. Moreover, for every $1 \leq i \leq n$, the curve $(\lambda_i + u_i \mu_i) \subset \partial Z_i$ defines a framing of the normal bundle of the core of $Z_i \hookrightarrow M$ which is even with respect to $\hat{\epsilon}(\vec{u})$.

The correspondence described in Proposition 3.45 can be further refined. Every relative spin^c structure σ on E is completely determined by its characteristic class $c(\sigma) \in G$. We can write

$$c(\sigma) = \mu^{\vec{\nu}}, \quad \nu \in \mathbb{Z}^n.$$

Proposition 3.42 and 3.43 imply that is $\vec{\nu}$ is, in the terminology of [116], a *charge*, i.e.

$$\vec{\nu} = \vec{e} + L\vec{e} \bmod 2 \iff \vec{\nu} = \boldsymbol{j}_2 \kappa_0 \bmod 2. \tag{3.31}$$

From now on, we will freely identify the spin^c structures on E with charges. For every $\vec{m} \in \mathbb{Z}^n$ we denote by $\mu^{\vec{m}} \in \mathbb{Z}[G]$ the monomial $\mu_1^{m_1} \dots \mu_n^{m_n}$.

E is equipped with a natural homology orientation (see [116]), and if $\mathcal{T}_{\mathcal{K},\sigma}$ is the sign-refined torsion of E corresponding to the spin^c structure σ then (see (3.5))

$$\mathcal{T}_{\mathcal{K},\sigma}(\mu_1^{-1}, \dots, \mu_n^{-1}) = (-1)^n \mu^{-\vec{v}} \mathcal{T}_{\mathcal{K},\sigma}(\mu_1, \dots, \mu_n). \tag{3.32}$$

The attaching solid tori are equipped with canonical spin^c structures and we get a surjection

$$\pi : \operatorname{Spin}^c(E, \partial E) \to \operatorname{Spin}^c(M), \quad \operatorname{Spin}^c(E, \partial E) \ni \vec{v} \mapsto \vec{v}^M \in \operatorname{Spin}^c(M).$$

The above correspondence is equivariant in the sense that if σ_1, σ_2 are two relative spin^c structures on E, $\sigma_2 = g \cdot \sigma_1$, $g = \mu^{\vec{\lambda}} \in G$, then

$$\frac{1}{2}(\vec{v}_2 - \vec{v}_1) = \vec{\lambda}, \quad v_2^M = \pi(g) \cdot v_1^M.$$

Moreover

$$c(\vec{v}^M) = \pi(c(\sigma)) \cdot \prod_i \boldsymbol{k}_i^{-1} = \pi\left(\mu^{\vec{v} - K\vec{e}}\right). \tag{3.33}$$

Observe that two charges $\vec{v}_1, \vec{v}_2$ induce identical spin^c structures on M if and only if

$$\frac{1}{2}(\vec{v}_1 - \vec{v}_2) \in \operatorname{Im}(P).$$

An *integral characteristic vector* of the surgery presentation (see [37, Def. 5.7.19]) is a vector $\vec{c} \in \mathbb{Z}^n$ such that, for all i,

$$p_i \equiv p_i c_i + \sum_{j \neq i} \ell_{ij} c_j q_j \bmod 2 \iff P\vec{c} \equiv D_{\vec{p}} \vec{e} \bmod 2. \tag{3.34}$$

Proposition 3.46. *Suppose σ is a spin^c structure on E with charge $\vec{v}$. Then the following conditions are equivalent.*

(i) *The spin^c structure $\vec{v}^M$ on M is induced by a spin structure.*

(ii) *There exists a $\vec{u} \in \mathbb{Z}^n$ such that $\vec{v} = K\vec{e} + P\vec{u}$.*

(iii) *There exists an integral characteristic vector $\vec{c} \in \mathbb{Z}^n$ such that $\vec{u} = \vec{c} + \vec{e}$ satisfies $\vec{v} = P(\vec{u}) + K\vec{e}$.*

Proof. (i) $\iff$ (ii) The spin^c structure σ_M is induced by a spin structure iff $c(\sigma_M) = 1$, iff $\pi(\mu^{\vec{v} - K\vec{e}}) = 1 \in H$, iff $\vec{v} - K\vec{e} \in \operatorname{Im}(P)$.

(ii) $\Longrightarrow$ (iii) Suppose there exists $\vec{u} \in \mathbb{Z}^n$ such that

$$\vec{v} = K\vec{e} + P\vec{u}.$$

We want to prove that $\vec{u} - \vec{\boldsymbol{e}}$ is an integral characteristic vector. Since $\vec{\nu}$ is a charge

$$\vec{\nu} - \vec{\boldsymbol{e}} \equiv L\vec{\boldsymbol{e}} \bmod 2.$$

We then have the following mod 2 equalities.

$$P(\vec{u} - \vec{\boldsymbol{e}}) = \vec{\nu} - K\vec{\boldsymbol{e}} - P\vec{\boldsymbol{e}} = \vec{\boldsymbol{e}} + L\vec{\boldsymbol{e}} - K\vec{\boldsymbol{e}} - P\vec{\boldsymbol{e}}$$

(use the identities (3.26), (3.28))

$$= \vec{\boldsymbol{e}} + L\vec{\boldsymbol{e}} + L(D_{\vec{\beta}} + D_{\vec{q}})\vec{\boldsymbol{e}} + D_{\vec{\alpha}}\vec{\boldsymbol{e}} + D_{\vec{p}}\vec{\boldsymbol{e}}.$$

At this point we recall the conditions (3.27) which imply

$$D_{\vec{\alpha}}\vec{\boldsymbol{e}} = \vec{\boldsymbol{e}} \bmod 2, \quad (D_{\vec{\beta}} + D_{\vec{q}})\vec{\boldsymbol{e}} \equiv \vec{\boldsymbol{e}} \bmod 2.$$

We deduce that

$$P(\vec{u} - \vec{\boldsymbol{e}}) \equiv D_{\vec{p}}\vec{\boldsymbol{e}} \bmod 2,$$

i.e. $(\vec{u} - \vec{\boldsymbol{e}}) \bmod 2$ is a characteristic vector, proving the desired implication. The implication (iii) $\Longrightarrow$ (ii) is trivial. □

We denote by Char_P the space of integral characteristic vectors, and set $\mathrm{Char}_P^2 :=$ $\mathrm{Char}_P \bmod 2$. We denote by $\mathcal{N}_P$ the space of charges $\vec{\nu}$ such that the spin^c structure $\vec{\nu}^M$ on M is induced by a spin structure. We have a natural equivalence relation on $\mathcal{N}_P$

$$\vec{\nu}_1 \sim \vec{\nu}_2 \iff \vec{\nu}_1^M = \vec{\nu}_2^M \iff \frac{1}{2}(\vec{\nu}_1 - \vec{\nu}_2) \in \mathrm{Im}(P).$$

Assume now that M *is a* $\mathbb{Q}$*-homology sphere*. Then for every $\vec{\nu} \in \mathcal{N}_P$ the element $\vec{u}$ postulated in Proposition 3.46(iii) is unique

$$\vec{\nu} - K\vec{\boldsymbol{e}} = P\vec{u} \iff \vec{u} = P^{-1}(\vec{\nu} - K\vec{\boldsymbol{e}}),$$

and thus we have a well defined bijection

$$\Psi_P = \Psi_{P,\vec{\alpha},\vec{\beta}}\colon \mathrm{Char}_P \to \mathcal{N}_P, \quad \vec{c} \mapsto K\vec{\boldsymbol{e}} + P(\vec{x} + \vec{\boldsymbol{e}}) = (K + P)\vec{\boldsymbol{e}} + P\vec{c}.$$

Lemma 3.47. $\Psi = \Psi_P$ *descends to a bijection*

$$\Psi_P\colon \mathrm{Char}_P^2 \to \mathcal{N}_P/\sim.$$

Proof. Since M is a rational homology sphere, the above spaces of equivalence classes are finite, so it suffices to prove only that Ψ descends to a well defined injection.

Clearly if $\vec{c}_1 = \vec{c}_2 \bmod 2$ there exists $\vec{x} \in \mathbb{Z}^n$ such that $\vec{c}_2 - \vec{c}_1 = 2\vec{x}$. Then

$$\frac{1}{2}\big(\Psi_P(\vec{c}_2) - \Psi(\vec{c}_1)\big) = P\vec{u} \Longrightarrow \Psi_P(\vec{c}_1) \sim \Psi_P(\vec{c}_2).$$

The opposite implication $\Psi_P(\vec{c}_1) \sim \Psi_P(\vec{c}_2) \Longrightarrow \vec{c}_1 = \vec{c}_2 \bmod 2$ is proved similarly. □

A priori, this identification could depend on the choice of $\vec{\alpha}, \vec{\beta}$ satisfying (3.27). We now prove that this is not the case.

Lemma 3.48. *The correspondence*

$$\Psi = \Psi_{P,\vec{\alpha},\vec{\beta}} \colon \operatorname{Char}^2_P \to \mathcal{N}_P / \sim .$$

is independent of $\vec{\alpha}$, $\vec{\beta}$ *satisfying* (3.27).

Proof. Suppose we have two pairs $(\vec{\alpha}', \vec{\beta}')$ and $(\vec{\alpha}'', \vec{\beta}'')$ satisfying (3.27). Then there exists a vector $\vec{k} \in \mathbb{Z}^n$ such that

$$D_{\vec{\alpha}''} = D_{\vec{k}} \cdot D_{\vec{p}} + D_{\vec{\alpha}''}, \quad D_{\vec{\beta}''} = D_{\vec{k}} \cdot D_{\vec{q}} + D_{\vec{\beta}'}.$$

Moreover,

$$D_{\vec{\alpha}''} - D_{\vec{\alpha}''} \equiv (D_{\vec{\beta}''} - D_{\vec{\beta}'}) \equiv 0 \bmod 2.$$

Since $(p_i, q_i) = 1$ we deduce that all the components of $\vec{k}$ must be even so that

$$\vec{k}_0 := \frac{1}{2}\vec{k} \in \mathbb{Z}^n.$$

Suppose $\vec{x} \in \operatorname{Char}_P$. Then

$$\begin{aligned}\Psi_{\vec{\alpha}'',\vec{\beta}''}(\vec{x}) - \Psi_{\vec{\alpha}',\vec{\beta}'}(\vec{x}) &= L(D_{\vec{\beta}''} - D_{\vec{\beta}'})\vec{e} + (D_{\vec{\alpha}''} - D_{\vec{\alpha}''})\vec{e} \\ &= LD_{\vec{k}} \cdot D_{\vec{q}}\vec{e} + D_{\vec{k}} \cdot D_{\vec{p}}\vec{e} = LD_{\vec{q}} \cdot D_{\vec{k}}\vec{e} + D_{\vec{p}}D_{\vec{k}} \cdot \vec{e}.\end{aligned}$$

Set $\vec{u}_0 = D_{\vec{k}_0}\vec{e}$. We can rewrite the above equality as

$$\Psi_{\vec{\alpha}'',\vec{\beta}''}(\vec{x}) - \Psi_{\vec{\alpha}',\vec{\beta}'}(\vec{x}) = 2(LD_{\vec{q}} + D_{\vec{p}})\vec{u}_0 = 2P\vec{u}_0.$$

This proves that

$$\Psi_{\vec{\alpha}'',\vec{\beta}''}(\vec{x}) \sim \Psi_{\vec{\alpha}',\vec{\beta}'}(\vec{x}) \quad \text{in } \mathcal{N}_P,$$

and completes the proof of the lemma. □

In view of Proposition 3.45 we can identify $\mathcal{N}_P / \sim$ with the space of spin structures on M,

$$\Phi \colon (\mathcal{N}_P / \sim) \to \operatorname{Spin}(M).$$

We have thus proved that the surgery presentation P of the rational homology sphere M produces an *explicit, canonical* identification

$$\Xi_P \colon \operatorname{Char}^2_P \xrightarrow{\Psi} (\mathcal{N}_P / \sim) \xrightarrow{\Phi} \operatorname{Spin}(M).$$

The Euclidean spin structure on $S^1 \times D^2$ is the restriction of the unique spin structure on the handlebody $D^2 \times D^2$. We deduce form the computation in Example 3.41 that the above identification is precisely the identification described in [37, §5.7] between the space Char^2_P and the space $\operatorname{Spin}(M)$ of spin structures on M.

There is a simple way to represent the integral characteristic vectors on a surgery diagram using colors,[5] or on the plumbing graph, using ●'s and o's. If $\vec{c} \in \mathrm{Char}_P$ then the components of $\mathcal{K}$ corresponding to $c_i = 1 \bmod 2$ will be colored in black, and the corresponding vertex of the plumbing graph will be indicated by o, while the components corresponding to $c_i = 0 \bmod 2$ will be colored in grey and the corresponding vertex of the plumbing graph will be indicated by a (grey) ●. As explained in [37, §5.7], the colors of the vertices left after a slam-dunk stay the same. We define a *surgery spin diagram* to be a surgery diagram with a characteristic vector indicated by coloring of the vertices by the rule explained above.

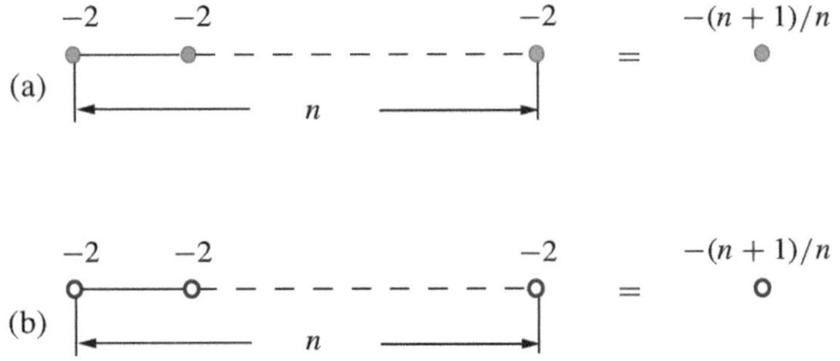

Figure 3.11. The two spin structures on the boundary of the A_n plumbing, $n \equiv 1 \bmod 2$.

Consider for example the boundary of the A_n-plumbing. Its associated plumbing graph is a"bamboo" of length n which can be iteratively slam-dunked to a single point with surgery coefficient $-(n+1)/n$ as in Figure 3.11.

Suppose $n+1$ is even. The boundary of this plumbing has two spin structures corresponding to the two characteristic vectors $\vec{c}_0 = (0, \dots, 0)$, $\vec{c}_1 = (1, \dots, 1)$. The first spin structure is depicted Figure 3.11 (a), and the second is depicted in Figure 3.11 (b). On the right hand side we depicted the equivalent diagrams obtained after iterated slam-dunks.

The importance of the above abstract results to torsion computations is best grasped on a concrete example.

Example 3.49. Consider again the plumbed rational homology sphere discussed in Example 3.29, page 142. We continue to use the same notations as in that example. We have

$$P := \begin{bmatrix} 3 & 1 & 0 & 0 & 1 & 1 \\ 1 & 3 & 1 & 1 & 0 & 0 \\ 0 & 1 & 2 & 0 & 0 & 0 \\ 0 & 1 & 0 & 2 & 0 & 0 \\ 1 & 0 & 0 & 0 & 2 & 0 \\ 1 & 0 & 0 & 0 & 0 & 2 \end{bmatrix}.$$

[5] ● = the spin structure *extends* over the corresponding handle, whence the *full grey* disk, o = the spin structure *does not extend* over the corresponding handle, whence the *holed black* disk.

In this case we can choose $K = -1$. A vector $\vec{c} \in \mathbb{Z}^6$ is characteristic iff the following mod 2 equalities hold:

$$\begin{cases} 1 = c_1 + c_2 + c_5 + c_6 \\ 1 = c_1 + c_2 + c_3 + c_4 \\ 0 = c_2 \\ 0 = c_1. \end{cases}$$

Hence $\vec{c} = (0, 0, c_3, c_4, c_5, c_6)$ where $c_3 + c_4 = c_5 = c_6 = 1 \bmod 2$. In particular the vector $\vec{c}_0 = (0, 0, 1, 0, 1, 0)$ is characteristic. Denote by ϵ_0 the associated spin structure. It is depicted using •'s and ∘'s in Figure 3.9. A charge on the complement of the plumbing link $\mathcal{K}$ which induces this spin structure is

$$\vec{v}_0 = (K + P)\vec{e} + P\vec{c}_0 = \begin{bmatrix} 5+1 \\ 5+1 \\ 2+2 \\ 2 \\ 2+2 \\ 2 \end{bmatrix}.$$

Denote by σ_0 the relative spin^c structure on E corresponding to $\vec{v}_0$. We deduce that there exists $\varepsilon = \pm 1$ and $\vec{k} \in \mathbb{Z}^4$ such that

$$\mathfrak{T}_{E,\sigma_0} = \varepsilon \mu^{\vec{k}} (\mu_1 - 1)^2 (\mu_2 - 1)^2,$$

where the vector $\vec{k}$ is uniquely determined by the condition (3.32)

$$\mu^{-\vec{k}} (\mu_1^{-1} - 1)^2 (\mu_2^{-1} - 1)^2 = \mu^{\vec{k} - \vec{v}_0} (\mu_1 - 1)^2 (\mu_2 - 1)^2.$$

This means

$$\mu_1^{-2} \mu_2^{-2} = \mu^{2\vec{k} - \vec{v}_0} \implies \vec{k} = (3, 3, 2, 1, 2, 1).$$

Thus

$$\mathfrak{T}_{E,\sigma_0} = \varepsilon \mu_1^3 \mu_2^3 \mu_3^2 \mu_4 \mu_5^2 \mu_6 (\mu_1 - 1)^2 (\mu_2 - 1)^2.$$

As in Example 3.29 suppose now that $\chi \in \hat{H}$ is such that $v = 1$ but $u^2 \neq 1$. Then

$$\hat{\mathfrak{T}}_{M,\epsilon_0}(\chi) = \varepsilon \chi^{-1} \big(\mu_1^3 \mu_2^3 \mu_3^2 \mu_4 \mu_5^2 \mu_6 \big) \lim_{t \to 1} F_P(t *_{\vec{w}} \chi).$$

$$= -\varepsilon z_1^{-3} z_2^{-3} z_3^{-2} 2 z_4^{-1} z_5^{-2} z_6^{-1} \frac{3}{(u-1)^2}.$$

We now use the identities (3.19) to conclude that

$$\hat{\mathfrak{T}}_{M,\epsilon_0}(\chi) = \varepsilon \frac{3u}{(u-1)^2}.$$

Observe that the last equality confirms the symmetry relation

$$\hat{\mathfrak{T}}_{M,\epsilon_0}(\chi) = \hat{\mathfrak{T}}_{M,\epsilon_0}(\bar{\chi}),$$

expected for the torsion associated to a spin-structure. We are left with one last ambiguity, the sign ε. In the next section we will explain how to remove it. □

§3.10 The torsion of rational homology 3-spheres. Part 2.

The identity (3.8) of §3.6 can be used to remove the sign ambiguity in the surgery formula of Theorem 3.23. We will explain how to achieve this for a special class of rational homology spheres, namely those described by a *nondegenerate, rational plumbing* along trees. We will follow closely the strategy in [75, Appendix A].

Consider a tree Γ, and denote by $\mathcal{V}$ its set of vertices. For each vertex $v \in \mathcal{V}$ we denote by d_v its degree. As in §3.8, page 141 we associate to Γ a link

$$\mathcal{K}_\Gamma = (\mathcal{K}_v)_{v\in\mathcal{V}}$$

whose components are unknots. Denote $E = E_\Gamma$ the exterior of this link, and by L the linking matrix of this link (or equivalently, the incidence matrix of Γ). Set $G := H_1(E)$.

Define $\mathcal{S} = \mathcal{S}_\Gamma \subset \mathbb{Q}^\mathcal{V}$ as the set consisting of all possible choices of surgery coefficients $\vec{r} = (r_v = p_v/q_v)_{v\in\mathcal{V}}$ so that the corresponding presentation matrix

$$P = P_{\vec{r}} = LD_{\vec{q}} + D_{\vec{p}}$$

is nondegenerate. We need to fix a convention. In the sequel we will assume that

$$q_v > 0 \quad \text{and} \quad \gcd(p_v, q_v) = 1, \quad \forall v \in \mathcal{V}.$$

Suppose $\mathcal{S} \neq \emptyset$. In this case $\mathcal{S}$ is an *open* subset of $\mathbb{Q}^\mathcal{V}$. For every $\vec{r} \in \mathcal{S}$ we denote by $M_{\vec{r}}$ the rational homology sphere obtained by Dehn surgery along Γ with surgery coefficients $\vec{r}$. We denote by Z_v the solid torus attached to the boundary of the v-component of $\mathcal{K}_\Gamma$. We set $H = H_{\vec{r}} = H_1(M_{\vec{r}})$, and we denote by $\pi : G \to H$ the natural projection. Set $\nu := |\det P|$ so that $|H| = \nu$.

For each $\vec{r}$ fix $\vec{\alpha}, \vec{\beta} \in \mathbb{Z}^\mathcal{V}$ as in (3.27) of §3.9, page 157, and then define a $\mathcal{V} \times \mathcal{V}$-matrix $K = K_{\vec{r}}$ as in (3.28)

$$K = LD_{\vec{\beta}} + D_{\vec{\alpha}}.$$

The equality $D_{\vec{p}}D_{\vec{\beta}} - D_{\vec{q}}D_{\vec{\alpha}} = 1$ implies

$$PD_{\vec{\beta}} - KD_{\vec{q}} = 1 \quad \text{and} \quad KD_{\vec{p}} - PD_{\vec{\alpha}} = L.$$

The v-th column of K, which we will denote by K_v, defines homology class $\boldsymbol{j}(\alpha_v\mu_v + \beta_v\lambda_v)$ in G, which we will continue to denote by K_v. Its image in H is the homology class of the axis of the attaching solid torus Z_v.

Fix a relative spin^c structure σ on E_γ. Using the surgery trick (2.19) on page 90 we deduce that there exists $\varepsilon_\Gamma \in \{\pm 1\}$ and $\vec{k}_\sigma \in \mathbb{Z}^{\mathcal{V}}$ such that

$$\mathcal{T}_{E,\sigma} = \varepsilon_\Gamma \mu^{\vec{k}_\sigma} \prod_{v\in\mathcal{V}} (\mu_v - 1)^{d_v - 1}.$$

Define the meromorphic function

$$F_P = F_{\Gamma,\vec{r}} \colon \tilde{G} \dashrightarrow \mathbb{C}, \quad \frac{\prod_{v\in\mathcal{V}}\big(\chi^{-1}(\mu_v) - 1\big)^{d_v-1}}{\prod_{v\in\mathcal{V}}\big(1 - \chi^{-1}(K_v)\big)}, \qquad \tilde{G} = \mathrm{Hom}(G, \mathbb{C}^*).$$

Theorem 3.23 states that F_P is holomorphic near each $\chi \in \hat{H}_{\vec{r}}$, and there exists $\varepsilon(\Gamma, \vec{r}) = \pm 1$ such that

$$\hat{\mathcal{T}}_{M_{\vec{r}},[\sigma]}(\chi) = \varepsilon(\Gamma, \vec{r})\chi^{-1}(\mu^{\vec{k}_\sigma}) F_P(\chi). \tag{3.35}$$

Above we have denoted by $[\sigma]$ the spin^c structure on M induced by σ. The ambiguous $\vec{k}_\sigma$ can be determined using the strategy outlined in the previous section.

To determine the sign $\varepsilon(\Gamma, \vec{r})$ note first that it depends continuously on $\vec{r}$. Thus by slightly changing the surgery coefficients we can arrange that $H_{\vec{r}}$ contains elements of order > 2. Since the curves K_v define a generating subset of H we deduce that there exists at least one $v_0 \in \mathcal{V}$ such that πK_{v_0} has order > 2 in $H_{\vec{r}}$. Write $\mathcal{T}_{M,[\sigma]}$ as a function

$$\mathcal{T}_M \colon H \to \mathbb{Q}.$$

Using the identity (3.8) on page 130 we deduce

$$\mathcal{T}_M(1) - \mathcal{T}_M(K_{v_0}) - \mathcal{T}_M(h) + \mathcal{T}_M(hK_{v_0}) = -\,\boldsymbol{lk}_M(K_{v_0}, h) \bmod \mathbb{Z}.$$

Using the Fourier inversion formula, and (3.35) we deduce that for every $h \in H$ we have

$$\frac{\varepsilon(\vec{r})}{|H|}\sum_{\chi}{}' F_P(\chi)\big(1 - \chi^{-1}(K_{v_0})\big)\big(1 - \chi(h)\big) = \boldsymbol{lk}_M(K_{v_0}, h) \bmod \mathbb{Z}, \tag{3.36}$$

for all $h \in H$, where $\sum'_\chi$ denotes summation over all the *nontrivial* characters of H.

The nontrivial terms in the above sum correspond to characters $\chi \in \hat{H}$ such that $\chi(\pi K_{v_0}) \neq 1$, i.e. $v_0 \in S_\chi$, where as in §3.8 S_χ denotes the support of the character χ. Denote by X_0 the set of such characters.

The integral basis $(\mu_v)_{v\in\mathcal{V}}$ of G defines a natural Euclidean inner product $(\bullet, \bullet)$ on $G \otimes \mathbb{Q}$. Define $\vec{\ell}_0 \in \check{G} := \mathrm{Hom}(G, \mathbb{Q})$ by

$$\langle \vec{\ell}_0, \mu_v \rangle = (\mu_{v_0}, P^{-1}\mu_v), \quad \forall v \in \mathcal{V}.$$

More intuitively, $\vec{\ell}_0$ is described by the v_0-th row of P^{-1}. There exists a positive integer ν_0 such that $\mathrm{Im}(\vec{\ell}_0) = \frac{1}{\nu_0}\mathbb{Z}^{\mathcal{V}}$. Set $\vec{w}_0 = \nu_0\vec{\ell}_0$, so that $\vec{w}_0 \in \check{G}$. For simplicity we set

$$m_v := \langle \vec{w}_0, \mu_v \rangle, \quad k_v := \langle \vec{w}_0, K_v \rangle.$$

From the identity $PD_{\vec{\beta}} - KD_{\vec{q}} = 1$ we deduce

$$-q_v k_v = m_v - \nu_0 \beta_v \delta_{vv_0}, \tag{3.37}$$

where δ_{uv} denotes the Kronecker symbol. The weight $\vec{w}_0$ is admissible for all the characters in X_0, and thus we can perform the regularization along this weight for all the characters in X_0. Using (3.15) we deduce that for every $\chi \in X_0$ we have

$$F_P(\chi) = \lim_{t\to 1} \frac{\prod_{v\in\mathcal{V}}\big(t^{m_v}\chi^{-1}(\mu_v) - 1\big)^{d_v-1}}{\prod_{v\in\mathcal{V}}\big(1 - t^{k_v}\chi^{-1}(K_v)\big)}, \quad \forall \chi \in X_0. \tag{3.38}$$

Denote by E_0 the manifold obtained from E_Γ attaching only the solid tori E_v, $v \neq v_0$. Set

$$G_{v_0} := H_1(E_{v_0}, \mathbb{Z}), \quad \mathfrak{S}_0 := \sum_{g\in \mathrm{Tors}(G_{v_0})} g \in \mathbb{Z}[G_0].$$

As explained in §B.2 we have

$$\mathrm{Tors}(G_{v_0}) \cong \langle K_{v_0}\rangle^{\perp} := \{h \in H;\ \boldsymbol{lk}_M(h, K_{v_0}) = 0\}$$

so that

$$|\,\mathrm{Tors}(G_{v_0})| = \frac{|H|}{\mathrm{ord}_H(K_{v_0})} = \frac{\nu}{\nu_0}.$$

Set $\tilde{G}_{v_0} = \mathrm{Hom}(G_{v_0}, \mathbb{C}^*)$, $\check{G}_{v_0} = \mathrm{Hom}(G_{v_0}, \mathbb{Z})$, and denote by $\tilde{G}^0_{v_0}$ the identity component of $\tilde{G}_{v_0}$. The complex Fourier transform of $\mathfrak{S}_0$ is

$$\hat{\mathfrak{S}}_0(\chi) = |\,\mathrm{Tors}(G_{v_0})| \cdot 1_{\tilde{G}^0_{v_0}}(\chi) = \frac{\nu}{\nu_0} \cdot \begin{cases} 1 & \text{if } \chi \in \tilde{G}^0_{v_0} \\ 0 & \text{if } \chi \notin \tilde{G}^0_{v_0}. \end{cases}$$

We have a natural projection $\pi_{v_0}\colon G \twoheadrightarrow G_{v_0}$, and thus inclusions $\tilde{G}_{v_0} \hookrightarrow \tilde{G}$, $\check{G}_{v_0} \hookrightarrow \check{G}$. The weight $\vec{w}_0$ generates the image of $\check{G}_{v_0} \hookrightarrow \check{G}$. There exist *exactly two* isomorphisms $\tilde{G}^0_{v_0} \to \mathbb{C}^*$, and the weight $\vec{w}_0$ fixes one such isomorphism. Via this isomorphism, the tautological action of $\tilde{G}^0_{v_0}$ on $\tilde{G}_{v_0}$ can be written as

$$\mathbb{C}^* \times \tilde{G}_{v_0} \ni (t, \chi) \mapsto \chi_t := t *_{\vec{w}_0} \chi \in \tilde{G}_{v_0}.$$

Define $R_{v_0}\colon \tilde{G}_{v_0} \setminus \{1\} \to \mathbb{C}$ by

$$\begin{aligned} R_{v_0}(\chi) &= \big(1 - \chi^{-1}(K_{v_0})\big)F_P(\chi) \\ &= \big(\chi^{-1}(\mu_{v_0}) - 1\big)^{d_{v_0}-1} \frac{\prod_{v\neq v_0}\big(\chi^{-1}(\mu_v) - 1\big)^{d_v-1}}{\prod_{v\neq v_0}\big(1 - \chi^{-1}(K_v)\big)}. \end{aligned}$$

Up to a sign, this is the complex Fourier transform of the torsion of E_0, and thus it is a holomorphic function.

Fix $g \in G_{v0}$ such that $\langle \vec{w}_0, g\rangle = 1$. According to (3.7) in §3.6 there exists $A \in \mathbb{Z}[G_{v0}]$ and $\varepsilon = \pm 1$ such that R_{v0} is the complex Fourier transform of $A + \varepsilon \frac{\mathfrak{S}_0}{1-g} \in \mathfrak{N}_1(G_{v0})$. Hence

$$R_{v0}(\chi)\big(1 - \chi^{-1}(g)\big) - \hat{A}(\chi)\big(1 - \chi^{-1}(g)\big) = \varepsilon \hat{\mathfrak{S}}_0(\chi)$$

We deduce

$$\varepsilon \frac{\nu}{\nu_0} = \lim_{t\to 1} R_{v0}(1_t)(1-t) = \lim_{t\to 1}(1-t)(t^{m_{v0}} - 1)^{d_{v0}-1} \prod_{v \neq v_0} \frac{(t^{m_v} - 1)^{d_v - 1}}{(1 - t^{k_v})}.$$

Thus the sign ε coincides with the sign of the limit on the right hand side. To determine this sign it is convenient to introduce the following notation. If $f(t)$ is a meromorphic function of the complex variable t then the notation $f(t) \sim \varepsilon(t-1)^n$, $\varepsilon = \pm 1, n \in \mathbb{Z}$, signifies that the function $g(t) = (t-1)^{-n} f(t)$ is holomorphic in a neighborhood of 1, $g(1) \in \mathbb{R}$, $\varepsilon g(1) > 0$. Observe that

$$(t^m - 1)^k \sim \operatorname{sign}(m)^k (t-1)^k.$$

Since the plumbing graph is a tree we deduce

$$\sum_v (\delta_v - 2) = -2 \times \text{Euler characteristic of } \Gamma = -2.$$

This identity shows that if $m_{v0} = 0$, then $R_{v0}(1_t)(1-t)$ would have a pole of order $d_{v0} - 1$ at $t - 1$. Since this function has a limit at $t = 1$ we deduce that when $m_{v0} = 0$ we must have $d_{v0} = 1$, in which case the term $(\chi^{-1}(\mu_{v0}) - 1)^{d_{v0}-1}$ has no contribution to the torsion. Hence, if we set $\operatorname{sign}(0) := 1$ we deduce that for every $v \in \mathcal{V}$ we have

$$(t^{m_v} - 1)^{d_v - 1} \sim \operatorname{sign}(m_v)^{\delta_v - 1}(t-1)^{d_v - 1}.$$

Moreover, for $v \neq v_0$ we have

$$(1 - t^{k_v}) \sim -\operatorname{sign}(k_v)(t-1) \overset{(3.37)}{\sim} \operatorname{sign}(m_v) \cdot (t-1).$$

We conclude that

$$R_{v0}(1_t) \sim (-1)^{d_{v0}-1} - \operatorname{sign}(m_{v0}) \prod_{v \in \mathcal{V}} \operatorname{sign}(k_v)^{d_v - 2} := \epsilon(v_0). \tag{3.39}$$

Thus

$$R_{v0}(\chi) = \hat{A}(\chi) + \epsilon(v_0)\frac{\hat{\mathfrak{S}}_0}{1 - \chi^{-1}(g)}.$$

Set $\Delta := A(1-g) + \epsilon_0 \mathfrak{S}_0 \in \mathbb{Z}[G_{v_0}]$ and $\hat{\Delta}(t) := \hat{\Delta}(1_t)$. Note that

$$R_{v_0}(\chi) = \frac{\hat{\Delta}(\chi)}{1-\chi^{-1}(g)}, \quad \hat{\Delta}(1) = \epsilon(v_0)\frac{\nu}{\nu_0}.$$

The equality (B.4) in Appendix §B.2, page 224, implies

$$\boldsymbol{lk}_M(K_{v_0}, g) = \langle \vec{\ell}_0, g\rangle \bmod \mathbb{Z} = \frac{1}{\nu_0}\langle \vec{w}_0, g\rangle \bmod \mathbb{Z} = \frac{1}{\nu_0}.$$

In particular, since the order ν_0 of K_{v_0} is > 2 we deduce

$$\boldsymbol{lk}_M(K_{v_0}, g) \neq -\boldsymbol{lk}_M(K_{v_0}, g) \bmod \mathbb{Z}. \tag{3.40}$$

Using (3.36) we deduce

$$\varepsilon(\vec{r})\frac{1}{\nu}\lim_{t\to 1}{\sum_{\chi}}' R_{v_0}(\chi_t)\big(1-\chi_t(g)\big) = \frac{1}{\nu_0} \bmod \mathbb{Z}. \tag{3.41}$$

To compute the expression in the left hand side we need an algebraic digression.

Lemma 3.50.

$$\frac{1}{|H|}{\sum}'_{\chi\in\hat{H}}\hat{P}(\chi_t) = -\frac{1}{|H|}\hat{P}(1_t) \bmod \mathbb{Z}, \tag{3.42}$$

for all $P \in \mathbb{Z}[G_{v_0}]$, $t \in \mathbb{C}^*$.

Proof. Denote by $\mathbf{p}_0 \colon G_{v_0} \twoheadrightarrow H$ the natural projection. We can then write

$$P = \sum_{\substack{n\in\mathbb{Z}\\ h\in \mathrm{Tors}(G_{v_0})}} p_{n,h} h g^n, \quad p_n \in \mathbb{Q}.$$

For every $\chi \in \hat{H}$ we have

$$\hat{P}(\chi_t) = \sum_{\substack{n\in\mathbb{Z}\\ h\in \mathrm{Tors}(G_{v_0})}} p_{n,h}\bar{\chi}(\mathbf{p}_0(hg^n))t^n.$$

Now observe that

$$\frac{1}{|H|}{\sum_{\chi\in\hat{H}}}'\bar{\chi}(\mathbf{p}_0(hg^n)) = -\frac{1}{|H|} \bmod \mathbb{Z}$$

The equality (3.42) follows by summing over h and n. □

Observe now that

$$R_{v_0}(\chi) = \frac{\hat{\Delta}(\chi)}{1-\chi^{-1}(g)} = \chi(g)\frac{\hat{\Delta}(\chi)}{\chi(g)-1}$$

so that

$$R_{v_0}(\chi)(1-\chi(g)) = -\chi(g)\hat{\Delta}(\chi).$$

Since $\Delta \in \mathbb{Z}[G_v]$, Lemma 3.50 implies

$$\frac{1}{\nu}\sum_{\chi}{}' R_{v_0}(\chi_t)\big(1-\chi_t(g)\big) = \frac{1}{\nu}1_t(g)\hat{\Delta}(t)$$

Hence

$$\varepsilon(\vec{r})\frac{1}{\nu}\lim_{t\to 1}\sum_{\chi}{}' R_{v_0}(\chi_t)\big(1-\chi_t(g)\big) = \frac{\varepsilon(\vec{r})\hat{\Delta}(1)}{\nu} = \frac{\varepsilon(\vec{r})\epsilon(v_0)}{\nu_0}.$$

Using (3.40) and (3.41) we deduce

$$\varepsilon(\vec{r}) = \epsilon(\vec{v}_0) = (-1)^{d_{v_0}-1}\operatorname{sign}(m_{v_0})^{d_{v_0}-1}\prod_{v\neq v_0}\operatorname{sign}(k_v)^{d_v-2}. \tag{3.43}$$

Remark 3.51. (a) If $P_0 := L + D_{\vec{r}}$ is negative definite then Proposition 3.26 implies $m_v < 0$, $\forall v$ and $k_v = -m_v/q_v > 0$ for all $v \neq v_0$. In this case we conclude that $\varepsilon(\vec{r}) = \epsilon(v_0) = 1$. This agrees with the conclusions in [75, Appendix A].

(b) A priori $\epsilon(v_0)$ depends on v_0 but formula (3.43) shows that this is not the case. If P is not negative definite it is not clear why such a fact should be true. Take for example the plumbing matrix

$$P := \begin{bmatrix} 2 & 1 & 0 \\ 1 & -3 & 1 \\ 0 & 1 & 5 \end{bmatrix}, \quad P^{-1} = \begin{bmatrix} \frac{16}{37} & \frac{5}{37} & -\frac{1}{37} \\ \frac{5}{37} & -\frac{10}{37} & \frac{2}{37} \\ -\frac{1}{37} & \frac{2}{37} & \frac{7}{37} \end{bmatrix}.$$

This corresponds to the plumbing

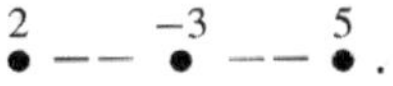

For example

$$\begin{aligned}
\epsilon(v_1) &= -\operatorname{sign}(16/37)\operatorname{sign}(5/37)\operatorname{sign}(-1/37) = 1,\\
\epsilon(v_2) &= -\operatorname{sign}(-10/37)\operatorname{sign}(5/37)\operatorname{sign}(2/37) = 1.\\
\epsilon(v_3) &= -\operatorname{sign}(7/37)^2\operatorname{sign}(-1/37) = 1.
\end{aligned}$$

□

Exercise 3.10. Suppose P is given by an *integral, negative definite* plumbing. Prove that the quantity

$$W(v_0) := m_{v_0}^{d_{v_0}-1}\prod_{v\neq v_0} m_v^{d_v-2}$$

is independent of the vertex v_0. □

The following result summarizes the facts proved so far.

Corollary 3.52.

$$\hat{\mathfrak{T}}_{M_{\vec{r}},[\sigma]} = -\operatorname{sign}\big((P^{-1}\mu_{v_0}, \mu_{v_0}\big) \cdot \prod_v \big(\mu_{v_0}, P^{-1}\mu_v)^{d_v-2}\big)$$
$$\times \chi^{-1}(\mu^{\vec{k}_\sigma}) \cdot \prod_v \frac{\big(\chi^{-1}(\mu_v) - 1\big)^{d_v-1}}{\big(1 - \chi^{-1}(K_v)\big)},$$

v_0 an arbitrary vertex of Γ. In the above formula the inner product $(\mu_{v_0}, P^{-1}\mu_v)$ is equal to the v_0v-entry in the matrix P^{-1}.

For example, the correct sign for the torsion computed in Example 3.49 in §3.8 is determined by inspecting an arbitrary row of the matrix P^{-1} in (3.18).

Suppose v_0 corresponds to the first row of P^{-1}. Note that the only terms that matter correspond to odd degree vertices v of the plumbing graph such the corresponding entry $(\mu_{v_0}, P^{-1}\mu_v)$ is negative. We deduce

$$\epsilon(v_0) = -\operatorname{sign}(2/3) \cdot \operatorname{sign}(-1/3)\operatorname{sign}(-1/3)\operatorname{sign}(-1/3) = 1.$$

If v_0 corresponds to the second row then

$$\epsilon(v_0) = -\operatorname{sign}(2/3)\operatorname{sign}(-1/3) \cdot \operatorname{sign}(-1/3)\operatorname{sign}(-1/3) = 1.$$

In our next example we illustrate how to compute the sign refined Reidemeister–Turaev torsion a rational homology sphere relevant in singularity theory.

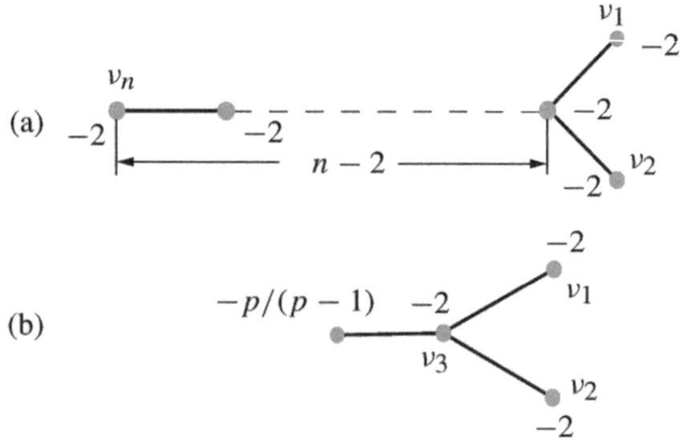

Figure 3.12. The D_n-plumbing.

Example 3.53 (The D_n plumbing.). The D_n plumbing is described in Figure 3.12(a). It consists of $n \geq 4$-vertices. The vector $(0, \dots, 0)$ is characteristic, and we have indicated this in the figure. After several slam-dunks it can transformed to the equivalent spin surgery diagram in Figure 3.12(b), where $p = n - 2$, and we have also indicated the result of the slam-dunks on the characteristic vector. More precisely the new characteristic vector is also the trivial vector. Assume for simplicity that n is even. We denote by M the boundary of this plumbing and define H and G as before. We denote by ϵ the spin structure described by the above characteristic vector.

The plumbing matrix is

$$P := \begin{bmatrix} -2 & 1 & 1 & p-1 \\ 1 & -2 & 0 & 0 \\ 1 & 0 & -2 & 0 \\ 1 & 0 & 0 & -p \end{bmatrix}, \quad P^{-1} = \frac{1}{4} \begin{bmatrix} -4p & -2p & -2p & 4-4p \\ -2p & -p-2 & -p & -2p+2 \\ -2p & -p & -p-2 & -2p+2 \\ -4 & -2 & -2 & -4 \end{bmatrix}.$$

Since $\det P = 4$ we deduce $H \cong \mathbb{Z}_4, \mathbb{Z}_2 \oplus \mathbb{Z}_2$. When p is even we have

$$H \cong \mathbb{Z}_2 \oplus \mathbb{Z}_2,$$

H is generated by $\mu_1, \mu_2, \mu_3, \mu_4$ subject to the relations

$$\mu_1 = \mu_2^2 = \mu_3^2, \quad \mu_1^{p-1} = \mu_4^p, \quad \mu_1^2 = \mu_2\mu_3\mu_4.$$

from which we deduce that $\mu_1 = 1$, and H is generated by the elements of order two μ_2, μ_3. As for the matrix K we can choose $\alpha_i = -1, \beta_i = 0, 1 \leq i \leq 3, \alpha_4 = p+1$, $\beta_4 = -p$ (recall that α_4 and β_4 must satisfy the congruences (3.27) on page 157) so that

$$K = \begin{bmatrix} -1 & 0 & 0 & -p \\ 0 & -1 & 0 & 0 \\ 0 & 0 & -1 & 0 \\ 0 & 0 & 0 & p+1 \end{bmatrix}.$$

Then

$$\hat{\mathfrak{T}}_{M,\epsilon}(\chi) = \varepsilon \chi^{-1}(\mu^{\vec{k}}) \frac{\left(\chi^{-1}(\mu_1) - 1\right)^2}{\prod_j \left(1 - \chi^{-1}(K_i)\right)}.$$

Corollary 3.52 shows that the sign $\varepsilon = 1$. The monomial $\mu^{\vec{k}}$ is determined using the arguments developed in §3.9. We can choose the trivial vector as integral lift of our characteristic vector. The corresponding charge is

$$\vec{v} = (K + P)\boldsymbol{e} = (-2, -2, -2, -2)^t.$$

The exponent $\vec{k}$ is determined from the equality

$$\mu^{-\vec{k}}(\mu_1^{-1} - 1)^2 = \mu^{\vec{k}-\vec{v}}(\mu_2 - 1)^2 \iff \mu^{2\vec{k}-\vec{v}-2\mu_1^2} = 1.$$

so that $2\vec{k} = (0, -2, -2, -2)$, i.e. $\vec{k} = (0, -1, -1, -1)$. Hence, for every $\chi \in \hat{H}\backslash\{1\}$ we have

$$\hat{\mathfrak{T}}_{M,\epsilon}(\chi) = \chi(\mu_2\mu_3\mu_4) \frac{\left(\chi^{-1}(\mu_1) - 1\right)^2}{\left(1 - \chi(\mu_1)\right)\left(1 - \chi(\mu_2)\right)\left(1 - \chi(\mu_3)\right)\left(1 - \chi(\mu_1^p \mu_4^{-p-1})\right)},$$

where the expression in the right hand side should be interpreted in the regularized sense, using admissible weights. In this case, the first row of $4P^{-1}$ produces such an admissible weight $\vec{w}$. The character χ is completely determined by the complex numbers $z_i := \chi(\mu_i) = \pm 1$, $i = 2, 3, 4$. Instead of χ we will write (z_2, z_3, z_4). Set $w_i := \langle \vec{w}, \mu_i \rangle$. Since $\mu_1 = \mu_2\mu_3\mu_4 = 1 \in H$ have

$$\hat{\mathfrak{T}}_{M,\epsilon}(\chi) = \lim_{t\to 1} \frac{(t^{w_1} - 1)}{\big(1 - t^{-w_2}z_2\big)\big(1 - t^{-w_3}z_3\big)\big(1 - t^{-pw_1+(p+1)w_4}z_4^{-(p+1)}\big)}$$
$$= \lim_{t\to 1} \frac{\big(t^{-4p} - 1\big)}{\big(1 - z_2t^{2p}\big)\big(1 - z_3t^{2p}\big)\big(1 - z_4t^4\big)}.$$

There are exactly three possibilities.

- $z_4 = 1$:

$$\hat{\mathfrak{T}}_{M,\epsilon}(-1, -1, 1) = \lim_{t\to 1} \frac{\big(t^{-4p} - 1\big)}{\big(1 + t^{2p}\big)\big(1 + t^{2p}\big)\big(1 - t^4\big)} = \frac{p}{4}.$$

- $z_3 = 1$:

$$\hat{\mathfrak{T}}_{M,\epsilon}(-1, 1, -1) = \lim_{t\to 1} \frac{\big(t^{-4p} - 1\big)}{\big(1 + t^{2p}\big)\big(1 - t^{2p}\big)\big(1 + t^4\big)} = \frac{1}{2}.$$

- $z_2 = 1$:

$$\hat{\mathfrak{T}}_{M,\epsilon}(1, -1, -1) = \lim_{t\to 1} \frac{\big(t^{-4p} - 1\big)}{\big(1 - t^{2p}\big)\big(1 + t^{2p}\big)\big(1 + t^4\big)} = \frac{1}{2}.$$

Hence

$$\mathfrak{T}_{M,\epsilon}(1) = \frac{1}{4}\Big(\frac{p}{4} + 1\Big) = \frac{p+4}{16}.$$

Similarly,

$$\mathfrak{T}_{M,\epsilon}(\mu_2) = \frac{1}{4}\sum_{\chi} \mathfrak{T}_{M,\epsilon}(\chi)\chi(\mu_2) = \frac{1}{4}\Big(-\frac{p}{4} - \frac{1}{2} + \frac{1}{2}\Big) = -\frac{p}{16},$$
$$\mathfrak{T}_{M,\epsilon}(\mu_3) = \frac{1}{4}\sum_{\chi} \mathfrak{T}_{M,\epsilon}(\chi)\chi(\mu_3) = \frac{1}{4}\Big(-\frac{p}{4} + \frac{1}{2} - \frac{1}{2}\Big) - \frac{p}{16}$$
$$\mathfrak{T}_{M,\epsilon}(\mu_4) = \frac{1}{4}\sum_{\chi} \mathfrak{T}_{M,\epsilon}(\chi)\chi(\mu_4) = \frac{1}{4}\Big(\frac{p}{4} - \frac{1}{2} - \frac{1}{2}\Big) = \frac{p-4}{16}.$$

The above computations have an interesting topological consequence. On the manifold M there are four spin structures $\{\epsilon_i;\ i = 1, \dots, 4\}$,

$$\epsilon_1 = \epsilon, \quad \epsilon_i = \mu_i \cdot \epsilon_1, \quad i = 2, 3, 4.$$

If f is an orientation preserving diffeomorphism of M such that $f^*(\epsilon_1) = \epsilon_i$, then we would have

$$\mathfrak{T}_{M,\epsilon}(1) = \mathfrak{T}_{M,\epsilon_i}(1) = \mathfrak{T}_{M,\epsilon}(\mu_i).$$

The above computations show that this is possible if and only if $\epsilon_i = \epsilon_1$, so that the spin structure ϵ_1 is a *topological invariant* of the oriented manifold M! The same is the case for ϵ_4.

We denote by Γ_M the group of components of the group of orientation preserving diffeomorphisms of M. Γ_M acts on $\mathrm{Spin}(M)$, and the above computations show that ϵ_1, ϵ_4 are fixed points of this action. Moreover these computations suggest that ϵ_2, ϵ_3 might belong to the same orbit of Γ_M. The exercise below shows that this is indeed the case. □

Exercise 3.11. (a) For each spin structure ϵ_i, $i = 1, 2, 3, 4$ indicate a characteristic vector representing it.

(b) Describe an orientation preserving diffeomorphism f of M such that

$$f^*\epsilon_2 = \epsilon_3.$$

□

Chapter 4
Alternative interpretations of the Reidemeister torsion

The Reidemeister torsion of an admissible 3-manifold can be given various equivalent interpretations. It is the goal of this final chapter to sketch three analytic methods of describing this invariant: gauge theoretic, Morse theoretic, and Hodge theoretic.

§4.1 A gauge theoretic interpretation: Seiberg–Witten invariants

Consider $(M, \sigma, \mathfrak{o}) \in \mathfrak{X}$. Recall that this means that M is an admissible 3-manifold, σ is a (relative) spin^c structure, and $\mathfrak{o}$ is an enhanced homology orientation. The attribute *enhanced* signifies that when $b_1 = 1$, the space $H_1(M, \mathbb{R})$ is oriented as well. We set $H := H_1(M)$, and we denote by $\mathbb{S}_\sigma$ the associated bundle of complex spinors.

If $\partial M \neq \emptyset$ we attach a semi-infinite cylinder $[0, \infty) \times \partial M$, and we equip the ensuing noncompact manifold $\hat{M}$ with a metric $\hat{g}$ which coincides with the cylindrical metric

$$dt^2 + g$$

on the cylindrical end, where g denotes a flat metric on a torus. Observe that if $\partial M \neq 0$ then the vector field V coincides with ∂_t along the cylindrical end.

The Seiberg–Witten equations depend on an additional deformation parameter. This is a co-closed 1-form η on $\hat{M}$ with the following properties.

P1. The restriction of η to the cylindrical end is nontrivial, t-covariant constant and harmonic. As such it defines a cohomology class $\eta|_{\partial M} \in H^1(\partial M, \mathbb{R})$.

P2. If $\partial M \neq \emptyset$ and $b_1(M) = 1$ then

$$\eta|_{\partial M} \in \operatorname{Im}\big(H^1(M, \mathbb{R}) \to H^1(\partial M, \mathbb{R})\big) \setminus \{0\}.$$

The configuration space of the Seiberg–Witten theory is

$$\mathcal{C}_\sigma := \Gamma(\mathbb{S}_\sigma) \times \mathcal{A}_\sigma,$$

where $\mathcal{A}_\sigma$ denotes the space of hermitian connections on the complex line bundle $\det \sigma := \det \mathbb{S}_\sigma \to \hat{M}$. When $\partial M \neq \emptyset$, so that $\hat{M}$ is noncompact, we require

$$\int_{\hat{M}} |F_A|^2 dv(\hat{g}) < \infty, \tag{E}$$

where F_A denotes the curvature of A. Observe that $\boldsymbol{PD}(c_1(\det\sigma)) = c(\sigma)$.

Every $A \in \mathcal{A}_\sigma$ defines a symmetric Dirac operator $\mathfrak{D}_A : \Gamma(\mathbb{S}_\sigma) \to \Gamma(\mathbb{S}_\sigma)$. A *finite energy monopole* is a configuration $\mathsf{C} = (\psi, A) \in \mathcal{C}_\sigma$ satisfying (E) and the *Seiberg–Witten equations*

$$\begin{cases} \mathfrak{D}_A\psi = 0 \\ \boldsymbol{c}(*F_A + \mathbf{i}\eta) = \frac{1}{2}q(\psi), \end{cases} \tag{SW}$$

where $q(\psi) \in \mathrm{End}(\mathbb{S}_\sigma)$ is defined by

$$\Gamma(\mathbb{S}_\sigma) \ni \phi \mapsto \langle\phi, \psi\rangle\psi - \frac{1}{2}|\psi|^2\phi \in \Gamma(\mathbb{S}_\sigma),$$

and $\boldsymbol{c}\colon \Lambda^*T^*M \to \mathrm{End}(\mathbb{S}_\sigma)$ denotes the Clifford multiplication "oriented" by the condition

$$\boldsymbol{c}(dv_M) = -1.$$

The group of *gauge transformations*

$$\mathfrak{G}_\sigma := C^\infty(M, S^1)$$

acts on the configuration space by

$$\mathfrak{G}_\sigma \times \mathcal{C}_\sigma \ni \big(\gamma, \mathsf{C} = (\psi, A)\big) \mapsto \big(\gamma \cdot \psi, A - 2d\gamma/\gamma\big) \in \mathcal{C}_\sigma.$$

It transforms monopoles to monopoles. The quotient $\mathcal{C}_\sigma/\mathfrak{G}_\sigma$ is equipped with a natural (Sobolev type) metric. We denote by $\mathfrak{M}_\sigma \subset \mathcal{C}_\sigma/\mathfrak{G}_\sigma$ the set of orbits of monopoles.

$\mathfrak{M}_\sigma$ is a compact subset of $\mathcal{C}_\sigma/\mathfrak{G}_\sigma$, and its infinitesimal deformations are described by an elliptic complex of index 0. For any $\hat{g}$ and a generic η these deformation complexes are acyclic showing that $\mathfrak{M}_\sigma$ is a finite set. By fixing a homology orientation of $H_*(M, \mathbb{R})$ we can associate a sign $\varepsilon(\mathsf{C}) = \pm 1$ to any orbit $[\mathsf{C}] \in \mathfrak{M}_\sigma$ (see [68]) and define

$$\boldsymbol{sw}_M(\sigma, \mathfrak{o}, \hat{g}, \eta) := \sum_{[\mathsf{C}]\in\mathfrak{M}_\sigma} \varepsilon(\mathsf{C}).$$

At this point we need to discuss separately three cases.

A. $\boldsymbol{b_1 > 1}$. A cobordism argument shows that $\boldsymbol{sw}_M(\sigma, \mathfrak{o}, \hat{g}, \eta)$ is independent of the choices $(\hat{g}, \eta)$. We denote this common value by $\boldsymbol{sw}_M(\sigma, \mathfrak{o})$, and we refer to it as the *Seiberg–Witten invariant*[1] of $(M, \sigma, \mathfrak{o})$. It has the following properties.

- $\boldsymbol{sw}_M(\sigma, \mathfrak{o}) = 0$ for all but finitely many σ's.
- If $\partial M = \emptyset$ then $\boldsymbol{sw}_M(\sigma) = \boldsymbol{sw}_M(\bar{\sigma})$, $\forall\sigma$.

[1] When working on closed manifolds, the homology orientation is the tautological one and we will not include it in the notation of the Seiberg–Witten invariant.

For every σ we can form the element

$$\boldsymbol{SW}_{M,\sigma,\mathfrak{o}} \in \mathbb{Z}[H], \quad \boldsymbol{SW}_{M,\sigma,\mathfrak{o}} = \sum_{h \in H} \boldsymbol{sw}_M(h^{-1}\sigma, \mathfrak{o})h.$$

Note that for every $h_0 \in H$ we have

$$\boldsymbol{SW}_{M,h_0\sigma,\mathfrak{o}} = h_0 \boldsymbol{SW}_{M,\sigma,\mathfrak{o}}.$$

Moreover, when $\partial M = \emptyset$

$$\boldsymbol{SW}_{M,\sigma} = c(\sigma)\boldsymbol{SW}_{M,\bar{\sigma}} = c(\sigma)\overline{\boldsymbol{SW}}_{M,\sigma}.$$

In this case for any spin structure ϵ we have $\det(\sigma(\epsilon)) = 1$ so that

$$\boldsymbol{SW}_{M,\sigma(\epsilon)} = \overline{\boldsymbol{SW}}_{M,\sigma(\epsilon)}.$$

For simplicity we set $\boldsymbol{SW}_{M,\epsilon} := \boldsymbol{SW}_{M,\sigma(\epsilon)}$.

B. $b_1 = 1$. This situation is a bit more delicate. We discuss separately the two cases $\partial M = \emptyset$ and $\partial M \neq \emptyset$.

B.1. $\partial M = \emptyset$. Fix a metric g. The enhanced homology orientation defines an orientation on $H \otimes \mathbb{R}$. Choose a harmonic 1-form ω_g which induces the chosen orientation on $H \otimes \mathbb{R}$, and $\|\omega_g\|_{L^2(g)} = 1$. Note that this orientation also produces a surjection

$$\deg = \deg_{\mathfrak{o}} \colon H \to H/\operatorname{Tors}(H) = \mathbb{Z}.$$

We identify $H^2(M)$ with H via the Poincaré duality, and for every complex line bundle $L \to M$, we set

$$\deg L := \deg(\boldsymbol{PD}\, c_1(L)).$$

For $\sigma \in \operatorname{Spin}^c(M)$ denote by $\mathcal{P}_\sigma(g)$ the space of co-closed 1-forms η such that

$$w_{\mathfrak{o}}(\sigma, \eta) := \int_M \big(\eta - 2\pi * c_1(\det\sigma)\big) \wedge *\omega_g \neq 0.$$

The *wall*

$$\mathcal{W}_\sigma := \{\eta \in \mathcal{P}_\sigma(g);\ w(\sigma, \eta) = 0\}$$

decomposes $\mathcal{P}_\sigma(g)$ into two chambers

$$\mathcal{P}^{\pm}_{\sigma,\mathfrak{o}}(g) = \{\eta \in \mathcal{P}_\sigma(g);\ \pm w_{\mathfrak{o}}(\sigma, \eta) > 0\}.$$

For generic $\eta \in \mathcal{P}_\sigma(g)$ we denote by $\boldsymbol{sw}_M(\sigma, \mathfrak{o}, \eta)$ the signed count of (σ, g, η)-monopoles. It is known that

$$\boldsymbol{sw}_M(\sigma, \mathfrak{o}, \eta) = \boldsymbol{sw}_M(\bar{\sigma}, \mathfrak{o}, \eta),$$

$\boldsymbol{sw}_M(\sigma, \mathfrak{o}, \eta) = 0$ for all but finitely many σ's, and

$$\boldsymbol{sw}_M(\sigma, \mathfrak{o}, \eta_1) = \boldsymbol{sw}_M(\sigma, \mathfrak{o}, \eta_2),$$

if $w_{\mathfrak{o}}(\sigma, \eta_1) \cdot w_{\mathfrak{o}}(\sigma, \eta_2) > 0$. We set

$$\boldsymbol{sw}^{\pm}_M(\sigma, \mathfrak{o}) := \boldsymbol{sw}_M(\sigma, \mathfrak{o}, \eta),$$

where $\pm w_{\mathfrak{o}}(\sigma, \eta) > 0$. The wall crossing formula (see [61]) states that

$$\boldsymbol{sw}^{+}_M(\sigma, \mathfrak{o}) - \boldsymbol{sw}^{-}_M(\sigma, \mathfrak{o}) = \frac{1}{2}\deg(\det \sigma).$$

We define

$$\boldsymbol{SW}_{M,\sigma,\mathfrak{o},\eta} = \sum_{h\in H} \boldsymbol{sw}_M(h^{-1}\sigma, \eta)h \in \mathbb{Z}[H],$$

$$\boldsymbol{SW}^{+}_{M,\sigma,\mathfrak{o}} = \sum_{h\in H} \boldsymbol{sw}^{+}_M(h^{-1}\sigma)h \in \mathbb{Z}[[H]].$$

Suppose we pick $\sigma_0 = \sigma(\epsilon)$, where ϵ is a spin structure, and $\eta = \eta_0$ such that $\int_M \eta_0 \wedge *\omega_g$ is a very small positive number. Fix $T \in H$ such that $\deg(T) = 1$. We deduce that

$$w_{\mathfrak{o}}(h^{-1} \cdot \sigma_0, \eta_0) > 0 \iff \int_M \eta_0 \wedge *\omega_g > -2\pi \deg_{\mathfrak{o}} h \iff \deg_{\mathfrak{o}} h \geq 0$$

Hence

$$(g, \eta_0) \in \mathcal{P}^{-}_{h^{-1}\cdot\sigma_0,\mathfrak{o}}(g) \iff \deg_{\mathfrak{o}}(h) < 0.$$

We can rephrase the wall crossing formula in the more compact form using the wall crossing term W_M introduced in §3.6. More precisely

$$\boldsymbol{SW}^{+}_{M,\sigma(\epsilon)} = \boldsymbol{SW}_{M,\sigma(\epsilon),\eta_0} + \sum_{h\in H} \deg^{+}(h^{-1})h = \boldsymbol{SW}_{M,\sigma(\epsilon),\eta_0} + \frac{\mathfrak{S}_M T}{(1-T)^2},$$

where

$$\deg^{+} = \max(\deg, 0), \quad \mathfrak{S}_M = \sum_{h\in \mathrm{Tors}(H)} h \in \mathbb{Z}[H].$$

Observe that

$$\boldsymbol{SW}^{0}_{M,\sigma(\epsilon)} := \boldsymbol{SW}^{+}_{M,\sigma(\varepsilon)} - W_M = \boldsymbol{SW}_{M,\sigma(\epsilon),\eta_0} \in \mathbb{Z}[H] \tag{4.1}$$

is a *topological invariant, independent on the orientation* $\mathfrak{o}$ *on* $H \otimes \mathbb{R}$, which satisfies the symmetry condition

$$\boldsymbol{SW}^{0}_{M,\sigma(\epsilon)} = \overline{\boldsymbol{SW}}^{0}_{M,\sigma(\epsilon)},$$

and the equivariance property

$$\boldsymbol{SW}^{0}_{M,\sigma(h_0\epsilon)} = h_0\boldsymbol{SW}^{0}_{M,\sigma(\epsilon)}, \quad \forall h_0 \in \mathrm{Tors}_2(H).$$

We will refer to $\boldsymbol{SW}^{0}_{M,\sigma(\epsilon)}$ as *the modified Seiberg–Witten invariant of* M *associated to the spin structure* ϵ.

Example 4.1 (The Seiberg–Witten invariants of $S^1 \times S^2$). Suppose $M = S^1 \times S^2$, and g_0 is the Cartesian product of the round metrics on S^1 and S^2. The scalar curvature of g_0 is a positive constant s_0. $H_1(M)$ is generated by the fiber S^1, and thus M carries a natural enhanced homology orientation $\mathfrak{o}$. We denote by $\omega_0 = d\varphi$ the angular form along S^1. It defines a generator of $H^1(M, \mathbb{R})$ which is positive with respect to the above enhanced homology orientation. ω_0 is a harmonic 1-form. After a possible rescaling of the metric g_0 we can assume it has L^2-norm 1. Choose $\eta_0 = c\omega_0$ where c is a very small positive number.

Note that for every spin^c structure σ we have $\deg(\det \sigma) \in 2\mathbb{Z}$. For every $n \in \mathbb{Z}$, we denote by σ_n the unique spin^c structure on M such that $\deg(\det \sigma_n) = 2n$. Observe now that

$$\eta_0 \in \begin{cases} \mathcal{P}^+_{\sigma_n}(g) & \text{if } n \leq 0 \\ \mathcal{P}^-_{\sigma_n}(g) & \text{if } n > 0. \end{cases} \tag{4.2}$$

We want to prove that for any $n \in \mathbb{Z}$ there are no (σ_n, g_0, η_0)-monopoles so that

$$\boldsymbol{sw}_M(\sigma_n, \eta_0) = 0, \quad \forall n \in \mathbb{Z}.$$

Indeed, suppose $\mathsf{C} = (\psi, A)$ is a σ_n-monopole. Using the first equation in (SW) we deduce $\mathfrak{D}_A^2 \psi = 0$. The Weitzenböck formula now yields

$$(\nabla^A)^* \nabla^A \psi + \frac{s_0}{4}\psi + \frac{1}{2}\boldsymbol{c}(F_A)\psi = 0.$$

The second equation in (SW) implies

$$\boldsymbol{c}(F_A) = \boldsymbol{c}(*F_A) = \frac{1}{2}q(\psi) - \boldsymbol{c}(\mathbf{i}\eta).$$

Hence

$$\Big((\nabla^A)^* \nabla^A + \frac{s_0}{4} - \frac{1}{2}\boldsymbol{c}(\mathbf{i}\eta_0) + \frac{1}{4}q(\psi)\Big)\psi = 0.$$

Taking the L^2-inner product of the last equality with ψ, and then integrating by parts we get

$$\int_M \Big(|\nabla^A \psi|^2 + \frac{s_0}{4}|\psi|^2 - \frac{1}{2}(\boldsymbol{c}(\mathbf{i}\eta_0)\psi, \psi) + \frac{1}{16}|\psi|^4\Big) dv_M = 0.$$

Since $\eta_0 = c\omega_0$ and $0 < c \ll 1$ we deduce that all the terms in the above equality must be zero so that $\psi \equiv 0$. Using this information in the second equation of (SW) we deduce

$$F_A = -\mathbf{i} * \eta_0 \Longrightarrow \frac{\mathbf{i}}{2\pi}\int_M F_A \wedge *\omega_0 = \frac{c}{2\pi}\int_M \omega_0 \wedge *\omega_0 = \frac{c}{2\pi}.$$

Hence

$$2\pi \int_M c_1(\det \sigma_n) \wedge *\omega_0 = c.$$

This equality is impossible when

$$0 < c < \min\{2\pi\langle c_1(\det\sigma_n), \omega_0\rangle_{L^2};\ n \in \mathbb{Z}\setminus\{0\}\},$$

which confirms our claim. We deduce that

$$\boldsymbol{sw}_M(\sigma_n, \eta_0) = 0, \quad \forall n.$$

Using (4.2) we conclude that

$$\boldsymbol{sw}_M^+(\sigma_n) = \boldsymbol{sw}_M(\sigma,\eta_0) = 0, \quad \forall n \le 0.$$

The wall crossing formula implies that for all $n > 0$ we have

$$\boldsymbol{sw}_M^+(\sigma_n) = n + \boldsymbol{sw}_M^-(\sigma_n) = n + \boldsymbol{sw}_M^-(\sigma_n, \eta_0) = n.$$

Now interpret $H_1(M)$ as a multiplicative group, and denote by T the generator of H satisfying $\deg T = 1$. We deduce

$$\boldsymbol{SW}_{M,\sigma_0}^+ = \sum_{n\in\mathbb{Z}} \boldsymbol{sw}_M^+(\sigma_{-n})T^n - \sum_{n<0} nT^n = \sum_{n>0} nT^{-n} = \frac{T^{-1}}{\left(1-T^{-1}\right)^2} = \frac{T}{(1-T)^2}.$$

□

B.2. $\partial M \cong T^2$. Using the orientation[2] of $H_1(M,\mathbb{R})$ we can choose a cycle $\Gamma_+ \in H_1(M,\mathbb{Z})$ inducing the positive orientation on 1-dimensional real space $H_1(M,\mathbb{R})$. For each Riemann metric g pick a harmonic 1-form ω_g on M such that along the end $[0,\infty)\times\partial M$ the restriction $\omega_g|_{t\times\partial M}$ converges exponentially as $t\to\infty$ to a nontrivial harmonic 1-form on ∂M and

$$\int_{\Gamma_+} \omega_g = 1.$$

The space of parameters (g,η) decomposes into

$$\mathcal{P}_\sigma^\pm = \{(g, s\omega_g + \varphi);\ \pm s > 0,\ \varphi \in \Omega^2(M),\ d^*\varphi = 0,\ \varphi \text{ has compact support}\}.$$

Then one can show that $\boldsymbol{sw}_M(\sigma,\hat{g},\eta)$ is independent of $(\hat{g},\eta)\in\mathcal{P}_+$. We denote this common value by $\boldsymbol{sw}_M^+(\sigma)$. $\boldsymbol{sw}_M^-(\sigma)$ is defined similarly. Now define

$$\boldsymbol{SW}_M^+ \colon \operatorname{Spin}^c(M,\partial M) \to \mathbb{Z}[[H_1(M)]], \quad \sigma \mapsto \sum_{h\in H_1(M)} \boldsymbol{sw}_M^+(h^{-1}\sigma)h. \tag{sw}$$

This is an $H_1(M)$-equivariant map and (see [68])

$$\boldsymbol{SW}_M^+(\sigma) \in \mathfrak{N}_1(H_1(M)), \quad \forall\sigma\in\operatorname{Spin}^c(M,\partial M).$$

[2]If $\partial M \neq \emptyset$ then the homology orientation on $H_*(M,\mathbb{R})$ induces an orientation of $H_1(M,\mathbb{R})$ which will be the one we choose.

Example 4.2. (a) ([3, 80]) If M is the total space of a degree d circle bundle over a Riemann surface Σ of genus g then for any spin^c structure σ on M we have

$$\mathbf{sw}_M(\sigma) \sim (1-t)^{2g-2} \in \mathbb{Z}[H_1(M)]$$

where t denotes the homology class of the fiber.

(b) If N_1 and N_2 are closed, oriented 3-manifolds such that $b_1(N_i) > 1$ then

$$\mathbf{sw}_{N_1 \# N_2} = 0. \qquad \square$$

The above examples and the computations in §2.1 suggest that the Reidemeister torsion and the Seiberg–Witten invariant $\mathbf{sw}_M$ could be related. We have the following more accurate statement.

Theorem 4.3 (Meng–Taubes [68], Turaev [115]). *The Seiberg–Witten invariant* $\mathbf{sw}_M$, $b_1(M) > 0$, *satisfies the Axioms* 1–3 *in* §3.5 *and thus coincides with the Reidemeister–Turaev torsion* (*up to a sign*).

Remark 4.4. For a general outline of the proof (based in essence on gauge theoretic gluing results) we refer to [68, 115]. As explained in Remark 3.19, the proof reduces to proving surgery formulæ for the Seiberg–Witten invariant of admissible 3-manifolds, and verifying that in the simplest case of a solid torus this invariant equals the torsion, $(1-t)^{-1}$. This is very easily deduced from the surgery formulæ, and the computation in Example 4.1.

Most of the analytical work needed to prove the gauge theoretic surgery results can be found in the recent paper [108]. The topological counterparts of these analytical gluing formulæ which are required in the proof can be found in [112].

D. Salamon outlines in [95] a different, more geometric approach in the special case of 3-manifolds which fiber over S^1. In [21] S. Donaldson describes yet another approach to the case $b_1 = 1$ of Meng–Taubes theorem based on Topological Quantum Field Theory. This approach was recently extended to any $b_1 > 0$ by T. Mark in [66]. □

Exercise 4.1. Assuming that the Seiberg–Witten invariant satisfies the excision formula in §3.7, prove that in the case of the canonical spin^c structure on the solid torus $Z = S^1 \times D^2$ it is equal, up to a sign, with the Reidemeister–Turaev torsion. □

We can eliminate the sign ambiguity in Theorem 4.3 at least when $b_1(M) = 1$ and $\partial M = \emptyset$. In §3.6 we defined the modified Reidemeister–Turaev torsion of closed 3-manifolds such that $b_1(M) = 1$ by the equality

$$\mathcal{T}^0_{M,\epsilon} := \mathcal{T}_{M,\epsilon} - W_M \in \mathbb{Z}[H], \quad \forall \epsilon \in \mathrm{Spin}(M). \tag{4.3}$$

We observed that

$$\mathcal{T}^0_{M,\epsilon} = \overline{\mathcal{T}}^0_{M,\epsilon}.$$

Proposition 4.5. *If $b_1(M) = 1$ and $\partial M = \emptyset$ then*

$$\boldsymbol{SW}^+_{M,\epsilon} = \mathfrak{T}_{M,\epsilon}, \quad \epsilon \in \mathrm{Spin}(M).$$

Proof. We have an equality

$$\boldsymbol{SW}^+_{M,\epsilon} = \pm\mathfrak{T}_{M,\epsilon} \iff (\boldsymbol{SW}^0_{M,\epsilon} + W_M) = \pm(\mathfrak{T}^0_{M,\epsilon} + W_M)$$

To prove that the correct choice of signs is "+" we argue by contradiction. Suppose

$$\boldsymbol{SW}^+_{M,\epsilon} = -\mathfrak{T}_{M,\epsilon}.$$

Then this implies that

$$\mathfrak{T}_{M,\epsilon} - W_M = -\boldsymbol{SW}^0_{M_\epsilon} - 2W_M \notin \mathbb{Z}[H]$$

which contradicts (4.3). □

Remark 4.6. We see that we can remove the sign ambiguity in the case $b_1(M) = 1$, $\partial M = \emptyset$ by relying on the wall cross formula for the Seiberg–Witten invariant and the structural equality (3.6) for the torsion. It is natural to expect that we could similarly remove the sign ambiguity in the case $b_1(M) = 1$, $\partial M \cong T^2$, by using the structural identity (3.7) and an as yet non existent wall crossing formula for the Seiberg–Witten invariants of M. Things are more delicate in this case. Lim's proof in [61] of the wall crossing formula in the closed case does not extends to the cylindrical end situation for a simple reason. In the noncompact case the Fredholm property of a partial differential operator is not decided by the symbol of the operator alone, as is the case in the closed case. For manifolds with cylindrical ends this is decided by global objects which are no longer invariant under lower order deformations of the operator. □

Proposition 4.7. *If M is a closed, homologically oriented 3-manifold such that $b_1(M) = 1$ then*

$$\widehat{\mathfrak{T}}^0_M(1) = \widehat{\boldsymbol{SW}}^0_M(1) = \frac{1}{2}\Delta''_M(1),$$

where $\Delta_M \in \mathbb{Z}[[T^{1/2}, T^{-1/2}]]$ denotes the symmetrized Alexander polynomial of M normalized such that $\Delta_M(1) = |\operatorname{Tors}(H_1(M))|$.

Proof. Set $H := H_1(M)$ and

$$\mathfrak{S}_M := \sum_{h \in \operatorname{Tors} H} h \in \mathbb{Z}[H].$$

The projection $\deg_{\mathfrak{o}} \colon H \twoheadrightarrow \mathbb{Z}$ defined by the homology orientation $\mathfrak{o}$ induces the augmentation morphism

$$\mathfrak{aug} \colon \mathbb{Z}[[H]] \to \mathbb{Z}[[t, t^{-1}]].$$

Fix $T \in H$ such that $\deg_{\mathfrak{o}} T = 1$. The symmetrized Alexander polynomial Δ_M is uniquely determined by the condition

$$\mathfrak{aug}\, \mathcal{T}_{M,\epsilon} = T^{k/2} \frac{\Delta_M(T)}{(1-T)^2},$$

for some $k \in \mathbb{Z}$. Using Theorem 4.9(a) we deduce

$$\begin{aligned} T^{k/2} \frac{\Delta_M(T)}{(1-T)^2} &= \mathfrak{aug}\, \boldsymbol{SW}_M = \mathfrak{aug}\, \boldsymbol{SW}_M^0 + \mathfrak{aug}\,(\mathfrak{S}_M) \frac{T}{(1-T)^2} \\ &= \mathfrak{aug}\, \boldsymbol{SW}_M^0 + |\operatorname{Tors}(H)| \frac{T}{(1-T)^2}. \end{aligned}$$

We conclude that

$$T^{k/2-1} \Delta_M(T) = (T - 2 + T^{-1}) \mathfrak{aug}\, \boldsymbol{SW}_M^0(T) + |\operatorname{Tors}(H)|.$$

The symmetry of $\boldsymbol{SW}^0$ implies $\boldsymbol{SW}_M^0(T) = \boldsymbol{SW}_M^0(T^{-1})$, and since Δ_M satisfies a similar symmetry we conclude $k/2 - 1 = 0$. Hence

$$\Delta_M(T) = (T - 2 + T^{-1}) \mathfrak{aug}\, \boldsymbol{SW}_M^0(T) + |\operatorname{Tors}(H)|.$$

Differentiating the above equality twice at $T = 1$ we deduce

$$\Delta_M''(1) = 2\mathfrak{aug}\, \boldsymbol{SW}_M(1) = 2\widehat{\boldsymbol{SW}}^0(1). \qquad \square$$

Remark 4.8. Observe a nice "accident". Suppose M is as in Proposition 4.7. Then

$$W_M = \mathfrak{S}_M \sum_{n \geq 1} n T^{-n}.$$

Formally

$$\begin{aligned} \widehat{W}_M(1) &= \widehat{\mathfrak{S}}_M(1) \sum_{n \geq 1} n = |\operatorname{Tors}(H)| \sum_{n \geq 1} n \\ &= |\operatorname{Tors}(H)| \zeta(-1) = -\frac{1}{12} |\operatorname{Tors}(H)|, \end{aligned}$$

where $\zeta(s)$ denotes Riemann's zeta function. In particular

$$\widehat{\boldsymbol{SW}}_M(1) = \widehat{\boldsymbol{SW}}_M^0(1) + \widehat{W}_M(1) = \frac{1}{2}\Delta_M''(1) - \frac{1}{12}|\operatorname{Tors}(H)|.$$

The expression in the right-hand-side is precisely the Casson–Walker–Lescop invariant of M, [58]. □

C. Suppose now that $b_1(M) = 0$, i.e. M is a rational homology sphere. Fix $\sigma \in \mathrm{Spin}^c(M)$. In this case the signed count of (σ, g, η)-monopoles depends on (g, η) in a more complicated way. To produce a topological invariant we need to add a correction to this count. For simplicity, we describe this correction only when $\eta = 0$.

The line bundle $\det \sigma = \det \mathbb{S}_\sigma$ admits a unique equivalence class of flat connections. Pick one such flat connection A_σ and denote by $\mathfrak{D}_{A_\sigma}$ the Dirac operator on $\mathbb{S}_\sigma$ determined by the twisting connection A_σ. We denote its eta invariant by $\eta_{\,\mathrm{dir}}(g, \sigma)$. Also, denote by $\eta_{\,\mathrm{sign}}(g)$ the eta invariant of the odd signature operator determined by g. Finally define the *Kreck–Stolz invariant* of (g, σ) by

$$KS(g, \sigma) = 4\eta_{\,\mathrm{dir}}(g, \sigma) + \eta_{\,\mathrm{sign}}(g).$$

Define *the modified Seiberg–Witten invariant* of (M, σ) by

$$\boldsymbol{sw}^0_M(\sigma) = \frac{1}{8} KS(g, \sigma) + \boldsymbol{sw}_M(\sigma) \in \mathbb{Q}.$$

As shown in [61], the above quantity is independent of the metric, and it is a topological invariant. Set

$$\boldsymbol{SW}^0_{M,\sigma} := \sum_{h \in H} \boldsymbol{sw}^0_M(h^{-1}\sigma)h \in \mathbb{Q}[H].$$

If $\sigma = \sigma(\epsilon)$ we have

$$\boldsymbol{SW}^0_{M,\sigma(\epsilon)} = \overline{\boldsymbol{SW}}^0_{M,\sigma(\epsilon)}.$$

To establish a relationship between the Reidemeister torsion and the Seiberg–Witten invariants for rational homology spheres we need to define a more sophisticated modification of the torsion.

Suppose $b_1(M) = 0$. We denote by $\mathrm{CW}_M \in \mathbb{Q}$ the *Casson–Walker invariant* of M (see [58, 118] for more information about this invariant) and define

$$\mathfrak{T}^0_{M,\epsilon} = \mathfrak{T}_{M,\epsilon} - \frac{1}{2}\mathrm{CW}_M\,\Theta_M.$$

Observe that $\hat{\mathfrak{T}}^0_{M,\epsilon}(1) = \mathfrak{aug}\mathfrak{T}^0_{M,\epsilon} = \frac{1}{2}|H|\,\mathrm{CW}_M =$ Lescop invariant of M (see [58, p. 80]).

We we will refer to the quantities $\mathfrak{T}^0_{M,\epsilon}$ for $b_1(M) = 0, 1$ the *modified Reidemeister–Turaev torsion of M*. The Seiberg–Witten invariant and the modified Reidemeister torsion are related. More precisely we have the following result.

Theorem 4.9. (a) [60, 65, 85]. $\widehat{\boldsymbol{SW}}^0_M(1) = \hat{\mathfrak{T}}^0_M(1)$ *if* $b_1(M) = 0$.

(b) [81]. $\boldsymbol{SW}^0_M = \mathfrak{T}^0_M$ *if M is a lens space.*[3]

[3] We have to warn the reader of a sign convention in [118, Prop. 6.2], where the lens space $L(p, q)$ is defined as the p/q-surgery on the unknot. However, the lens space $L(p, q)$ as defined in §2.7 is the $-p/q$-surgery on the unknot, [37, p.158], [46, p. 65-66].

Part (b) of the above theorem can be slightly strengthened to

$$\boldsymbol{SW}^0_M = \mathcal{T}^0_M, \quad \text{if } M \text{ is a connected sum of lens spaces.} \tag{4.4}$$

This equality follows from the vanishing of the torsion under connected sums, the additivity[4] of the Casson–Walker invariant, and the additivity of the Kreck–Stolz invariant.

Theorem 4.10 ([83]).

$$\boldsymbol{SW}^0_M = \mathcal{T}^0_M$$

for any rational homology 3*-sphere* M.

Outline of the proof. The first temptation would be to prove an extension of the Uniqueness Theorem 3.17 which is limited to manifolds satisfying $b_1(M) > 0$. There is however a major obstacle.

The Uniqueness Theorem 3.17 involves invariants of admissible manifolds with boundary, and for such manifolds the functional set-up for the Seiberg–Witten equations requires perturbation parameters satisfying the nondegeneracy condition **P2** described at the beginning of this section. If we glue along their boundaries two manifolds $M_i, i = 0, 1$, such that $b_1(M_i) = 1, \partial M_i = T^2$, aiming to produce a rational homology sphere $N = M_0 \#_{T^2} M_1$, then we would have to use perturbation parameters ν_i matching along the boundaries. These would produce a non-exact closed 2-form $*\nu_0 \# * \nu_1$ on N which would contradict the condition $b_1(N) = 0$. Thus, whatever uniqueness statement we would attempt to prove, it cannot involve the torsion of manifolds with boundary.

On the analytical side, the way out of this quandary is to work with a different class of allowable perturbations, the compactly supported ones. This creates a serious problem. It substantially changes the structure of the deformation complex for the Seiberg–Witten equations on manifolds with boundary. Its index will no longer of be zero, as desired. However this choice of perturbation is forced upon us, and cannot be avoided. We must renounce all the analytical achievements in [68, 108] and prove new gluing formulæ.

All is not lost. The analytical results in [62, 65, 85] can produce gluing results for the modified Seiberg–Witten invariant, albeit much weaker than the ones in [108]. These new formulæ for the Seiberg–Witten invariant do not involve manifolds with boundary.

On the topological side, one has to renounce the multiplicative gluing formulæ discussed so far and use additive ones [116, §6]. We outline below some the difficulties of this approach and the method we propose to get out of trouble.

We denote by D_M the difference $\boldsymbol{SW}^0_M - \mathcal{T}^0_M$. Proving the equality of these two invariants is equivalent to showing that $D_M \equiv 0$. At this point the harmonic analysis in §1.6 comes in extremely handy. For example, the Marcolli–Wang result [65] translates

[4]This follows from the very general surgery results for eta invariants in [52].

into $\widehat{D}_M(1) = 0$, for all rational homology spheres. The *true nature* of the surgery formulæ is *best displayed* in the Fourier picture. To explain the gist of these formulæ consider a 3-manifold N with $b_1 = 1$ and boundary T^2. N can be thought of as the complement of a knot in a $\mathbb{Q}HS$. Pick two simple closed curves c_1, c_2 on ∂N with nontrivial intersection numbers with the longitude $\lambda \in H_1(\partial N, \mathbb{Z})$.

By Dehn surgery with c_i as attaching curves we obtain two rational homology spheres M_1, M_2 and two knots $K_i \hookrightarrow M_i$, $i = 0, 1$. Let $H_i := H_1(M_i, \mathbb{Z})$, $G := H_1(N, \partial N; \mathbb{Z})$. Denote by j the inclusion induced morphism $H_1(\partial N) \to H_1(N)$. The knot K_i determines a subgroup $K_i^\perp \subset \hat{H}_i$, consisting of the characters vanishing on K_i. These subgroups are naturally isomorphic to the group of characters χ of $H_1(N)$ with the property that the composition

$$H_1(\partial N) \xrightarrow{j} H_1(N) \xrightarrow{\chi} \mathbb{C}^*$$

is trivial (see §B.2). We thus have a natural isomorphism

$$f\colon K_1^\perp \to K_2^\perp.$$

Putting together the gluing formulæ in [62, 65, 85] and the additive gluing formulaæ in [116, §6] we get a gluing formula for $\hat{D}_M$ of the form[5]

$$\langle \lambda, c_2 \rangle \widehat{D}_{M_1}(\chi) = \langle \lambda, c_1 \rangle \widehat{D}_{M_2}\big(f(\chi)\big) + |G| \mathcal{K}, \quad \forall \chi \in K_1^\perp$$

where $\langle \bullet, \bullet \rangle$ denotes the intersection pairing on $H_1(\partial N, \mathbb{Z})$, and $\mathcal{K}$ is a universal correction term which depends only on the divisibility m_0 of the longitude and the $\mathrm{SL}_2(\mathbb{Z})$-orbit of the pair (c_1, c_2) with respect to the obvious action of this group on the space of pairs of primitive vectors in a 2-dimensional lattice (see §B.2). We will thus write $\mathcal{K}_{m_0;[c_1,c_2]}$, and call the triplet $(m_0; [c_1, c_2])$ the arithmetic type of the surgery. The results of [85] prove that

$$\mathcal{K}_{1;[c_1,c_2]} \equiv 0, \quad \forall [c_1, c_2].$$

We call surgeries with $m_0 = 1$ *primitive*, and the surgeries with trivial correction term, *admissible*. We denote by $\mathfrak{R}$ the class of rational homology spheres M such that $\widehat{D}_M \equiv 0$. Both the family of admissible surgeries and the family $\mathfrak{R}$ are "time dependent" families, and during our proof we gradually produce larger and larger classes of surgeries/manifolds inside these families.

The class $\mathfrak{R}$ is closed under connected sums and certain primitive surgeries. Using this preliminary information and basic Kirby calculus one can show[6] that all homology lens spaces belong to $\mathfrak{R}$. As a bonus, we can include many more arithmetic types of Dehn surgeries in the class of admissible surgeries.

[5]The reader should compare this description of the surgery formula with the ones in [65, 85] to truly appreciate the amazing simplifying power of the Fourier transform.

[6]I learned this fact from Nikolai Saveliev.

Loosely speaking, the homology lens spaces have the simplest linking forms. We take this idea seriously and define an appropriate notion of complexity of a linking form. The proof then proceeds by induction, including in $\mathfrak{X}$ manifolds of larger and larger complexity. This process also increases the class of admissible surgeries, which can be used at the various inductive steps. Such a proof is feasible if we can produce a large supply of complexity reducing Dehn surgeries. Fortunately, this can be done using elementary arithmetic. We refer for more details to [83]. □

Using Theorem 4.10 we can now establish a relationship between the maps $\mathfrak{q}_{\text{tors}}$ and $\mathfrak{q}_{\text{top}}$ introduced in §3.9.

Proposition 4.11.

$$\mathfrak{q}_{\text{tors}} = \mathfrak{q}_{\text{top}}.$$

Proof. Since both maps are $\hat{H}$-equivariant it suffices to prove that $\mathfrak{q}_{\text{top}}(\epsilon) = \mathfrak{q}_{\text{tors}}(\epsilon)$ for some spin structure ϵ on M. Using Theorem 4.10 we deduce that

$$\mathfrak{q}_{\text{tors}}(\epsilon)(h) = \frac{1}{8}\big(KS(\epsilon) - KS(h\cdot\epsilon)\big) \bmod \mathbb{Z}, \quad \forall h \in H.$$

To compute $\mathfrak{q}_{\text{top}}(\epsilon)$ consider a simply connected spin four-manifold $(\hat{M}, \hat{\epsilon})$ such that (see [37])

$$\partial(\hat{M}, \hat{\epsilon}) = (M, \epsilon).$$

Next, choose $\hat{h} \in H^2(\hat{M}, \mathbb{Z})$ such that $[\hat{h}] := \hat{h}|_{\partial M} \in \hat{H}$ is the Poincaré dual of h, i.e.

$$[\hat{h}](\bullet) = \boldsymbol{lk}_M(h, \bullet).$$

Set $\hat{\sigma}_h := \hat{h}\cdot\hat{\epsilon}$.

Fix a metric g on M and extend it to a metric $\hat{g}$ on $\hat{M}$ which is a product near the boundary. Denote by $\hat{\mathbb{S}}_h$ the $\mathbb{Z}_2$-graded of spinors associated to $\hat{\sigma}_h$. Extend the flat connection A_σ on $\det\sigma$ to a connection $\hat{A}_h$ on $\det\hat{\mathbb{S}}_h^+$. We get in this fashion an operator $\mathfrak{D}_h$ on $\hat{\mathbb{S}}_h$. Using the Atiyah–Patodi–Singer index theorem [2] for $\mathfrak{D}_h$ and the signature operator on $\hat{M}$ we deduce

$$-\frac{1}{24}\int_{\hat{M}} p_1(\hat{M}, \hat{g}) + \frac{1}{2}\check{Q}_{\hat{M}}(\hat{h}, \hat{h}) = \frac{1}{2}\eta_{\,\text{dir}}(h\epsilon, g) \bmod \mathbb{Z},$$

and

$$\frac{1}{24}\int_{\hat{M}} p_1(\hat{M}, \hat{g}) - \frac{1}{8}\,\text{sign}(\hat{M} = \frac{1}{8}\eta_{\,\text{sign}} \bmod \mathbb{Z}.$$

Thus

$$\frac{1}{8}KS(h\epsilon) = \frac{1}{2}\check{Q}_{\hat{M}}(\hat{h}, \hat{h}) - \frac{1}{8}\,\text{sign}(\hat{M}) \bmod \mathbb{Z}.$$

Hence

$$\mathfrak{q}_{\text{tors}}(h) = \frac{1}{8}\big(KS(\epsilon) - KS(h\cdot\epsilon)\big) \bmod \mathbb{Z} = -\frac{1}{2}\check{Q}_{\hat{M}}(\hat{h}, \hat{h}) \bmod \mathbb{Z} = \mathfrak{q}_{\text{top}}(\epsilon)(h).$$ □

Remark 4.12. While the book was in print, F. Deloup and G. Massuyeau have given a purely topological proof of Proposition 4.11. For details we refer to their preprint `math.GT/0207/188`. This result plays a key role in the investigations in [75]. □

§4.2 A Morse theoretic interpretation

The Reidemeister torsion resembles in many respects the Euler characteristic. According to the classical Poincaré–Hopf theorem, we can interpret the Euler characteristic as counting stationary points of smooth vector fields.

In the early 80's, D. Fried [33, 34] has shown that the Reidemeister torsion of a smooth manifold X can too be interpreted as counting closed orbits of nowhere vanishing Morse–Smale vector fields, i.e. gradient vector fields associated to smooth maps $X \to S^1$ without critical points (see also [28] for earlier results of this nature). M. Hutchings and Y. Lee have recently extended Fried's result to any generic map $\alpha\colon X \to S^1$. The goal of this subsection is to formulate this result and loosely explain it. For details and proof we refer to [49] which served as our main source of inspiration. For different approaches we refer to [47, 48, 87].

Suppose X is a closed, compact, oriented, smooth manifold such that $\chi(X) = 0$. Fix a Riemann metric g on X and a smooth function

$$\alpha\colon X \to S^1$$

such that the pair (α, g) is *admissible*, i.e.

- The critical points of α and the closed orbits of $\nabla\alpha$ are nondegenerate.
- The ascending and descending manifolds of the critical points of α intersect transversally.

A *closed orbit* of $\nabla\alpha$ is a nonconstant map

$$u\colon S^1 \to X$$

such that

$$\frac{du}{dt} = -\lambda \nabla\alpha(u)$$

for some positive constant λ. Two orbits are considered equivalent if they differ by a reparametrization. We denote by $\mathcal{O}$ the set of equivalence classes of closed orbits. Each equivalence class of closed orbits u determines an unique homology class $[u] \in H_1(X)$.

The *period* of a closed orbit u is the largest positive integer p such that u factors through a p-fold covering

$$S^1 \to S^1, \quad z \mapsto z^p.$$

Equivalently, the period is the largest positive integer p such that

$$\frac{1}{p}[u] \in H_1(X).$$

A closed orbit u defines a return map

$$\phi_u \colon \langle \dot{u}(0) \rangle^{\perp} \to \langle \dot{u}(0) \rangle^{\perp},$$

where $\langle \dot{u}(0) \rangle^{\perp}$ denotes a neighborhood of the origin in the orthogonal complement of $\dot{u}(0)$ in $T_{u(0)}X$. The orbit is called *nondegenerate* if

$$\det(1 - D\phi_u) \neq 0.$$

In this case we define the Lefschetz sign

$$\varepsilon(u) := \operatorname{sign}\det(1 - D\phi_u).$$

The function α defines a cohomology class

$$\omega = \omega_\alpha := \frac{1}{2\pi} f^*(d\theta) \in H^1(X, \mathbb{Z}).$$

Define the *Novikov ring*

$$\Lambda_\alpha := \big\{ s = \textstyle\sum_h s_h h \in \mathbb{Z}[[H_1(X)]];\ \forall C \in \mathbb{R},\ \#\{h,\ s_h \neq 0,\ \omega(h) < C\} < \infty \big\},$$

and the *zeta function* of α

$$\zeta_\alpha := \exp\Big(\sum_{u \in \mathcal{O}} \frac{\varepsilon(u)}{p(u)} [u] \Big) \in \Lambda_\alpha.$$

It is not a priori clear that ζ_α is a Laurent series with *integral* coefficients. This follows from the equivalent description (see [34, 48, 95])

$$\zeta_\alpha = \prod_{u \in \mathcal{O}^*} \big(1 - (-1)^{i_-([u])}[u]\big)^{(-1)^{i_0([u])}},$$

where $\mathcal{O}^*$ denotes the set of primitive orbits (period 1), $i_{-/0}([u])$ denotes the number of real eigenvalues of the return map in the intervals $(-\infty, -1)$ and respectively $(-1, 1)$.

The Morse–Novikov complex associated to α and g is a chain complex $(\underline{C}, \partial) = (\underline{C}(\alpha; g), \partial)$ of *free* Λ_α-modules defined as follows (see [47, 84, 86] for more details).

• C_k is the free Λ_α module generated by the critical points of f of index k. Denote by $\tilde{X}$ the universal abelian cover and by $\tilde{\alpha}$ the induced smooth function

$$\tilde{\alpha} \colon \tilde{X} \to \mathbb{R}.$$

Choose a lift $\hat{x} \in \tilde{X}$ of any critical point x of f. Each such lift is a critical point of $\tilde{\alpha}$. Then the collection $\{\hat{x};\ \mathrm{ind}(x) = k\}$ is a Λ_α-basis of C_k. A choice of Euler structure uniquely specifies such a lift. Here, we prefer to think of an Euler structure as a 1-chain γ on X such that

$$\partial\gamma = \sum_{\nabla\alpha(x)=0} (-1)^{\mathrm{ind}(x)} x.$$

• The boundary map $\partial\colon C_k \to C_{k-1}$ is defined by

$$\partial\hat{x} = \sum_{\mathrm{ind}\,(y)=k-1} \langle x, y\rangle y,$$

where

$$\langle x, y\rangle = \sum_{h\in H_1(X)} \langle x, y\rangle_h \cdot h.$$

We denote by $\tau_g(\alpha)$ the torsion of the Novikov complex $(\underline{C}(\alpha, g), \partial)$, $\tau_g(\alpha) \in Q(\Lambda_\alpha)/\pm$. Denote by i_α the natural morphism

$$Q(H_1(X)) \to Q(\Lambda_\alpha).$$

We have the following result.

Theorem 4.13 (Hutchings–Lee, Pajitnov [49, 87]). *For any admissible pair* (α, g) *and for any Euler structure* σ *we have*

$$i_\alpha(\tau(X, \sigma)) = I_{\alpha,g}(\sigma) := \zeta_\alpha \tau_g(\alpha, \sigma).$$

For a proof of this result we refer to [47, 48, 49, 66, 87]. The strategy is easy to describe. One constructs a homotopy equivalence between $(\underline{C}(\alpha, g), \partial)$ and the cellular chain complex of $\tilde{X}$ with coefficients in Λ_α. This is not a simple homotopy equivalence, and its torsion is precisely the zeta function of the flow determined by α. The theorem is then a consequence of the multiplicativity properties of the torsion.

We will however sketch an argument of M. Hutchings [47] showing that the above identity follows immediately if we assume $I_{\alpha,g}$ is a topological invariant. More precisely, we have an apparently weaker result.

Theorem 4.14 ([47, 87]). *The invariant* $I_{\alpha,g}$ *depends only on the class* $\omega_\alpha \in H^1(X, \mathbb{Z})$.

We will show, following [47], that

$$\text{Theorem 4.14} \implies \text{Theorem 4.13}.$$

The proof will be carried out in two steps.

Step 1. *Theorem* 4.13 *is true when* $\omega_\alpha = 0$, *i.e.* α *lifts to a map* $f\colon X \to \mathbb{R}$. In this case $\zeta_\alpha = 1$ and the result is classical; see [72, Sec.9] or [86].

Step 2. *Reduce the general case to Step* 1. This can be achieved immediately using the following technical result, [47, 87].

Lemma 4.15 ([56]). *Fix an Euler structure σ on X and a smooth map $\alpha\colon X \to S^1$. Then there exist a metric g on X and a smooth map $f\colon X \to \mathbb{R}$ such that*

- *The pairs (α, g) and $(\beta = \alpha + \exp(\mathbf{i}f), g)$ is admissible.*
- *The vector field $\nabla^g \beta$ has no nontrivial periodic orbits.*
- *There is a canonical isomorphisms of chain complexes*

$$(\underline{C}(f, g), \partial) \otimes \Lambda_\alpha \to (\underline{C}(\beta, g), \partial).$$

Proof. The result is obviously true when α is homotopically trivial. Thus we only need to consider the case when $\omega_\alpha \neq 0 \in H^1(X, \mathbb{Z})$. By eventually perturbing α within its homotopy class we can find a metric g so that (α, g) is admissible. Pick a smooth function $\xi : X \to \mathbb{R}$. We look for f of the form $n\xi$, where n is a *very large* positive integer. This assumption will guarantee the existence of a bijection between the zeroes of ∇f and

$$\nabla \beta_n = n\nabla f + \nabla\alpha,$$

and the corresponding stable and unstable manifolds of these vector fields. Let us now show that for large n the vector field $\nabla\beta_n$ will have no nontrivial periodic orbits.

We argue by contradiction. Suppose that for every $n \gg 0$ there is such an orbit γ_n. We denote by $2\pi s_n \in 2\pi\mathbb{Z}$ its principal period. Then $[\gamma_n]$ has infinite order in $H_1(X, \mathbb{Z})$ since

$$\int_{\gamma_n} d\beta_n = \int_0^{2\pi s_n} |d\beta_n|^2 dt > 0, \quad (d\beta_n := \beta_n^*(d\theta) = 2\pi\omega_{\beta_n}).$$

In fact, since $\omega_{\beta_n} = \omega_\alpha \in H^1(X, \mathbb{Z})$ there exists $c > 0$ such that

$$\langle d\beta_n, [\gamma_n]\rangle \geq c, \quad \forall n.$$

Isolate the critical set $\operatorname{Crit}(f)$ of f in a tiny neighborhood U_r consisting of geodesic balls of radii $r > 0$ centered at the critical points of f. We denote by $L = L_r$ the minimum distance between two distinct components of U_r. Denote by $\lambda_{n,r}$ the part of the path γ_n outside U_r and by $\mu_{n,r}$ the part inside U_r. For each $0 < r \ll 1$ we can find $N = N(r) > 0$ such that

$$\operatorname{Crit}(\beta_n) \subset U_r, \quad \forall n \geq N(r).$$

$\lambda_{n,r}$ and $\mu_{n,r}$ consist of the same number of components

$$\lambda_{n,r,k}, \quad \mu_{n,r,k}, \quad k = 1, \dots, \nu(n).$$

We label the components so that $\mu_{n,r,k}$ follows $\lambda_{n,r,k}$. Using the Morse Lemma for $\frac{1}{n}\beta_n|_{U_r}$ we deduce that the length of each component of $\mu_{n,r}$ is $O(r)$. We need to distinguish several cases.

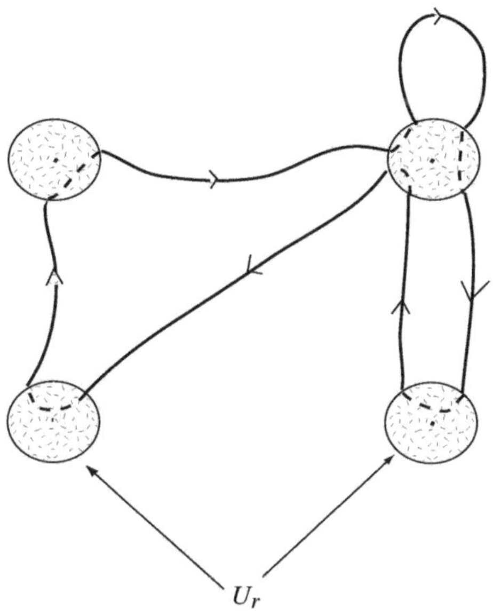

Figure 4.1. A periodic orbit.

Case 1. *The path γ_n intersects several of the components of U_r; (Figure 4.1).* We reparametrize

$$\gamma_n(t) := \gamma_n(t/n)$$

so that

$$\dot{\gamma}_n = V_n := \nabla\xi + \frac{1}{n}\nabla\alpha_n.$$

Assume first that $\lambda_{n,r}$ has no "loops", i.e. components which start and end at the same component of U_r. We deduce that there exists $C > 0$ such that

$$\inf |d\lambda_{n,r}(t)| \geq \inf_{X\setminus U_r} |\nabla\xi + \frac{1}{n}\nabla\alpha| \geq Cr - \frac{1}{Cn}.$$

Hence

$$\begin{aligned} 0 = \int_{\gamma_n} d\xi &= \int_{\lambda_{n,r}} |d\lambda_{n,r}|^2 - \frac{1}{n} g(d\lambda_{n,r}, d\alpha)ds + \int_{\mu_{n,r}} g(d\xi, d\mu_{n,r})ds \\ &= \sum_{k=1}^{\nu(n)} \left\{ \int_{\lambda_{n,r,k}} \left(|d\lambda_{n,r,k}|^2 - \frac{1}{n} g(d\lambda_{n,r,k}, d\alpha) \right) + \int_{\mu_{n,r,k}} g(d\xi, d\mu_{n,r})ds \right\} \qquad (*) \\ &\geq \nu(n) C \Big(L_r \Big(r - \frac{1}{Cn} \Big) - r^2 \Big) = C\nu(n) \Big(-r^2 + rL_r - \frac{L_r}{nC} \Big) \end{aligned}$$

If we now choose

$$n > \max \Big(N(r), \frac{C}{rL_r(L_r - r)} \Big), \qquad 0 < r < L_r$$

we obtain a contradiction in (*).

If $\lambda_{n,r}$ contains "loops", we can shortcut them away, by connecting the initial point and the final point of such a loop by a geodesic segment inside U_r. We obtain a new closed path $\hat{\gamma}_n$ which is tangent to V_n outside U_r. This leads as above to a contradiction.

Case 2. *γ_n intersects a single component of U_r so that $\lambda_{n,r}$ consists only of loops.* Suppose the component of U_r in question is a ball of radius r centered at the critical point x_0 of f. We assume r is considerably smaller than the injectivity radius R_0 of X at x_0,

$$0 < r \ll R_0.$$

Let us observe that γ_n cannot be included in any contractible open set of X because it carries a nontrivial homology class. Thus, one of the components, say $\lambda_{n,r,1}$, must go out of the geodesic ball of radius $R_0/2$ centered at x_0, and in particular

$$\text{length}(\lambda_{n,r,1}) > \frac{R_0}{2}.$$

Now form a closed path $\hat{\gamma}_n$ by joining the endpoints of $\lambda_{n,r,1}$ by a path of length $O(r)$ inside U_r. The equality

$$\int_{\hat{\gamma}_n} d\xi = 0$$

leads as in the previous case to a contradiction. This concludes the proof of the lemma. □

Remark 4.16. The Meng–Taubes–Turaev theorem in the previous section shows that the Morse invariant I_α of a closed 3-manifolds coincides with the Seiberg–Witten invariant. At this moment there is no proof which directly identifies these two invariants. However, the work of C.H. Taubes, [104, 105, 106, 107], on the invariants of (degenerate) symplectic manifolds suggests one explanation. We consider for simplicity the case when X is a 3-manifold which fibers over a circle

$$\begin{array}{ccc} F & \overset{j}{\hookrightarrow} & X \\ & & \downarrow \pi \\ & & S^1 \end{array}$$

where F is a compact, oriented Riemann surface of genus g. We can find a metric g_0 on X such that $\omega_0 := \pi^*(d\theta)$ is harmonic. Then the 4-manifold

$$M := S^1 \times X$$

is symplectic. Indeed, if $d\varphi$ denotes the angular form on the component S^1 of M and $g := d\varphi^2 + g_0$ then

$$\omega := d\varphi \wedge \omega_0 + *_{g_0}\omega_0$$

is a symplectic form.

The Seiberg–Witten invariant of M coincides with the Seiberg–Witten invariant of X. On the other hand, the closed trajectories $\gamma\colon S^1 \to X$ of the gradient flow of π, which contribute to the Morse invariant I_π lead to symplectically embedded tori

$$1 \times \gamma\colon S^1 \times S^1 \to S^1 \times X.$$

According the work of Taubes, the Seiberg–Witten invariants of M count precisely such tori. The equality between the Seiberg–Witten invariant and the Morse invariant of X implies that the correspondence

$$\text{closed orbits of the gradient flow} \longrightarrow \text{symplect tori} \longrightarrow \text{monopoles}$$

is in some sense a bijection. The work of Taubes explains why the second arrow above is a bijection. The work of D. Salamon [95] offers strong evidence that the first arrow is a bijection as well.

§4.3 A spectral interpretation: the Ray–Singer analytic torsion

Like the Euler characteristic, the Reidemeister torsion too has a Hodge theoretic interpretation. Suppose X is a closed, connected, compact, oriented, smooth manifold of *odd* dimension $n = 2m + 1$.

A morphism $\rho\colon \pi_1(X) \to S^1$ defines a pair (L_ρ, A_ρ) consisting of a hermitian line bundle $L_\rho \to X$ and a *flat* hermitian connection A_ρ on it.

We denote by $\Omega^k(L)$ the space of smooth L-valued degree k-differential forms on X, i.e. sections of the bundle $L \otimes \Lambda^k T^*X$. Since A is flat we obtain a *co*-chain complex

$$0 \to \Omega^0(L) \xrightarrow{d_A} \Omega^1(L) \xrightarrow{d_A} \cdots \xrightarrow{d_A} \Omega^n(L) \to 0.$$

For consistency reasons, we will think of it as a *chain* complex

$$(\underline{C}(\rho), \partial), \quad C_k := \Omega^{n-k}(L).$$

This is an *infinite dimensional* complex.

Suppose $(\underline{C}(\rho), \partial)$ is acyclic. We would like to define a notion of torsion for this complex. Clearly the definitions we have used so far are useless. However, the formula (A.1) in §A.1 will provide a way out of this trouble.

A Riemann metric g on X induces a Hermitian metric on $\Omega^*(L)$ and Laplacians

$$\Delta_j(\rho) = d_A^* d_A + d_A d_A^*\colon \Omega^j(L) \to \Omega^j(L).$$

The equality (A.1) shows that, provided we can make some rigorous sense of $\det \Delta_i(\rho)$, then we can define the torsion of this complex by the equality

$$\tau(\rho)^2 = \frac{\prod_{j \text{ odd}} \det(\Delta_j(\rho))^j}{\prod_{k \text{ even}} \det(\Delta_k(\rho))^k}.$$

Since the operator Δ_j is elliptic, selfadjoint and positive, its possibly nonexistent determinant ought to be positive. We can pass to logarithms and obtain

$$\log|\tau(\rho)| = \frac{1}{2}\sum_{k=0}^{n}(-1)^{k+1}k \log\det \Delta_k(\rho).$$

If Δ were a positive symmetric matrix, then we would have

$$\begin{aligned}\log\det\Delta &= \sum_{\lambda\in\sigma(\Delta)} \log\lambda \ (\sigma(\Delta) = \text{the spectrum of } \Delta, \text{ multiplicities included})\\ &= -\frac{d}{ds}|_{s=0}\Big(\sum_{\lambda\in\sigma(\Delta)}\lambda^{-s}\Big) = -\frac{d}{ds}|_{s=0}\operatorname{Tr}(\Delta^{-s}).\end{aligned}$$

Using the classical formula

$$\Gamma(s)\lambda^{-s} = \int_0^\infty t^{s-1}e^{-\lambda t}dt$$

we can further write

$$\log\det\Delta = -\frac{d}{ds}|_{s=0}\Big(\frac{1}{\Gamma(s)}\int_0^\infty t^{s-1}\operatorname{Tr} e^{-t\Delta}dt\Big).$$

Fortunately, the last expression makes sense in infinite dimensions as well. We have the following result, going back to H. Weyl.

Theorem 4.17 ([39], Chap. 1). *Suppose Δ is a second order, selfadjoint, positive, elliptic operator on a closed, compact Riemannian manifold (X, g) of dimension n. Then, the operator $e^{-t\Delta}$ is of trace class, the integral*

$$\int_0^\infty t^{s-1}\operatorname{Tr} e^{-t\Delta}dt$$

converges for all $s\in\mathbb{C}$ $|s|\gg 0$, and the function

$$s\mapsto \zeta_\Delta(s) = \frac{1}{\Gamma(s)}\int_0^\infty t^{s-1}\operatorname{Tr} e^{-t\Delta}dt$$

admits an extension to $\mathbb{C}$ as a meromorphic function with only simple poles located at $s=\frac{n+2-j}{2}$, $j=0,1,2,\dots$. In particular, if n is odd, then $\zeta_\Delta(s)$ is analytic at $s=0$.

Thus we can define the *Ray–Singer analytic torsion* of the acyclic representation $\rho\colon \pi_1(X)\to S^1$

$$RS(\rho) = RS_X(\rho, g) := \frac{1}{2}\sum_{k=0}^{n}(-1)^k k\zeta'_{\Delta_k(\rho)}(0). \tag{4.5}$$

The above heuristic argument suggests that $\exp(RS(\rho))$ is a natural candidate for the torsion of $\underline{C}(\rho)$.

Example 4.18 (The analytic torsion of the circle). Suppose $X = S^1$. All the lines bundles on S^1 are trivial. The flat connections on the trivial bundle $\underline{\mathbb{C}} \to X$ are parameterized by $a \in \mathbb{R} \mod \mathbb{Z}$. Given $a \in [0, 1)$ we can form the connection

$$\nabla^a = d - \mathbf{i}ad\theta : \Omega^0(\underline{\mathbb{C}}) \to \Omega^1(\underline{\mathbb{C}})$$

with holonomy

$$\rho_a(\boldsymbol{t}) = \exp(2\pi \mathbf{i} a),$$

where $\boldsymbol{t}$ denotes the canonical generator of $\pi_1(S^1)$.

We have two Laplacians

$$\Delta_0(a) = \Delta_1(a) = \Delta_a : C^\infty(\underline{\mathbb{C}}) \to C^\infty(\underline{\mathbb{C}}),$$

$$\Delta_a = -\frac{d^2}{d\theta^2} + 2\mathbf{i}\frac{d}{d\theta} + a^2 = \Big(-\mathbf{i}\frac{d}{d\theta} - a\Big)^2$$

with identical spectra

$$\{\lambda_n = (n-a)^2;\ n = 0, 1, \dots\}.$$

We see that the representation ρ_a is acyclic if and only if $a = 0$. The eigenvalue λ_n has an one-dimensional eigenspace generated by $\mathbf{e}_n := \exp(n\theta\mathbf{i})$. The zeta function of Δ_a is

$$\begin{aligned}\zeta_{\Delta_a}(s) &= \sum_{n\in\mathbb{Z}} \frac{1}{(n-a)^{2s}} = \sum_{n\geq 0} \frac{1}{(n+1-a)^{2s}} + \sum_{n\geq 0} \frac{1}{(n+a)^{2s}} \\ &= \zeta(2s; 1-a) + \zeta(2s; a),\end{aligned}$$

where $\zeta(s; a)$ denotes the Riemann–Hurwitz function, [122, Chap. XIII]. Thus

$$\begin{aligned}RS(\rho_a) &= -\frac{1}{2}\frac{d}{ds}|_{s=0}\big(\zeta(2s; a) + \zeta(2s; 1-a)\big) \\ &= -(\zeta'(0; a) + \zeta'(0; 1-a).\end{aligned}$$

To proceed further we need the Lerch identity [122, §13.21],

$$\zeta'(0; a) = \log\Gamma(a) - \frac{1}{2}\log(2\pi).$$

Thus

$$RS(\rho_a) = -\log\Gamma(a)\Gamma(1-a) + \log(2\pi) = -\log\Big(\frac{\pi}{\sin(\pi a)}\Big) + \log(2\pi)$$

$$= -\log(2\sin(\pi a)) = \log\left(\frac{1}{|1-\rho_a(\boldsymbol{t})|^2}\right).$$

We now interpret the holonomy function $\boldsymbol{t} \mapsto \rho_a(\boldsymbol{t})$ as a character of $H_1(S^1)$ so that the last identity can be rephrased as

$$\exp(RS(\rho_a)) = \big|\hat{\tau}_{S^1}(\rho_a)\big|^2. \qquad \square$$

The identity proved in the above example is no accident. In fact, we have the following remarkable results.

Theorem 4.19 (Ray–Singer, [89]). *Suppose (X, g) is a compact, oriented Riemann manifold and ρ is a nontrivial character of $H_1(X)$. The quantity $RS_X(\rho, g)$ is independent of the metric so that it is a topological invariant of the pair (X, ρ).*

Theorem 4.20 (Cheeger–Müller, [4, 9, 14, 74]).

$$\exp(RS_X(\rho)) = \left|\hat{\mathfrak{T}}_X(\rho)\right|^2.$$

(Observe that the $\pm t^n$ multiplicative ambiguity of $\mathfrak{T}_X$ does not affect the value of $|\hat{\mathfrak{T}}_X(\rho)|$.)

We refer to [22, 23] for a more conceptual interpretation of these results in terms of metrics on determinant lines. Also, we want to mention that the proof in [4] directly identifies the Ray–Singer analytic torsion to the Morse theoretic description of the torsion.

Appendix A
Algebra

§A.1 Formal Hodge theory

Suppose $\mathbb{K}$ is a field of characteristic $\neq 2$.

Definition A.1. A *formal metric* on a $\mathbb{K}$-vector space V is a bilinear, symmetric, map

$$g: V \times V \to \mathbb{K}.$$

such that the induced map

$$D_g: V \to V^*, \quad v \mapsto g(v, \bullet).$$

is an isomorphism. D_g is called the *metric duality*. A *metric $\mathbb{K}$-vector space* is a $\mathbb{K}$-space equipped with a formal metric. □

If T is a linear operator between two metric $\mathbb{K}$-spaces

$$T: (V_0, g_0) \to (V_1, g_1)$$

then its *formal metric adjoint* is the operator

$$T^{\sharp}: (V_1, g_1) \to (V_0, g_0)$$

defined by the commutative diagram

$$\begin{array}{ccc} V_1 & \xrightarrow{T^{\sharp}} & V_0 \\ {\scriptstyle D_{g_1}}\downarrow & & \downarrow{\scriptstyle D_{g_0}} \\ V_1^* & \xrightarrow{T^*} & V_0^* \end{array}$$

where T^* denotes the adjoint.

If (V, g) is a metric $\mathbb{K}$-space and $U \subset V$ is a subspace, then the orthogonal complement $U^{\perp}$ of U in V is defined in the usual fashion.

Lemma A.2. *Suppose $(V, g) = \langle \bullet, \bullet \rangle$ is a metric $\mathbb{K}$-space, and $U \hookrightarrow V$ is a metric subspace, i.e. the restriction of g to U is a metric. Then*

$$U \cap U^{\perp} = 0, \quad V = U + U^{\perp}$$

Proof. Denote by i the inclusion $U \hookrightarrow V$. The equality

$$U \cap U^{\perp} = 0$$

follows from the fact that U is a metric subspace. Thus

$$U + U^{\perp} \cong U \oplus U^{\perp}.$$

If $v \in V$ then

$$i^{\sharp}(v) \in U, \quad v - i^{\sharp}(v) \in U^{\perp}.$$

Indeed, the statement $i^{\sharp}(v) \in U$ is tautological. The second follows from

$$\langle v - i^{\sharp}(v), u\rangle = \langle v, u\rangle - \langle v, i(u)\rangle = 0, \forall u \in U.$$

Thus

$$v = i^{\sharp}(v) + (v - i^{\sharp}(v)) \in U + U^{\perp}. \qquad \square$$

Corollary A.3. *If U is a metric subspace of a metric space V and i denotes the inclusion $U \to V$ then $i^{\sharp}$ is the orthogonal projection onto U. Moreover, there exists a natural isomorphism*

$$U^{\perp} \cong V/U.$$

Proposition A.4 (Formal Hodge theorem). *Consider a length n chain complex of finite dimensional $\mathbb{K}$-vector spaces*

$$(\underline{C}, \partial):\quad 0 \to C_n \xrightarrow{\partial} C_{n-1} \xrightarrow{\partial} \cdots \xrightarrow{\partial} C_1 \xrightarrow{\partial} C_0 \to 0.$$

equipped with formal metrics $\langle \bullet, \bullet \rangle$ such that both $\mathrm{Range}(\partial)$ *and* $\ker(\partial)$ *are metric subspaces. Then there exist natural isomorphisms*

$$H_{\mathrm{odd}}(\underline{C}, \partial) \to \ker\big(\partial + \partial^{\sharp} : \underline{C}_{\mathrm{odd}} \to \underline{C}_{\mathrm{even}}\big),$$

and

$$H_{\mathrm{even}}(\underline{C}, \partial) \to \ker\big(\partial + \partial^{\sharp} : \underline{C}_{\mathrm{even}} \to \underline{C}_{\mathrm{odd}}\big).$$

Proof. Using the above corollary we deduce

$$H_*(\underline{C}, \partial) \cong \mathrm{Range}(\partial)^{\perp} \cap \ker \partial.$$

Suppose $u \in \ker(\partial + \partial^{\sharp})$. Thus

$$0 = \partial^{\sharp}\partial u + (\partial^{\sharp} u)^2 = \partial^{\sharp}\partial u$$

so that

$$\langle \partial u, \partial v \rangle = 0, \quad \forall v.$$

Thus

$$\partial u \in \mathrm{Range}(\partial)^{\perp}.$$

Since $\mathrm{Range}(\partial)^{\perp} \cap \mathrm{Range}(\partial) = 0$ we deduce $\partial u = 0$. Thus

$$u \in \ker(\partial + \partial^{\sharp}) \iff \partial u = \partial^{\sharp} u = 0.$$

The condition $\partial u = 0$ implies $u \in \ker \partial$ while the condition $\partial^{\sharp} u = 0$ implies that

$$\forall v \;\; 0 = \langle \partial^{\sharp} u, v \rangle = \langle u, \partial v \rangle \iff u \in \mathrm{Range}(\partial)^{\perp}$$

so that

$$\ker(\partial + \partial^{\sharp}) \subset \mathrm{Range}(\partial)^{\perp} \cap \ker \partial.$$

The opposite inclusion is immediate. □

Corollary A.5. *Suppose $(\underline{C}, \partial)$ is an acyclic complex of $\mathbb{K}$-spaces. Then there exists an algebraic contraction.*

Proof. Fix a formal metric on $\underline{C}$ such that both $\mathrm{Range}(\partial)$ and $\ker \partial$ are metric subspaces. Since the complex is acyclic, we deduce from the formal Hodge theorem that

$$\partial + \partial^{\sharp} : \underline{C} \to \underline{C}$$

is an *odd* isomorphism. In particular, the *even map*

$$\Delta := (\partial + \partial^{\sharp})^2 = \partial\partial^{\sharp} + \partial^{\sharp}\partial$$

is a selfadjoint isomorphism which commutes with ∂. The map

$$\eta := \Delta^{-1}\partial^{\sharp}$$

is a contraction satisfying $\eta^2 = 0$. □

The last results admits the following generalization.

Corollary A.6. *Assume $(\underline{C}, \partial)$ is a finite length chain complex of finite dimensional $\mathbb{K}$-spaces. Then there exists a subcomplex $\underline{X} \subset \underline{C}$ which is **perfect** i.e.*

$$\underline{X} \subset \ker \partial,$$

*and also an **algebraic deformation retract**. The last condition means that there exist maps*

$$p : C_k \to X_k, \quad \eta : C_j \to C_{j+1}$$

such that

$$p \circ i = \mathbf{1}_{\underline{X}}, \quad 1 - i \circ p = \partial\eta + \eta\partial, \quad \eta^2 = 0$$

where i denotes the inclusion $\underline{X} \hookrightarrow \underline{C}$. In particular, i induces an isomorphism

$$i_*\colon \underline{X} \to H_*(\underline{C}, \partial).$$

Proof. Pick a formal metric on $\underline{C}$ such that both $\ker\partial$ and $\mathrm{Range}\,\partial$ are metric subspaces. Define

$$\underline{X} := \ker(\partial + \partial^\sharp) = (\mathrm{Range}\,\partial)^\perp \cap \ker\partial.$$

$\underline{X}$ is a metric subspace and we can define

$$p := i^\sharp.$$

Corollary A.3 now implies $p \circ i = \mathbf{1}_{\underline{X}}$.

Set

$$\Delta := (\partial\partial^\sharp + \partial^\sharp\partial) \quad \text{and} \quad J := ii^\sharp + \Delta.$$

The finite dimensionality of the complex implies that J is selfadjoint, invertible, commutes with ∂ and for every $x \in X$ we have

$$Jx = x.$$

The last equality implies

$$Jii^\sharp = J^{-1}ii^\sharp = ii^\sharp.$$

Now define

$$\eta := J^{-1}\partial^\sharp.$$

Then

$$\partial\eta + \eta\partial = J^{-1}\Delta = J^{-1}(J - ii^\sharp) = 1 - J^{-1}ii^\sharp = 1 - ii^\sharp. \qquad \square$$

Definition A.7. A *generalized contraction* of a chain complex of $\mathbb{K}$-vector spaces $(\underline{C}, \partial)$ is a degree one map

$$\eta\colon C_k \to C_{k+1}$$

such that $\eta^2 = 0$, the chain morphism

$$P := (\partial\eta + \eta\partial)$$

is a projector ($P^2 = P$) and $\mathrm{Range}(P)$ is a perfect subcomplex. $\square$

According to Corollary A.6, every finite dimensional complex of vector spaces admits a generalized contraction. Observe that any contraction of a necessarily acyclic complex is a generalized contraction.

Suppose now that $(\underline{C}, \partial)$ is an acyclic complex of finite dimensional $\mathbb{K}$-vector spaces and $\underline{c}$ is a basis of $\underline{C}$. It determines a canonical metric $g = g(\underline{c})$ on $\underline{C}$ by requiring that $\underline{c}$ is an *orthonormal* basis. Suppose moreover that $\ker \partial$ is a metric subspace of $(\underline{C}, g(\underline{c}))$. Then

$$\mathcal{T}(\underline{C}, \underline{c}) = \det\left(\partial + \eta \colon \underline{C}_{\text{even}} \to \underline{C}_{\text{odd}}\right), \quad \eta := \Delta^{-1}\partial^{\sharp}.$$

We deduce

$$\begin{aligned}\mathcal{T}(\underline{C})^2 &= \det\big((\partial^{\sharp} + \eta^{\sharp})(\partial + \eta) \colon \underline{C}_{\text{even}} \to \underline{C}_{\text{even}} \big)\\ &= \det\left(\partial^{\sharp}\partial + \Delta^{-2}\partial\partial^{\sharp} \colon \underline{C}_{\text{even}} \to \underline{C}_{\text{even}}\right)\end{aligned}$$

To compute this determinant we decompose

$$C_j := K_j \oplus B_j,$$

where

$$K_j = \ker(\partial \colon C_j \to C_{j-1}), \quad B_j = K_j^{\perp} = \text{Range}(\partial^{\sharp}; C_{j-1} \to C_j).$$

Observe that the Laplacian is compatible with these splittings, i.e.

$$\Delta(K_j) \subset K_j, \quad \Delta(B_j) \subset B_j,$$

and

$$\Delta|_{K_j} = \partial\partial^{\sharp}, \quad \Delta|_{B_j} = \partial^{\sharp}\partial.$$

Set

$$k_j := \det(\Delta|_{K_j}), \quad b_j := \det(\Delta|_{B_j}), \quad \delta_j := \det(\Delta|_{C_j}) = k_j b_j.$$

Using the decompositions

$$\underline{C}_{\text{even}} = B_0 \oplus K_2 \oplus B_2 \oplus \cdots$$

we deduce that $\partial^*\partial + \Delta^{-2}\partial\partial^*$ has the diagonal block decomposition

$$\partial^*\partial + \Delta^{-2}\partial\partial^* = \begin{bmatrix} \Delta & 0 & 0 & \dots \\ 0 & \Delta^{-1} & 0 & \dots \\ 0 & 0 & \Delta & \dots \\ \vdots & \vdots & \vdots & \ddots \end{bmatrix}.$$

Thus

$$\mathcal{T}(\underline{C})^2 = \frac{b_0 b_2 \dots b_{2j} \dots}{k_2 k_4 \dots k_{2j} \dots}.$$

Now observe that since ∂ induces bijections $B_j \to K_{j+1}$ and ∂^* induces bijections $K_{j+1} \to B_j$ we have

$$\begin{aligned} b_j &= \det(\partial^{\sharp}\partial \colon B_j \to B_j) = \det(\partial^{\sharp} \colon K_{j+1} \to B_j) \det(\partial \colon B_j \to K_{k+1}k_{j+1})\\ &= \det(\partial\partial^{\sharp} \colon K_{j+1} \to K_{j+1}) = k_{j+1},\end{aligned}$$

so that

$$\mathcal{T}(\underline{C})^2 = \frac{\prod_{i \text{ even}} b_i}{\prod_{j \text{ odd}} b_j},$$

and

$$b_j b_{j+1} = \delta_{j+1} \iff b_{j+1} = \frac{\delta_{j+1}}{b_j} \iff b_{j+1} = \frac{\prod_{i\geq 0} \delta_{j+1-2i}}{\prod_{i\geq 0} \delta_{j-2i}},$$

where $\delta_k = 1$ if $k < 0$. If $n > 0$ denotes the length of the chain complex ($C_k = 0$, $\forall k \geq n$ and $C_{n-1} \neq 0$) then we deduce

$$\mathcal{T}(\underline{C}, \underline{\boldsymbol{c}})^2 = \frac{\prod_{k \text{ even}} \delta_k^{n-k}}{\prod_{j \text{ odd}} \delta_j^{n-j}}. \tag{A.1}$$

The terms on the right hand side depend on the metric $g(\underline{\boldsymbol{c}})$. Let us point out that this formula holds for *any metric*, not just $g(\underline{\boldsymbol{c}})$.

§A.2 Determinants and zeta functions

Suppose $\mathbb{K}$ is a field of characteristic $\neq 2$, U is a finite dimensional $\mathbb{K}$-vector space, and $A\colon U \to U$ is an endomorphism. The *characteristic polynomial* of A is defined by

$$p_A(t) := \det(t - A).$$

Using the identity

$$\det(1 - A) = \sum_{j\geq 0} (-1)^j \operatorname{Tr}(\Lambda^j A), \tag{A.2}$$

where $\Lambda^j A$ denotes the linear map $\Lambda^j U \to \Lambda^j U$ induced by A, we deduce

$$p_A(t) = t^{\dim U} \det(\mathbf{1} - t^{-1} A) = t^{\dim U} \sum_{j\geq 0} \Big(\frac{-1}{t}\Big)^j \operatorname{tr}\big(\Lambda^j A\big).$$

The characteristic polynomial is intimately related to the zeta function of A, defined by

$$\zeta_A(s) := \exp\Big(\sum_{\nu\geq 1} \operatorname{Tr}(A^\nu)\frac{s^\nu}{\nu}\Big).$$

We define $\zeta(A) := \zeta_A(1)$ so that, formally

$$\zeta_A(s) = \zeta(sA).$$

To explain the relationship between ζ_A and p_A we use the elementary identity

$$-\log(1 - (as)) = \sum_{\nu\geq 1} \frac{(as)^\nu}{\nu} \iff \frac{1}{1 - as} = \exp\Big(\sum_{\nu\geq 1} \frac{(as)^\nu}{\nu}\Big),$$

which implies that

$$\zeta_A(s) = \frac{1}{\det(\mathbf{1} - sA)} \iff \zeta_A(t^{-1}) = \frac{t^r}{p_A(t)}.$$

If we denote by $S^k V$ the k-th symmetric power of a vector space V, then for every endomorphism $B\colon V \to V$ we have

$$\frac{1}{\det(\mathbf{1} - B)} = \sum_{k\geq 0} \operatorname{tr}\big(S^k B\big). \tag{A.3}$$

Hence

$$\zeta_A(t) = \sum_{k\geq 0} t^k \operatorname{tr}\big(S^k A)\big).$$

If U is a superspace, $U = U_{\text{even}} \oplus U_{\text{odd}}$, and A is an *even* endomorphism

$$A = A_{\text{even}} \oplus A_{\text{odd}}$$

then we can define the *s-characteristic polynomial*

$$\hat{p}_A(t) = \det{}_s(t - A) := \frac{\det(t - A_{\text{even}})}{\det(t - A_{\text{odd}})}.$$

We deduce that for any even endomorphism A of U we have

$$\hat{p}_A(t) = t^{\chi(U)}\Big(\sum_{j\geq 0}\Big(\frac{-1}{t}\Big)^j \operatorname{tr}(\Lambda^j A_{\text{even}})\Big)\Big(\sum_{k\geq 0}\Big(\frac{1}{t}\Big)^k \operatorname{tr}(S^k A_{\text{odd}})\Big), \tag{A.4}$$

where the Euler characteristic of U is the s-trace of the identity map $\mathbf{1}_U$

$$\chi(U) := \operatorname{tr}_s \mathbf{1}_U := \dim_{\mathbb{K}} U_{\text{even}} - \dim_{\mathbb{K}} U_{\text{odd}}.$$

There is a super-version of the zeta function

$$\hat{\zeta}_A(t) := \exp\Big(\sum_{\nu\geq 1} \operatorname{tr}_s(A^\nu)\frac{t^\nu}{\nu}\Big),$$

where tr_s denotes the s-trace. We then have the identities

$$\hat{\zeta}_A(t^{-1}) \cdot \hat{p}_A(t) = t^{\chi(U)} \iff \hat{\zeta}_A(t) = \frac{1}{\det_s(1 - tA)} = \frac{\det(1 - tA_{\text{odd}})}{\det(1 - tA_{\text{even}})}. \tag{*}$$

Using (A.2) and (A.3) we deduce

$$\begin{aligned}\hat{\zeta}_A(t) &= \Big(\sum_{k\geq 0} \operatorname{Tr}(S^k A_{\text{even}})\Big) \cdot \Big(\sum_{j\geq 0} (-1)^j t^j \operatorname{Tr}(\Lambda^j A_{\text{odd}})\Big)\\ &= \sum_{d\geq 0} t^d \Big(\sum_{j=0}^{d} (-1)^j \operatorname{Tr}(\Lambda^j A_{\text{odd}}) \cdot \operatorname{Tr}(S^{d-j} A_{\text{even}})\Big).\end{aligned} \tag{**}$$

If X is a compact, closed, oriented smooth manifold and $f\colon X \to X$ is a smooth map, then its *Lefschetz number* is

$$L(f) := \operatorname{tr}_s\bigl(f_*\colon H_*(X;\mathbb{R}) \to H_*(X;\mathbb{R})\bigr).$$

The celebrated *Lefschetz fixed point theorem* states that (see [57, VIII.5]) if all the fixed points of f are nondegenerate, i.e.

$$\delta_x(f) := \det\bigl((\mathbf{1} - df)\colon T_xX \to T_xX\bigr) \neq 0, \quad \forall f(x) = x$$

then

$$L(f) = \sum_{x\in \operatorname{Fix}(f)} L(f,x)$$

where the *local Lefschetz number* $L(f,x)$ is defined by

$$L(f,x) = \operatorname{sign} \delta_x(f).$$

The *zeta function* of f is defined by

$$\zeta_f(t) := \exp\Bigl(\sum_{k\geq 1} \frac{t^k}{k} L(f^k)\Bigr).$$

Using the identities (*) and (**) we deduce

$$\zeta_f(t) = \frac{\det\bigl(\mathbf{1} - tH_{\text{odd}}(f)\bigr)}{\det\bigl(\mathbf{1} - tH_{\text{even}}(f)\bigr)}$$

$$= \sum_{d\geq 0} t^d \Bigl(\sum_{j=0}^{d} (-1)^j \operatorname{Tr}\bigl(\Lambda^j H_{\text{odd}}(f)\bigr) \cdot \operatorname{Tr}\bigl(S^{d-j} H_{\text{even}}\bigr)\Bigr).$$

The last sum can be expressed in terms of the symmetric powers of X. S^nX is defined as the quotient of the Cartesian product of X^n modulo the natural action of the symmetric group S^n. Then (see [63])

$$H_{\text{odd}}(S^nX) = \bigoplus_{j\ odd} \Lambda^j H_{\text{odd}}(X;\mathbb{R}) \otimes S^{n-j} H_{\text{even}}(X;\mathbb{R}),$$

$$H_{\text{even}}(S^nX) = \bigoplus_{j\ even} \Lambda^j H_{\text{odd}}(X;\mathbb{R}) \otimes S^{n-j} H_{\text{even}}(X;\mathbb{R}).$$

Hence

$$L(S^d f) = \sum_{j=0}^{d} (-1)^j \operatorname{Tr}\bigl(\Lambda^j H_{\text{odd}}(f)\bigr) \cdot \operatorname{Tr}\bigl(S^{d-j} H_{\text{even}}(f)\bigr),$$

so that

$$\zeta_f(t) = \sum t^d L(S^d f).$$

If for example X is a Riemann surface of genus g, then $H_0(f) = H_2(f) = 1$. If we denote by $A \in \mathrm{Sp}_{2g}(\mathbb{Z})$ a symplectic matrix representing $H_1(f)$ then we deduce

$$\zeta_f(t) = \frac{\det(\mathbf{1} - tA)}{(1-t)^2}.$$

§A.3 Extensions of Abelian groups

In this section we survey some basic facts concerning extensions of Abelian groups needed in surgery theory. We will denote by $\mathbb{T}$ the rational circle $\mathbb{Q}/\mathbb{Z}$ and, for any Abelian group A we will denote by $\hat{A}$ its dual

$$\hat{A} := \mathrm{Hom}(A, \mathbb{T}).$$

Suppose A and C are Abelian groups. An *extension* of A by C is a short exact sequence of Abelian groups

$$0 \to A \to B \to C \to 0.$$

Two extensions $0 \to A \to B_i \to C \to 0$, $i = 0, 1$ are isomorphic if there exists an isomorphism $f\colon B_0 \to B_1$ such that the diagram below is commutative.

$$\begin{array}{ccccccccc} 0 & \longrightarrow & A & \longrightarrow & B_0 & \longrightarrow & C & \longrightarrow & 0 \\ & & \Big\downarrow{\scriptstyle \mathbf{1}_A} & & \Big\downarrow{\scriptstyle f} & & \Big\downarrow{\scriptstyle \mathbf{1}_C} & & \\ 0 & \longrightarrow & A & \longrightarrow & B_1 & \longrightarrow & C & \longrightarrow & 0. \end{array}$$

We denote by $\mathrm{Ext}(C, A)$ the set of isomorphism classes of extensions.

Proposition A.8 ([64]). (a) *The correspondence* $\mathrm{Ext}(C, -)$ *is a covariant functor from the category of Abelian groups to the category of sets.*

(b) *The correspondence* $\mathrm{Ext}(-, A)$ *is a contravariant functor from Abelian groups to sets.*

Consider the group morphisms

$$\nabla^A\colon A \oplus A \to A, \quad a_1 \oplus a_2 \mapsto a_1 + a_2$$

and

$$\Delta_C\colon C \to C \oplus C, \quad c \mapsto c \oplus c.$$

Given two extensions $E_i \in \text{Ext}(C, A)$

$$E_i\colon\ 0 \to A \to B_i \to C \to 0, \quad i = 0, 1$$

we can construct in the obvious fashion $E_0 \oplus E_1 \in \text{Ext}(C \oplus C, A \oplus A)$. The *Baer sum* of the two extensions is

$$E_0 + E_1 := \nabla_*^A \Delta_C^* \in \text{Ext}(C, A)$$

where $\nabla_*^A \Delta_C^*$ is the composition

$$\text{Ext}(C \oplus C, A \oplus A) \xrightarrow{\Delta_C^*} \text{Ext}(C, A \oplus A) \xrightarrow{\nabla_*^A} \text{Ext}(C, A).$$

Proposition A.9 ([64, 93]). *The Baer sum introduces a structure of Abelian group on* $\text{Ext}(C, A)$. *The trivial (zero) extension is the split extension*

$$0 \to A \to A \oplus C \to C \to 0.$$

Moreover, for any short exact sequence of Abelian groups

$$0 \to X \to Y \to Z \to 0$$

and any Abelian group A we have the exact sequences

$$\begin{aligned} 0 \longrightarrow \text{Hom}(Z, A) &\longrightarrow \text{Hom}(Y, A) \longrightarrow \text{Hom}(X, A) \\ &\xrightarrow{\delta} \text{Ext}(Z, A) \longrightarrow \text{Ext}(Y, A) \longrightarrow \text{Ext}(X, A) \longrightarrow 0 \end{aligned}$$

and

$$\begin{aligned} 0 \longrightarrow \text{Hom}(A, X) &\longrightarrow \text{Hom}(A, Y) \longrightarrow \text{Hom}(A, Z) \\ &\xrightarrow{\delta} \text{Ext}(A, X) \longrightarrow \text{Ext}(A, Y) \longrightarrow \text{Ext}(A, Z) \longrightarrow 0. \end{aligned}$$

Example A.10. Since $\mathbb{Z}$ is a projective $\mathbb{Z}$-module we deduce that every extension

$$0 \to C \to B \to \mathbb{Z} \to 0$$

is split so that $\text{Ext}(\mathbb{Z}, C) = 0$ for any Abelian group C. □

The next result is particularly relevant in topology.

Proposition A.11. *Suppose C is a finite Abelian group. Given*

$$\lambda\colon C \to \mathbb{T} := \mathbb{Q}/\mathbb{Z}$$

we define

$$C_\lambda = \{q \in \mathbb{Q} \oplus C;\ \lambda(c) = q \bmod \mathbb{Z}\}.$$

(a) *The sequence* E_λ *defined by*

$$0 \to \mathbb{Z} \xrightarrow{\iota} C_\lambda \xrightarrow{\pi} C \to 0$$

is exact, where ι *is induced by the inclusions* $\mathbb{Z} \hookrightarrow \mathbb{Q} \oplus 0 \hookrightarrow C_\lambda$ *while* π *is induced by the natural projection* $\mathbb{Q} \oplus C \to C$.

(b) *The correspondence* $\psi : \hat{C} := \mathrm{Hom}(C, \mathbb{T}) \to \mathrm{Ext}(C, \mathbb{Z})$ *defined by*

$$\hat{C} \ni \lambda \mapsto E_\lambda \in \mathrm{Ext}(C, \mathbb{Z})$$

is a group isomorphism.

Proof. Part (a) is obvious. We will show that the correspondence in part (b) is a bijection. For simplicity we will confine ourselves to the special case when C is a cyclic group of order N.

Fix a generator t of C. Given an extension $E \in \mathrm{Ext}(C, \mathbb{Z})$

$$0 \to \mathbb{Z} \to B \to C \to 0$$

we can find $b_0 \in B$ which maps to t. Then Nb_0 maps to $0 \in C$ so that there exists $m \in \mathbb{Z} \hookrightarrow B$ such that

$$Nb_0 = m.$$

(Abusing notations we can write $b_0 = m/N$.) Now define

$$\lambda_E : C \to \mathbb{T}, \quad kt \mapsto km/N \bmod \mathbb{Z}.$$

The morphism λ_E does not depend on the choice of the generator t and the element $b_0 \in B$ mapping to t. We have thus constructed a map

$$\phi : \mathrm{Ext}(C, \mathbb{Z}) \to \mathrm{Hom}(C, \mathbb{T}), \quad E \mapsto \lambda_E.$$

We let the reader check that

$$E_{\lambda_E} = E \iff \psi \circ \phi = \mathbf{1},$$

and

$$\lambda_{E_\lambda} = \lambda \iff \phi \circ \psi = \mathbf{1}. \qquad \square$$

The result in the above proposition allows us to determine $\mathrm{Ext}(C, F)$ where F is a free Abelian group of rank m. More precisely, we have a natural isomorphism

$$\mathrm{Ext}(C, F) \to \mathrm{Hom}(C, F_{\mathbb{Q}}/F)$$

where $F_{\mathbb{Q}} := F \otimes \mathbb{Q}$. A morphism $\vec{\lambda} : C \to F_{\mathbb{Q}}/F \cong F \otimes \mathbb{T}$ defines the extension

$$C_{\vec{\lambda}} := \{(\vec{q}, c) \in (F_{\mathbb{Q}}) \oplus C;\ \vec{\lambda}(c) = \vec{q} \bmod F\}. \tag{A.5}$$

Example A.12. Consider the inclusions $i_k : \mathbb{Z}^2 \to \mathbb{Z}^2, k = 1, 2$, given by the matrices

$$A_1 := \begin{bmatrix} 4 & 0 \\ 0 & 1 \end{bmatrix}, \quad A_2 := \begin{bmatrix} 2 & 0 \\ 0 & 2 \end{bmatrix}.$$

These lead to the Abelian extensions

$$0 \to \mathbb{Z}^2 \xrightarrow{A_1} \mathbb{Z}^2 \to \mathbb{Z}_4 \to 0,$$

and

$$0 \to \mathbb{Z}^2 \xrightarrow{A_2} \mathbb{Z}^2 \to \mathbb{Z}_2 \oplus \mathbb{Z}_2 \to 0.$$

The first extension is given by the morphism

$$\lambda_1 : \mathbb{Z}_4 \to (\mathbb{T})^2, \quad 1 \mapsto (1/4, 0)$$

while the second is given by the morphism

$$\lambda_2 : \mathbb{Z}^2 \times \mathbb{Z}^2 \to (\mathbb{T})^2, \quad (1, 0) \mapsto (1/2, 0), \quad (0, 1) \mapsto (0, 1/2).$$

More generally, if G is a finite abelian group which admits a presentation

$$0 \to F \xrightarrow{A} F \to G \to 0$$

then this extension is classified by a morphism

$$G \to F_{\mathbb{Q}}/F.$$

This can be easily describes as follows. Consider the inverse

$$A^{-1} : F_{\mathbb{Q}} \to F_{\mathbb{Q}}.$$

Observe that $A^{-1}(A(F)) = F \subset F_{\mathbb{Q}}$ so that we have an induced map

$$G \cong F/(AF) \to F_{\mathbb{Q}}/F.$$

It is precisely the classifying map of the presentation. The map $G \to F_{\mathbb{Q}}/F$ is clearly an inclusion. Conversely, every inclusion

$$G \hookrightarrow F_{\mathbb{Q}}/F$$

produces a (class of) presentation(s) of G.

It is perhaps instructive to see how this works in a concrete situation and to point out a confusing fact. Suppose $F \cong \mathbb{Z}^n$ and A is described by an $n \times n$ matrix

$$Ae_j = \sum_i a_{ij} e_i, \quad i, j = 1, \ldots, n,$$

where e_i denotes the canonical integral basis of $\mathbb{Z}^n$. Suppose

$$A^{-1}e_j = \sum_i a'_{ij}e_i, \quad a'_{ij} \in \mathbb{Q},\ i, j = 1, \dots, n.$$

Set $G := \mathbb{Z}^n/A\mathbb{Z}^n$, and denote by $[e_i]$ the image of e_i in G. Then the morphism $G \to \mathbb{T}^n$ corresponding to the extension

$$0 \to \mathbb{Z}^n \xrightarrow{A} \mathbb{Z}^n \to G \to 0$$

is described by

$$[e_j] \mapsto \begin{bmatrix} a'_{1j} \\ \vdots \\ a'_{nj} \end{bmatrix} \bmod \mathbb{Z}.$$

The right-hand-side of the above equality is the j-th *column* of A^{-1}.

In topological applications the map $G \to \mathbb{T}^n$ is described by n characters of G. Since the dual $\hat{G}$ embeds in the dual of $\mathbb{Z}^n$ it is customary to describe the map $\hat{G} \to \mathbb{T}^n$ as a vector consisting of n characters of $\mathbb{Z}^n$. In this case they are

$$\vec{\lambda} = (\lambda_1, \dots, \lambda_n), \quad \lambda_i([e_j]) = a'_{ij} \bmod \mathbb{Z},\ i, j = 1, \dots, n.$$

In other words, λ_i is the character of $\mathbb{Z}^n$ described by the i-th *row* of A^{-1}. □

Appendix B
Topology

§B.1 How to compute the Alexander polynomial of a knot

In this section we will survey a few methods of computing the Alexander polynomial of a knot. As testing ground for each of these methods we will use the trefoil knot (see Figure B.2). For details and proofs we refer to [27, 92].

The Alexander polynomial of a knot $K \subset S^3$ is determined by the universal Abelian cover of its complement $X_K := S^3 \setminus K$. By Alexander duality we have $H := H_1(X_K, \mathbb{Z}) \cong \mathbb{Z}$ and we denote by $\hat{X}_K \to X_K$ the universal Abelian cover. Set $R := \mathbb{Z}[t, t^{-1}]$. The homology group $H_1(\hat{X}_K, \mathbb{Z})$ has a natural R-module structure induced by the deck transformations of the covering $\hat{X}_K \to X_K$.

The ring R is a unique factorization domain and the Alexander polynomial is by definition

$$\Delta_K := \operatorname{ord}_R\big(H_1(\hat{X}_K)\big).$$

It is an element of R uniquely determined up to a multiplicative term $\pm t^k$, $k \in \mathbb{Z}$. To concretely compute Δ_K we need to produce a presentation

$$\mathfrak{R} \to \mathfrak{G} \to H^1(\hat{X}_K) \to 0$$

where $\mathfrak{R}$ and $\mathfrak{G}$ are finitely generated free R-modules. We will present two algorithms for producing such presentations.

1. Seifert matrices. Consider an oriented Seifert surface $\Sigma \subset S^3$ such that $\partial\Sigma = K$ and a small tubular neighborhood N_K of $\Sigma \hookrightarrow S^3$. Then $N_K \setminus \Sigma$ consists of two components which we label $N_K^{\pm}$. (The orientation of Σ allows us to canonically label these components with $+$ or $-$.)

Σ is a genus g surface with boundary S^1 and thus $H_1(\Sigma, \mathbb{Z}) \cong \mathbb{Z}^{2g}$. Fix a set of generators $x_1, \dots, x_{2g}$ of this homology group represented by embedded circles. By pushing the circle x_i into N_K^+ we obtain a circle x_i^+. The *Seifert matrix* is the $2g \times 2g$ matrix V_K with entries defined by

$$v_{ij} := \boldsymbol{Lk}(x_i, x_j^+).$$

$\boldsymbol{Lk}(K_1, K_2)$ denotes the linking number of two disjoint knots and can be computed using the sign rules in Figure B.1. We refer to [59, 92] for more details.)

The matrix $A_K(t) := V_K^T - tV_K$ is called an *Alexander matrix* of the knot K and provides a presentation of $H_1(\hat{X}_K, \mathbb{Z})$

$$\mathbb{Z}[t, t^{-1}]^{2g} \xrightarrow{A(t)} \mathbb{Z}[t, t^{-1}]^{2g} \to H^1(\hat{X}_K, \mathbb{Z}) \to 0.$$

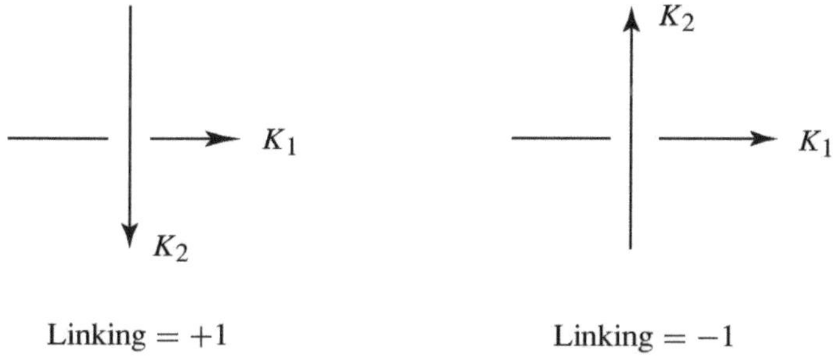

Figure B.1. Sign rules for linking numbers.

In particular, we have

$$\Delta_K(t) \sim \det A_K(t).$$

Example B.1. Consider as promised the case of the trefoil knot. In the second diagram in Figure B.2 we can clearly visualize a Seifert surface for the knot. It is obtained by joining two disjoint disks by three twisted bands. This Seifert surface has genus one and in Figure B.2 we describe a set of generators of H_1.

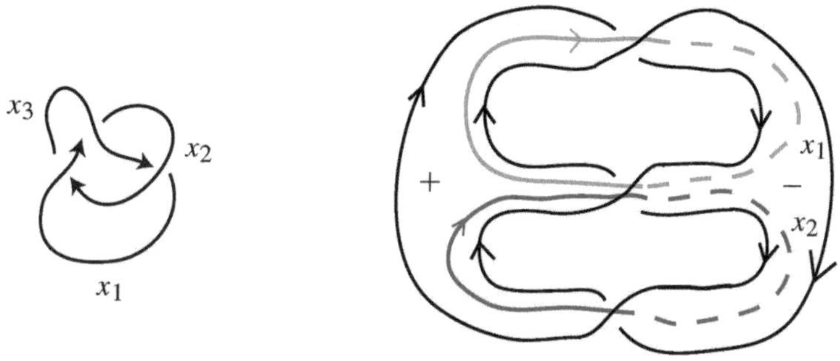

Figure B.2. Two equivalent diagrams for the trefoil knot.

The "+" and "−" signs in this picture fix the orientation of this *two sided surface*. It is clear that x_2 and x_1^+ do not link so that $v_{21} = 0$. As for the other entries of the Seifert matrix, they are described in Figure B.3. Thus

$$V = \begin{bmatrix} -1 & 1 \\ 0 & -1 \end{bmatrix}$$

so that

$$A(t) = \begin{bmatrix} t-1 & -t \\ 1 & t-1 \end{bmatrix}, \quad \Delta(t) \sim t^2 - t + 1 = \frac{(t^6-1)(t-1)}{(t^2-1)(t^3-1)}.$$

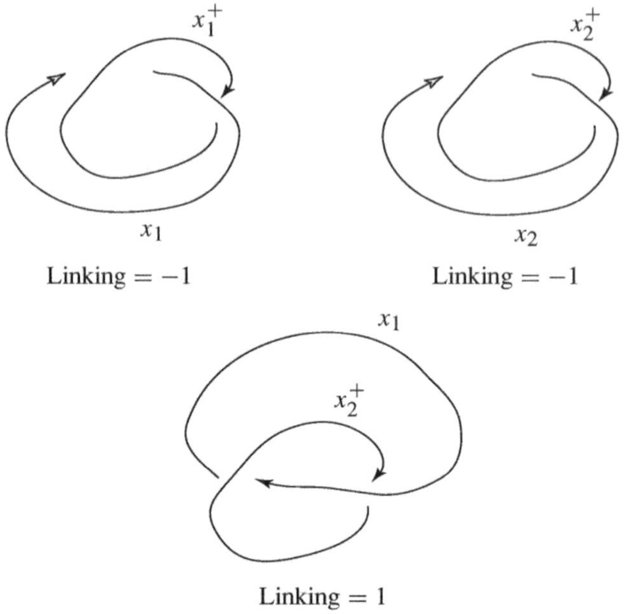

Figure B.3. Computing a Seifert matrix for the trefoil knot.

The formula we have just obtained is a special case of the more general description of the Alexander polynomial of a (p, q)-torus knot (see [96]). The trefoil is a (2, 3)-torus knot. □

2. Fox free differential calculus. R. Fox has developed [24, 25, 26, 109] an algebraic machinery of determining a presentation of the $\mathbb{Z}[t, t^{-1}]$-module $H_1(\hat{X}_K)$ once a presentation of the knot group $\pi_1(X_K)$ is given.

Suppose we are given a finite presentation $\langle x_1, \dots, x_n; R_1, \dots, R_m\rangle$ of a group G. Then there exist natural $\mathbb{Z}$-linear maps

$$D_1, \dots, D_n \colon \mathbb{Z}[G] \to \mathbb{Z}[G]$$

uniquely determined by the requirements

$$D_\bullet 1 = 0, \quad D_i x_j = \delta_{ij}$$

$$D_\bullet(u \cdot v) = D_\bullet u + u D_\bullet v, \quad \forall u, v \in G.$$

We can form the *Jacobian* of the presentation which is the $n \times m$ matrix $\mathfrak{J} = \mathfrak{J}(x_j; r_i)$ with entries in $\mathbb{Z}[G]$ described by

$$\mathfrak{J}_{ij} = D_i R_j.$$

If H denotes the abelianization of G, $H := G/D(G)$, $D(G) := [G, G]$ then we get a natural morphism $\psi: \mathbb{Z}[G] \to \mathbb{Z}[H]$ and correspondingly, an $n \times m$ matrix $J := \psi(\mathfrak{J})$ with entries in $\mathbb{Z}[H]$. The transpose of J defines a presentation the $\mathbb{Z}[H]$-module $D(G)/D^2(G) \oplus \mathbb{Z}[H]$,

$$\mathbb{Z}[H]^m \to \mathbb{Z}[H]^n \to D(G)/D^2(G) \oplus \mathbb{Z}[H] \to 0.$$

Suppose now that G is the fundamental group of the complement of a knot $K \hookrightarrow S^3$

$$G = G_K := \pi_1(X_K), \quad X_K = S^3 \setminus K.$$

If we are given a presentation of G_K with n generators and m relations then

$$H = H_1(X_K) \cong \mathbb{Z}, \quad D(G)/D^2(G) \cong H_1(\hat{X}_K),$$

where as before, $\hat{X}_K \to X_K$ denotes the universal Abelian cover. In this case we denote by J_K the abelianization of the Jacobian matrix $\mathfrak{J}$. Then the Alexander polynomial is the greatest common divisor of the set of $(n-1) \times (n-1)$ minors of J; see [27] or [59, Chap.11] for more details. Equivalently, and more invariantly, we can define the Alexander polynomial as a generator of the first Fitting ideal $F_1(H_1(\hat{X}_K) \oplus \mathbb{Z}[t, t^{-1}])$ which admits a presentation given by the transpose of J_K. Using Proposition 2.25 we deduce that $F_1(M) = F_0(H_1(\hat{X}_K))$.

One can obtain a presentation of G_K (called the *Wirtinger presentation*) from the diagram of K as follows.

• Orient the knot and mark the undercrossings in the order given by the orientation. We have thus divided K into n oriented arcs $x_1, \dots, x_n$ each connecting two consecutive undercrossings.
• G_K admits a presentation with $x_1, \dots, x_n$ as generators and one relation for each undercrossing, as described in Figure B.4. We can drop any one relation and still obtain a correct presentation. It is clear that all the generators of the Wirtinger presentations are mapped by the abelianization map into the generator t of $H_1(X_K) \cong \mathbb{Z}$.

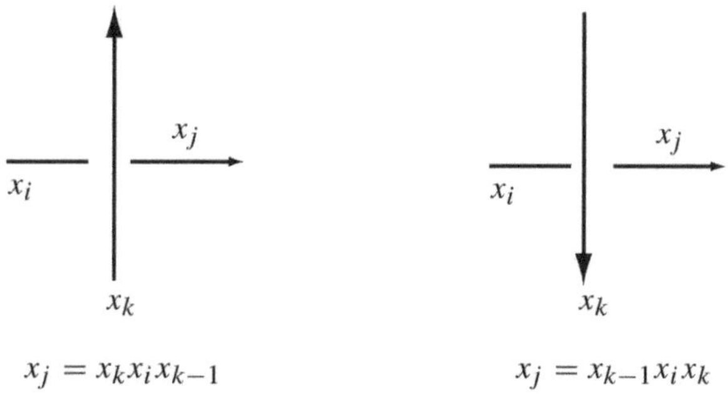

Figure B.4. Wirtinger relations.

Example B.2. We consider again the trefoil knot. From the top of Figure B.2 we read the following presentation

$$\langle x_1, x_2, x_3;\ x_2 = x_3^{-1}x_1x_3,\ x_3 = x_1^{-1}x_2x_1,\ x_1 = x_2^{-1}x_3x_2\rangle.$$

Observe that only two generators are important since $x_3 = x_1^{-1}x_2x_1$. We set $a = x_1$ and $b = x_2$ so that $x_3 = bab^{-1}$. We obtain the equivalent presentation

$$\langle a, b;\ a = baba^{-1}b^{-1},\ b = abab^{-1}a^{-1}\rangle \iff \langle a, b;\ bab = aba\rangle$$

If we set

$$R = baba^{-1}b^{-1}a^{-1}$$

then

$$D_aR = D_a(baba^{-1}b^{-1}) - (baba^{-1}b^{-1})a^{-1} = D_a(bab) - baba^{-1} - baba^{-1}b^{-1}a^{-1}$$

$$= D_a(ba) - baba^{-1} - baba^{-1}b^{-1}a^{-1} = b - baba^{-1} - baba^{-1}b^{-1}a^{-1}$$

$$D_bR = D_b(baba^{-1}b^{-1}) = D_b(baba^{-1}) - baba^{-1}b^{-1} = D_b(bab) - baba^{-1}b^{-1}$$

$$= D_b(ba) + ba - baba^{-1}b^{-1} = 1 + ba - baba^{-1}b^{-1}.$$

By passing to abelianization we get

$$D_aR = t - t^2 - 1, \quad D_bR = 1 + t^2 - t$$

so that

$$J_K = [-(t^2 - t + 1) \quad t^2 - t + 1].$$

This shows $\Delta_K(t) \sim t^2 - t + 1$. □

Remark B.3. The Fox free calculus works in the more general case of links with several components as well, with very few but obvious changes. The Wirtinger presentation is obtained in the similar fashion, and we obtain a presentation of $\pi_1(X_K)$ with the same number ν of generators and relations. A generator g of this presentation will represent in $H_1(X_K)$ the same homology class as the meridian of the component of the link to which g belongs. As in the case of knots, any relation can be dropped from the presentation. We obtain an exact sequence of $\mathbb{Z}[H]$-modules

$$\mathbb{Z}[H]^\nu \xrightarrow{J} \mathbb{Z}[H]^\nu \to H_1(\hat{X}_K) \oplus \mathbb{Z}[H]$$

from which we deduce that

$$\operatorname{ord} H_1(\hat{X}_K) = F_1(H_1(\hat{X}_K) \oplus \mathbb{Z}[H]).$$ □

3. Conway's skein relations. The Alexander polynomial of a knot is uniquely determined by the manner in which it changes when the crossing patterns of a diagram are changed; see Figure B.5. A formula describing such a change is called a *skein* relation. The Alexander polynomial of a knot K satisfies the symmetry property

$$\Delta_K(t) \sim \Delta_K(t^{-1}).$$

Any polynomial $Q \in \mathbb{Z}[t, t^{-1}]$ is equivalent to a polynomial $P \in \mathbb{Z}[t^{1/2}, t^{-1/2}]$ satisfying $P(t) = P(t^{-1})$. The polynomial P is unique *up to a sign*. E.g., the polynomial $(t-1) \sim t^{1/2} - t^{-1/2} \sim t^{-1/2} - t^{+1/2}$.

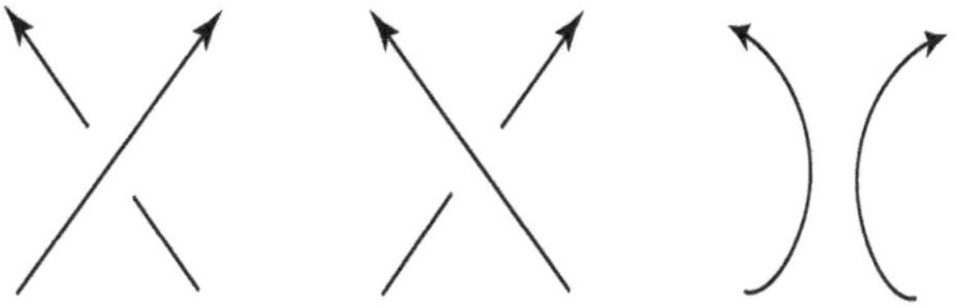

L_+ L_- L_0

Figure B.5. Changing the crossing patterns.

Remarkably, any *oriented* link L determines a polynomial $\Delta_L(t) \in \mathbb{Z}[t^{1/2}, t^{-1/2}]$ uniquely determined by the following conditions (see [59, Chap8]).

$$\Delta_L(t) = \Delta_L(t^{-1}).$$

$$\Delta_{\text{unknot}}(t) = 1.$$

and the Conway's skein relation (see Figure B.5)

$$\Delta_{L_+}(t) - \Delta_{L_-}(t) = (t^{-1/2} - t^{1/2})\Delta_{L_0}(t).$$

When L is a knot, Δ_L is equivalent to the Alexander polynomial. We see that this algorithm picks up a canonical representative for the Alexander polynomial of an *oriented* knot. This is called the *Conway normalized Alexander polynomial.*

Example B.4. We want to compute the (Conway normalized) Alexander polynomial of the trefoil knot, oriented as in Figure B.2. Look at the top crossing in the first diagram of Figure B.2. In the conventions of Figure B.5 it is a L_+. The corresponding L_- represents the unknot while L_0 is the *Hopf link H* depicted in Figure B.6.

$$\Delta_K(t) - 1 = (t^{-1/2} - t^{1/2})\Delta_H(t).$$

The crossing in the left hand side of Figure B.6 is a L_+. Changing it to a L_- transforms H into a pair of unlinked unknots, while the move $L_+ \to L_0$ transforms the Hopf link to an unknot. Since $\Delta_{L_-}(t) = 0$ we deduce

$$\Delta_H(t) = (t^{-1/2} - t^{1/2})$$

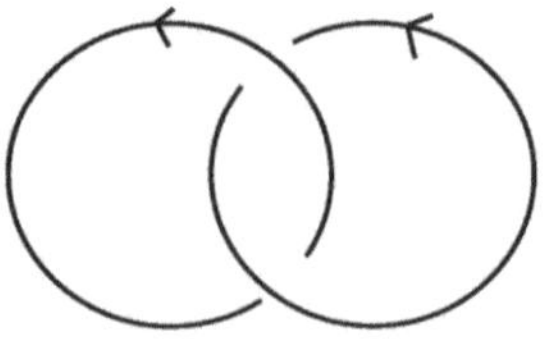

Figure B.6. The Hopf link.

so that

$$\Delta_K(t) = 1 + (t^{-1/2} - t^{-1/2})^2 = t - 1 + t^{-1} \sim t^2 - t + 1. \qquad \square$$

§B.2 Dehn surgery and linking forms

The existing literature on Dehn surgery can be quite confusing, especially as far as the various sign conventions are concerned. For the reader's convenience, we have decided to include several useful facts concerning this important concept, paying special attention to the many orientation conventions.

Suppose M^3 is a rational homology 3-sphere, $\mathcal{K} \hookrightarrow M^3$ is an oriented link in M^3 with components $\mathcal{K}_1, \dots, \mathcal{K}_m$, $U = \bigcup_{j=1}^m U_j$ is a small open tubular neighborhood of $\mathcal{K}$ and

$$M_{\mathcal{K}} := M \setminus U.$$

Set $T := \partial \bar{U}$ and $H_*(X) := H_*(X, \mathbb{Z})$ for any topological space X.

1. The homology of $M_{\mathcal{K}}$ as an extension of $H_1(M)$. Since

$$H_2(M) \cong H^1(M) \cong 0$$

we deduce from the long exact sequence of the pair $(M, M_{\mathcal{K}})$ that

$$0 \to H_2(M, M_{\mathcal{K}}) \to H_1(M_{\mathcal{K}}) \to H_1(M) \to H_1(M, M_{\mathcal{K}}) \to 0.$$

Using excision we can rewrite

$$0 \to H_2(\bar{U}, T) \to H_1(M_{\mathcal{K}}) \to H_1(M) \to H_1(\bar{U}, T) \to 0.$$

The long exact sequence of the pair $(\bar{U}, T)$ now implies

$$H_2(\bar{U}, T) \cong \ker\big(H_1(T) \to H_1(\bar{U})\big) \cong \mathbb{Z}^m$$

and $H_1(\bar{U}, T) \cong 0$. Moreover $\ker\big(H_1(T) \to H_1(\bar{U})\big)$ admits a natural basis $\vec{\mu} = (\mu_1, \dots, \mu_m)$ consisting of meridians. More precisely, μ_j is the generator of

$$\ker\big(H_1(T_j) \to H_1(\bar{U}_j)\big)$$

such that

$$\boldsymbol{Lk}_M(\mu_j, \mathcal{K}_j) = 1$$

where $\boldsymbol{Lk}_M$ is the $\mathbb{Q}$-valued linking number of two disjoint embedded circles. Denote by $\mathbb{Z}\langle\vec{\mu}\rangle$ the free Abelian group generated by the meridians. We thus have a short exact sequence

$$0 \to \mathbb{Z}\langle\vec{\mu}\rangle \cong \mathbb{Z}^m \stackrel{j}{\to} H_1(M_{\mathcal{K}}) \to H_1(M) \to 0,$$

where

$$j \colon H_1(T) \to H_1(M_{\mathcal{K}})$$

is the inclusion induced morphism. As explained in Appendix §A.3, it defines an element

$$\vec{\lambda}_{\mathcal{K}} \in \operatorname{Ext}(H_1(M), \mathbb{Z}^m) \cong \operatorname{Hom}(H_1(M), \mathbb{T}^m).$$

We claim that

$$\vec{\lambda}_{\mathcal{K}}(c) = \sum_{i=1}^{m} \boldsymbol{lk}_M(\mathcal{K}_i, c)\mu_i \in \mathbb{Q}\langle\vec{\mu}\rangle/\mathbb{Z}\langle\vec{\mu}\rangle, \quad \forall c \in H_1(M),$$

where

$$\boldsymbol{lk}_M \colon H_1(M) \times H_1(M) \to \mathbb{T}$$

is the linking form.

To see this, denote by ν the order of $H_1(M)$. If $c \in H_1(M)$ is represented by an embedded circle $C \subset M_{\mathcal{K}}$ then

$$\nu C = \sum_j k_j \mu_j \in H_1(M_{\mathcal{K}})$$

and thus

$$\vec{\lambda}_{\mathcal{K}}(C) = \sum \frac{k_j}{\nu}\mu_j.$$

On the other hand

$$\boldsymbol{lk}_M(C, \mathcal{K}_i) = \frac{1}{\nu}\sum_j k_j \, \boldsymbol{Lk}_M(\mu_j, \mathcal{K}_i) = k_i/\nu \bmod \mathbb{Z}.$$

Using the description (A.5) in §A.3 we deduce that

$$H_1(M_{\mathcal{K}}) \cong \left\{(\vec{\alpha}, c) \in \mathbb{Q}^m \times H_1(M);\ \alpha_j = \boldsymbol{lk}_M(\mathcal{K}_j, c) \bmod \mathbb{Z},\ \forall j \in \overline{1, m}\right\}. \quad (*)$$

2. The morphism $j: H_1(T) = H_1(\partial M_{\mathcal{K}}) \to H_1(M_{\mathcal{K}})$. We use the exact sequence of the pair $(M_{\mathcal{K}}, T)$ and we deduce

$$\begin{aligned} 0 &\to H_3(M_{\mathcal{K}}, T) \to H_2(T) \to H_2(M_{\mathcal{K}}) \to H_2(M_{\mathcal{K}}, T) \\ &\to H_1(T) \xrightarrow{j} H_1(M_{\mathcal{K}}) \to H_1(M_{\mathcal{K}}, T) \\ &\to H_0(T) \to H_0(M_{\mathcal{K}}) \to 0. \end{aligned} \tag{**}$$

To deal with the relative homology we use excision

$$H_*(M_{\mathcal{K}}, T) \cong H_*(M, \bar{U})$$

and then the long exact sequence of the pair $(M, \bar{U})$,

$$\begin{aligned} 0 &\to H_2(M, \bar{U}) \to H_1(U) \to H_1(M) \\ &\to H_1(M, U) \to H_0(\bar{U}) \to H_0(M) \to 0. \end{aligned}$$

We deduce that

$$H_2(M_{\mathcal{K}}, T) \cong \ker\big(H_1(\bar{U}) \to H_1(M)\big) \cong \mathbb{Z}^m$$

and that the boundary map

$$H_2(M_{\mathcal{K}}, T) \to H_1(\bar{U})$$

is $1-1$. The above morphism factors through the inclusion induced map

$$\imath: H_1(T) \to H_1(\bar{U})$$

so that the morphisms

$$H_2(M_{\mathcal{K}}, T) \to H_1(T) \quad \text{and} \quad \ker \boldsymbol{j} \subset H_1(T) \xrightarrow{\imath} H_1(\bar{U})$$

must be $1-1$. Using the sequence (**) we deduce that $H_2(M_{\mathcal{K}}, T) \cong \ker \boldsymbol{j}$ so that

$$\imath(\ker \boldsymbol{j}) \cong \ker\big(H_1(\bar{U}) \to H_1(M)\big). \tag{B.1}$$

If we denote by $\langle \mathcal{K} \rangle \subset H_1(M)$ the subgroup generated by the components of the link $\mathcal{K}$ we deduce that we have the short exact sequences

$$0 \to \imath(\ker \boldsymbol{j}) \to H_1(\bar{U}) \to \langle \mathcal{K} \rangle \to 0$$

and

$$0 \to H_1(M)/\langle \mathcal{K} \rangle \to H_1(M, \bar{U}) \to \mathbb{Z}^{m-1} \cong \ker(H_0(\bar{U}) \to H_0(M)) \to 0.$$

Hence,

$$H_1(M_{\mathcal{K}}, T) \cong H_1(M)/\langle \mathcal{K} \rangle \oplus \mathbb{Z}^{m-1},$$

and

$$\operatorname{coker} \boldsymbol{j} \cong H_1(M)/\langle \mathcal{K} \rangle. \tag{B.2}$$

3. Longitudes. Suppose $\mathcal{K}$ is a link with a single component, i.e. a knot. If M is not an integral homology sphere there is no canonical way of choosing an integral basis of $H_1(T)(\cong \mathbb{Z}^2)$. On the other hand, there is a natural way of choosing a $\mathbb{Q}$-basis of $H_1(T) \otimes \mathbb{Q}$. A *longitude* is a generator λ of

$$\ker \boldsymbol{j} = \ker\big(H_1(T) \to H_1(M_{\mathcal{K}})\big)$$

such that $\mu \cdot \lambda > 0$ where μ denotes a meridian and the intersection product is defined with respect to the orientation of T as boundary of U. Denote by r the order of $\mathcal{K}$ in $H_1(M)$. From the identity (B.1) we deduce that $\lambda = r\mathcal{K}$ in $H_1(\bar{U})$. In particular, this implies $\mu \cdot \lambda = r$. Moreover, since λ bounds in the complement of $\mathcal{K}$ we deduce that $\boldsymbol{Lk}_M(\lambda, \mathcal{K}) = 0$.

We can now conclude that any homology class c in $H_1(T)$ is uniquely determined by a pair $(\alpha, n) \in \mathbb{Q} \times \mathbb{Z}$ satisfying the conditions

$$n := \mu \cdot c \iff \imath(c) = n\mathcal{K} \in H_1(\bar{U}),$$

and

$$\alpha := \frac{1}{r}(c \cdot \lambda)\, \boldsymbol{Lk}_M(c, \mathcal{K}) = n\, \boldsymbol{lk}_M(\mathcal{K}, \mathcal{K}) \bmod \mathbb{Z}.$$

As an element of $H_1(T; \mathbb{Q})$ the cycle c has the decomposition

$$c = \alpha\mu + \frac{n}{r}\lambda.$$

Using the above basis of $H_1(T; \mathbb{Q})$ we can identify the inclusion $H_1(T) \hookrightarrow H_1(T; \mathbb{Q})$ with the inclusion $G \hookrightarrow \mathbb{Q}^2$ where G is the additive subgroup of $\mathbb{Q}^2$ defined by the conditions

$$G = \left\{ \left(\alpha, \frac{n}{r}\right) \in \mathbb{Q} \oplus \frac{1}{r}\mathbb{Z};\ \alpha = n\, \boldsymbol{lk}_M(\mathcal{K}, \mathcal{K}) \bmod \mathbb{Z} \right\}.$$

Observe that the longitude λ need not be a primitive element in $H_1(T)$. Indeed,

$$\frac{1}{m}\lambda \in G \iff n = \frac{r}{m}, \quad n\, \boldsymbol{lk}_M(\mathcal{K}, \mathcal{K}) \in \mathbb{Z}.$$

If we write

$$\boldsymbol{lk}_M(\mathcal{K}, \mathcal{K}) = \frac{\nu}{r}, \quad 0 \le \nu < r,$$

then the above conditions become

$$\frac{r}{m}, \frac{\nu}{m} \in \mathbb{Z},$$

so that we can choose $m = (r, \nu)$ (:=(g.c.d.(r, ν))). Hence

$$\lambda = (\nu, r)\lambda_0, \quad \lambda_0 \in H_1(T).$$

Remark B.5. We would like to discuss a rather subtle point. The pair $(r, \nu \bmod r\mathbb{Z})$ was determined by the order and the self-linking number of $\mathcal{K}$. We would like to show that the same information is algebraically encoded by the pair of cycles $\mu, \lambda \in H_1(T)$.

The Abelian group $F := H_1(T)$ is free of rank two, and in order to perform concrete computations we need to choose $\mathbb{Z}$-bases. This is involves non-canonical choices, and thus we need to be able to separate the invariant quantities from those which are not. Clearly, the coordinates of μ and λ with respect to some $\mathbb{Z}$-basis are not invariant quantities. The determination of numerical invariants of the pair (μ, λ) boils down to a group theoretic problem.

Describe the space $\mathcal{O}$ of orbits of the group $\mathfrak{G} :=$ Aut(F) acting diagonally on the space $\mathcal{P} \subset F \times F$ of pairs of linearly independent vectors.

Observe that $\pi := (\mu, \lambda) \in F \times F$, and the orbit

$$\{(T\mu, T\lambda);\ T \in \mathfrak{G}\}$$

corresponds to the different choices of bases of F. A pair $\pi = (\boldsymbol{e}_1, \boldsymbol{e}_2) \in \mathcal{P}$ defines an injection

$$j_\pi : \mathbb{Z}^2 \to F, \quad (n_1, n_2) \mapsto n_1\boldsymbol{e}_1 + n_2\boldsymbol{e}_2.$$

The extension

$$0 \to \mathbb{Z}^2 \xrightarrow{j_\pi} F \to G_\pi := F/j_\pi(\mathbb{Z}^2) \to 0$$

is a complete invariant of the orbit $(\mathfrak{G} \cdot \pi)$. It is completely characterized by the group G_π, and the characteristic element

$$\chi \in \mathrm{Hom}(G_\pi, \mathbb{T}^2) = \mathrm{Ext}(G_\pi, \mathbb{Z}^2).$$

In our special case, $\pi = (\mu, \lambda)$, G_π is a cyclic group of order r, namely the cyclic group generated by the knot $\mathcal{K} \hookrightarrow M$. The last statement implicitly assumed the existence of a canonical generator. This is indeed the case. Pick as generator the unique vector $\kappa \in F \cap \{x\mu + y\lambda;\ x, y \in [0, 1]\}$ such that $\mu \cdot \kappa = 1$. Geometrically, κ is the vertex of the Newton polygon of the cone $s\mu + t\lambda$, $s, t > 0$, closest to μ.

The characteristic element χ is given by the pair of characters $\chi_1, \chi_2 \in \hat{G}_\pi$.

$$\chi_1(\mathcal{K}) = 1/r \in \mathbb{T}, \quad \chi_2(\mathcal{K}) = \boldsymbol{lk}_M(\mathcal{K}, \mathcal{K}) = \nu/r \in \mathbb{T}.$$

This can be seen easily using the $\mathbb{Z}$-basis (μ, κ) of $H_1(T)$. In particular, $\lambda = -\nu\mu + r\kappa$. □

4. The morphism $j : H_1(T) \to H_1(M_\mathcal{K})$ revisited. Denote the components of $\mathcal{K}$ by $\mathcal{K}_j$, a small, open tubular neighborhood of $\mathcal{K}_j$ by U_j, the meridian of $\mathcal{K}_j$ by μ_j and the longitude of $\mathcal{K}_j$ by λ_j. Finally, denote by r_j the order of $\mathcal{K}_j$ in $H_1(M)$.

Observe that if $i \neq j$ we have

$$\ell_i^j := r_i\, \boldsymbol{Lk}_M(\mathcal{K}_i, \mathcal{K}_j) = \boldsymbol{Lk}_M(\lambda_i, \mathcal{K}_j) \in \mathbb{Z}.$$

Since λ_i bounds in $M_{\mathcal{K}_i}$ we deduce we deduce that we have the following equality in $H_1(M_{\mathcal{K}})$

$$\lambda_i = \sum_{j\neq i} \ell_i^j \mu_j.$$

Using (*) in 1. on page 217 we deduce

$$H_1(M_{\mathcal{K}}) = \big\{ \big(\textstyle\sum_j \alpha_j \mu_j, c\big) \in \mathbb{Q}\langle\vec{\mu}\rangle \times H_1(M);\ \alpha_j = \boldsymbol{lk}_M(c, \mathcal{K}_j) \bmod \mathbb{Z},\ \forall j\big\}.$$

Using the description of $H_1(T)$ in 3., we deduce that the morphism $\boldsymbol{j}$ acts according to the rule

$$\begin{aligned}\boldsymbol{j}\colon \alpha_j\mu_j + \frac{n_i}{r_i}\lambda_i &\mapsto \Big(\alpha_j\mu_j + \frac{n_i}{r_i}\sum_{s\neq i}\ell_i^s \mu_s, n_i\mathcal{K}_i\Big)\\ &= \Big(\alpha_j\mu_j + n_i \sum_{s\neq i} \boldsymbol{Lk}_M(\mathcal{K}_s, \mathcal{K}_i)\mu_s,\ n_i\mathcal{K}_i\Big), \quad i, j = 1, \dots, m.\end{aligned}$$

The natural map $H_1(M_{\mathcal{K}}) \to H_1(M)$ is given by

$$\Big(\sum_j \alpha_j\mu_j, c\Big) \mapsto c.$$

5. Mayer–Vietoris interpretation. The Mayer–Vietoris sequence associated to the decomposition $M = \bar{U} \cup M_{\mathcal{K}}$ leads to the Abelian group extension

$$0 \to \mathbb{Z}^{2m} \cong H_1(T) \xrightarrow{\imath \oplus \boldsymbol{j}} H_1(\bar{U}) \oplus H_1(M_{\mathcal{K}}) \to H_1(M) \to 0. \tag{B.3}$$

This extension is classified by a linear map

$$H_1(M) \to H_1(T) \otimes \mathbb{T}.$$

Arguing as in 1. we deduce that this classifying map is given by

$$c \mapsto \sum_j \boldsymbol{lk}_M(\mathcal{K}_j, c)\mu_j \in H_1(T) \otimes \mathbb{T}.$$

6. Dehn surgery. The manifold M can be described as a quotient space

$$\bar{U} \cup_{f_0} M_{\mathcal{K}}$$

where $f_0\colon \partial\bar{U} \to \partial M_{\mathcal{K}}$ is an orientation reversing diffeomorphism. Given any orientation *preserving* diffeomorphism

$$\vec{\gamma} = (\gamma_1, \dots, \gamma_m), \quad \gamma_j\colon \partial\bar{U}_j \to \partial\bar{U}_j$$

we can form a new manifold

$$M_{\vec{\gamma}} := \bar{U} \bigcup_{f_0 \circ \vec{\gamma}} M_{\mathcal{K}}$$

called the *Dehn surgery* determined by $\vec{\gamma}$. Clearly the diffeomorphism type of $M_{\vec{\gamma}}$ depends only on the isotopy type of $\vec{\gamma}$. We denote by Γ the group of these isotopy classes. Observe that

$$\Gamma \cong \prod_{j=1}^{m} \mathrm{SL}(H_1(T_j)).$$

We denote by Γ_0 the subgroup of Γ consisting of diffeomorphism which extend to $\bar{U}$. It is not difficult to see that $M_{\vec{\gamma}}$ depends only on the orbit $\vec{\gamma} \circ \Gamma_0 \in \Gamma / \Gamma_0$.

The set of orbits Γ / Γ_0 can be identified with the set of m-uples

$$\vec{c} := (c_1, c_2, \dots, c_m) \in \prod_{j=1}^{m} H_1(T_j)$$

such that c_j is a nontrivial primitive element of $H_1(T_j)$. A diffeomorphism $\vec{\gamma}$ belongs to the orbit labelled by $\vec{c}$ if and only if

$$\vec{\gamma}(\vec{\mu}) = \vec{c} \iff \gamma_j(\mu_j) = c_j.$$

For this reason, the Dehn surgery determined by $\vec{\gamma}$ is often denoted by $M_{\vec{c}}$.

Using the bases (μ_j, λ_j) of $H_1(T_j; \mathbb{Q})$ we can write

$$c_j := \alpha_j \mu_j + \frac{n_j}{r_j} \lambda_j, \quad \alpha_j = n_j\, \boldsymbol{lk}_M(\mathcal{K}_j, \mathcal{K}_j) \bmod \mathbb{Z}. \qquad (***)$$

It is often convenient to identify c_j with the pair of numbers (α_j, n_j). These are known as the *surgery coefficients*. When M is an integral homology sphere then the surgery coefficients (α, n) are both integers. In this case the (α, n) surgery is traditionally referred to as the *α/n-surgery*. Furthermore, if $n = 1$ then the surgery is called *integral*. Two natural question arise.

A. *Describe the homology of $M_{\vec{c}}$ in terms of the homology of M, invariants of the link $\mathcal{K} \hookrightarrow M$ and $\vec{c}$.*

B. *Describe the linking form of $H_1(M_{\vec{c}})$ in similar terms.*

7. The homology of $M_{\vec{c}}$. Arguing exactly as in 1., we obtain the extension

$$0 \to \mathbb{Z}\langle \vec{\mu} \rangle \stackrel{j \circ \gamma}{\longrightarrow} H_1(M_{\mathcal{K}}) \to H_1(M_{\vec{c}}) \to 0$$

or, equivalently,

$$0 \to \mathbb{Z}\langle \vec{c} \rangle \stackrel{j}{\longrightarrow} H_1(M_{\mathcal{K}}) \to H_1(M_{\vec{c}}) \to 0.$$

Using the description of $\boldsymbol{j}$ in 4. we deduce that $H_1(M_{\vec{c}})$ is the quotient of $H_1(M_{\mathcal{K}})$ modulo the subgroup generated by

$$\Big(\alpha_i\mu_i + n_i\sum_{j\neq i}\boldsymbol{Lk}_M(\mathcal{K}_j,\mathcal{K}_i)\mu_j,\, n_i\mathcal{K}_i\Big).$$

If we form the $m\times m$ matrix $P(\mathcal{K},\vec{c})$ with rational entries

$$p_{ji} = \begin{cases} n_i\,\boldsymbol{Lk}_M(\mathcal{K}_j,\mathcal{K}_i) & j\neq i\\ \alpha_i & j=i,\end{cases}$$

we deduce that $M_{\vec{c}}$ is a rational homology sphere if and only if

$$\det P(\mathcal{K},\vec{c})\neq 0.$$

We consider two extreme situations.

a. M is an integral homology sphere. Then $P = P(\mathcal{K},\vec{c})$ defines a presentation of $H_1(M_{\vec{c}})$,

$$0\to\mathbb{Z}^m\xrightarrow{P}\mathbb{Z}^m\to H_1(M_{\vec{c}})\to 0.$$

Alternatively, $M_{\vec{c}}$ can be given the Mayer–Vietoris description

$$0\to H_1(T)\xrightarrow{i\oplus \boldsymbol{j}\circ\gamma} H_1(\bar{U})\oplus H_1(M_{\mathcal{K}})\to H_1(M_{\vec{c}})\to 0.$$

b. $\mathcal{K}$ is consists of a single component, $m=1$. Set $\alpha:=\alpha_1$, $n:=n_1$ etc. α and n are constrained by

$$\alpha\equiv n\,\boldsymbol{lk}_M(\mathcal{K},\mathcal{K}) \bmod \mathbb{Z}.$$

Then

$$\begin{aligned} H_1(M_c) &= H_1(M_{\mathcal{K}})/\mathbb{Z}\langle c\rangle\\ &\cong \{(t\mu,\gamma)\in\mathbb{Q}\langle\mu\rangle\times H_1(M);\ t=\boldsymbol{lk}_M(\gamma,\mathcal{K}) \bmod \mathbb{Z}\}/\mathbb{Z}\langle\alpha\mu+n\mathcal{K}\rangle.\end{aligned}$$

8. Linking theory on $H_1(M_{\vec{c}})$. We will again consider two cases.

a. The manifold M is an integral homology sphere. In this case $H_1(M_{\vec{c}})$ admits a presentation of the form

$$0\to\mathbb{Z}\langle\vec{\mu}\rangle\xrightarrow{P(\mathcal{K},\vec{c})}\mathbb{Z}\langle\vec{\mu}\rangle\to H_1(M_{\vec{c}})\to 0.$$

According to the computations in Example A.12, this extension is classified by the map

$$\Omega := P(\mathcal{K},\vec{c})^{-1}\colon \mathbb{Z}\langle\vec{\mu}\rangle/P\mathbb{Z}\langle\vec{\mu}\rangle\to\mathbb{Q}\langle\vec{\mu}\rangle/\mathbb{Z}\langle\vec{\mu}\rangle.$$

On the other hand, according to the computation in **1.**, page 216, this classifying map can be described in terms of the linking theory on $H_1(M_{\vec{c}})$. We denote by $[\mathcal{K}_j]_{\vec{c}}$ the homology class of the core of U_j in $H_1(M_{\vec{c}})$. These classes define via the linking form of $M_{\vec{c}}$ a vector of n characters of $H_1(M_{\vec{c}})$ which classifies the above extension. Moreover, we have

$$\boldsymbol{lk}_{M_{\vec{c}}}([\mathcal{K}_i]_{\vec{c}}, [\mu_j]) = \Omega_{ij} \bmod \mathbb{Z}, \quad \forall i, j, \tag{B.4}$$

where $\Omega := (\Omega_{ij}) = P^{-1}$, that is

$$P^{-1}\mu_j = \sum_i \Omega_{ij}\mu_i.$$

If we denote by $(\bullet, \bullet)$ the natural inner product on $\mathbb{Q}\langle\mu\rangle$ defined by $(\mu_i, \mu_j) = \delta_{ij}$, then we can write

$$\Omega_{ij} = (\mu_i, P^{-1}\mu_j).$$

If we are "lucky enough", so that $[\mathcal{K}_j]_{\vec{c}}$ generate $H_1(M_{\vec{c}})$, then the above trick allows us to determine the linking form of $M_{\vec{c}}$. In fact, this is not a matter of luck.

Proposition B.6. *Suppose that the surgery coefficient of* $\mathcal{K}_i$ *is* $\frac{p_i}{q_i}$ *so that*

$$P(\mathcal{K}, \vec{c}) = (p_{ij})_{1\leq i,j\leq n}$$

where

$$p_{ij} = \begin{cases} q_j\, \boldsymbol{Lk}_M(\mathcal{K}_i, \mathcal{K}_j) & \text{if } i \neq j \\ p_i & \text{if } i = j. \end{cases}$$

Then the classes μ_i *generate* $H_1(M_{\vec{c}})$, *and we have the equalities*

$$q_i[\mathcal{K}_i]_{\vec{c}} = -\mu_i \quad \text{in } H_1(M_{\vec{c}}),\ \forall i.$$

Moreover,

$$\boldsymbol{lk}_{M_{\vec{c}}}(\mu_i, \mu_j) = -q_i(\mu_i, P^{-1}\mu_j) \bmod \mathbb{Z}.$$

Proof. The Dehn surgery is described by a family

$$\vec{\gamma} = (\gamma_1, \dots, \gamma_m) \in \prod_{j=1}^m \mathrm{SL}(2,\mathbb{Z}), \quad \gamma_j := \begin{bmatrix} p_j & \alpha_j \\ q_j & \beta_j \end{bmatrix}, \quad p_j\beta_j - \alpha_j q_j = 1.$$

These matrices describe the attaching rules

$$\boldsymbol{m}_i \mapsto \boldsymbol{j}(c_i) = \boldsymbol{j}(p_i\mu_i + q_i\lambda_i) = \boldsymbol{b}_i := p_i\mu_i + q_i\sum_{j\neq i}\ell_{ji}\mu_j,$$

$$\boldsymbol{l}_i \mapsto \boldsymbol{j}(\alpha_i\mu_i + \beta_i\lambda_i) = \boldsymbol{k}_i := \alpha_i\mu_i + \beta_i\sum_{j\neq i}\ell_{ij}\mu_j,$$

where $\boldsymbol{m}_i$ (resp. $\boldsymbol{l}_i$) denotes the meridian (resp. the longitude) of the i-th attaching solid torus and

$$\ell_{ij} = \boldsymbol{Lk}_M(\mathcal{K}_i, \mathcal{K}_j) \in \mathbb{Z}.$$

The group $H_1(M_{\vec{c}})$ is the quotient of $\mathbb{Z}\langle\vec{\mu}\rangle$ modulo the lattice spanned by the vectors $\boldsymbol{b}_i$. Using the identities

$$p_j\beta_j - \alpha_j q_j = 1$$

we deduce

$$\beta_i \boldsymbol{b}_i - q_i \boldsymbol{k}_i = \mu_i$$

which shows that $-q_i[\mathcal{K}_i]_{\vec{c}} = [\mu_i]$ in $H_1(M_{\vec{c}})$. The second statement in the proposition follows from the identity

$$-q_i\Omega_{ij} = -q_i\, \boldsymbol{lk}_{M_{\vec{c}}}([\mathcal{K}_i], [\mu_j]) = \boldsymbol{lk}_{M_{\vec{c}}}([\mu_i], [\mu_j]) \bmod \mathbb{Z}. \qquad \square$$

Remark B.7. Denote by P_0 the symmetric matrix defined by

$$p_{ij} = \begin{cases} \ell_{ij} & \text{if } i \neq j \\ \frac{p_i}{q_i} & \text{if } i = j. \end{cases}$$

Then $P(\mathcal{K}, \vec{c}) = P_0 \cdot \mathrm{diag}(q_1, \dots, q_n)$, and the (i, j)-th entry of the symmetric matrix P^{-1} is $q_i\Omega_{ij}$, that is

$$(\mu_i, P_0^{-1}\mu_j) = -q_i(\mu_i, P^{-1}\mu_j).$$

The linking form of $M_{\vec{c}}$ is completely characterized by P_0^{-1} via the equalities,

$$\boldsymbol{lk}_M(\mu_i, \mu_j) = -(\mu_i, P_0^{-1}\mu_j). \qquad \square$$

Example B.8. The arguments in the proof of the above proposition can be easily grasped in the following simple situation. Suppose $\mathcal{K}$ is a knot in $M := S^3$. The Dehn surgeries on $\mathcal{K}$ are determined by a pair of relatively prime positive integers (p, q). If $\mathcal{K}$ is the unknot in S^3 then the (p, q) surgery on $\mathcal{K}$ produces the lens space $L(p, -q)$. Denote by λ and μ the longitude and respectively the meridian of $\mathcal{K}$ such that $\mu \cdot \lambda = 1$. The complement $M_{\mathcal{K}}$ is a solid torus and μ is its core. μ is a generator of $H_1(M_{\mathcal{K}})$. The characteristic curve of the (p, q) surgery is $p\mu + q\lambda$.

The first homology group of the manifold $M_{p/q} := M_c$ admits the presentation

$$0 \to \mathbb{Z}\mu \xrightarrow{p\cdot} \mathbb{Z}\vec{\mu} \cong H_1(M_{\mathcal{K}}) \to H_1(M_{p/q}) \to 0.$$

We deduce that $H_1(M_{p,q})$ and the generator μ of $H_1(M_{\mathcal{K}})$ induces a generator of $H_1(M_{p/q})$. Moreover, according to Example A.12 the above extension is classified by the map

$$\mathbb{Z}_p \to \mathbb{T}, \quad 1 \bmod p\mathbb{Z} \mapsto 1/p \bmod \mathbb{Z}.$$

If we now denote by $[\bullet]_{p/q}$ the class in $H_1(M_{p/q})$ determined by a closed curve $\bullet$ we deduce

$$\boldsymbol{lk}_{M_{p/q}}([\mathcal{K}]_{p/q}, [\mu]_{p/q}) = 1/p.$$

Since $[\mu]_{p/q}$ is a generator of $H_1(M_{p/q})$ we can write

$$[\mathcal{K}]_{p/q} = x[\mu]_{p/q}, \quad x \in \mathbb{Z} \bmod p\mathbb{Z}.$$

Now observe that the gluing map γ of this Dehn surgery is described by a matrix

$$\gamma := \begin{bmatrix} p & \alpha \\ q & \beta \end{bmatrix} \in \mathrm{SL}(2, \mathbb{Z})$$

with inverse

$$\gamma^{-1} = \begin{bmatrix} \beta & -\alpha \\ -q & p \end{bmatrix}.$$

The meaning of the entries of this matrix are given by the attaching rules

$$\mu \mapsto c = p\mu + q\lambda \in H_1(\partial M_{\mathcal{K}}), \quad \mathcal{K} = \lambda \mapsto \alpha\mu + \beta\lambda \in H_1(\partial M_{\mathcal{K}}).$$

Since $\lambda = 0 \in H_1(M_{\mathcal{K}})$ deduce from these rules that

$$[\mathcal{K}]_{p/q} = \alpha[\mu]_{p/q}.$$

Since $\det\gamma = p\beta - \alpha q = 1$ we deduce $\alpha = -q^{-1} \bmod p\mathbb{Z}$. Hence

$$\boldsymbol{lk}_{M_{p/q}}(-q^{-1}[\mu]_{p,q}, [\mu]_{p,q}) = 1/p,$$

so that

$$\boldsymbol{lk}_{M_{p/q}}([\mu]_{p/q}, [\mu]_{p/q}) = -q/p \bmod \mathbb{Z}.$$ □

b. $\mathcal{K}$ has only one component but M may have nontrivial homology. In this case it is wiser to treat M and M_c "democratically", as equal partners. These two manifolds have something in common, the 3-manifold with boundary $M_{\mathcal{K}}$. This notation does not respect our "democracy rule" and we will set $N := M_{\mathcal{K}}$. The Dehn surgery process can now be described as attaching the solid torus

$$\bar{U} := S^1 \times D^2$$

to the boundary ∂N so that the curve $\mathbf{pt} \times \partial D^2$ is attached along a primitive curve $\mu \in H_1(\partial N)$. We denote the resulting closed manifold by N_μ. (In the old notations $M = N_\mu$.)

As an aside, note that the homologies of N and ∂N contain no information about the homology sphere M. This is determined by an additional internal data, namely the homology class in $H_1(T)$ carried by the meridian of the knot. We can loosely rephrase this by saying that the homology groups of N have no idea about the manifold M. There is however one element in $H_1(T)$ which carries this information.

As the complement of a knot in a rational homology sphere, the manifold N has several special topological features we would like to single out and rephrase in a language which makes no mention of M.

Observe first that $b_1(N) = 1$. Moreover, the boundary map

$$H_2(N, \partial N) \to H_1(\partial N)$$

is injective (see **2.**, page 218). Its image is a rank one subgroup of $H_1(\partial N)$ generated by the longitude λ. This subgroup is isomorphic to the kernel of the morphism

$$\boldsymbol{j} : H_1(\partial N) \to H_1(N).$$

λ need not be a primitive element of $H_1(\partial N)$ and we can write

$$\lambda = m_0\lambda_0, \quad m_0 > 0, \ \lambda_0 \in H_1(\partial N) \text{ is primitive.}$$

At **2.** we have shown that $H_1(N, \partial N)$ is a torsion group. Using the Universal Coefficients Theorem we obtain the split exact sequence

$$0 \to \mathrm{Ext}(H_1(N, \partial N), \mathbb{Z}) \to H^2(N, \partial N) \to \mathrm{Hom}(H_2(N, \partial N), \mathbb{Z}) \to 0.$$

The Poincaré duality now leads to the isomorphisms

$$\begin{aligned} H_1(N) \cong H^2(N, \partial N) &\cong \mathrm{Hom}(H_2(N, \partial N), \mathbb{Z}) \oplus \mathrm{Ext}(H_1(N, \partial N), \mathbb{Z}) \\ &\cong \mathrm{Hom}(H_2(N, \partial N), \mathbb{Z}) \oplus \mathrm{Hom}(H_1(N, \partial N), \mathbb{T}). \end{aligned}$$

This isomorphism is most conveniently expressed in intersection theoretic terms. Denote by $H_1^\tau(N)$ the torsion part of $H_1(N)$. The above isomorphisms implies that we have a nondegenerate *linking pairing*

$$\boldsymbol{lk}_N \colon H_1^\tau(N) \times H_1(N, \partial N) \to \mathbb{T}.$$

We obtain a bilinear map

$$\widehat{\boldsymbol{lk}}_N \colon H_1^\tau(N) \times H_1^\tau(N) \to \mathbb{T}, \quad \widehat{\boldsymbol{lk}}_N(c_1, c_2) = \boldsymbol{lk}_N(c_1, i(c_2))$$

where i denotes the morphism $H_1(N) \to H_1(N, \partial N)$. This is a symmetric, yet possibly degenerate form. If we identify $H_2(N, \partial N)$ with $\ker \boldsymbol{j} = \mathbb{Z}\langle\lambda\rangle \subset H_1(\partial N)$ we see that we have a map

$$L \colon H_1(N) \to \mathrm{Hom}(\mathbb{Z}\langle\lambda\rangle, \mathbb{Z}) \cong \mathbb{Z} \iff \boldsymbol{Lk} \colon H_1(N) \times \mathbb{Z}\langle\lambda\rangle \to \mathbb{Z}$$

which can be described geometrically as the linking with λ. λ bounds a chain Λ in N and we define

$$L(c) = \boldsymbol{Lk}_N(c, \lambda) := \#(c \cap \Lambda), \quad \forall c \in H_1(N).$$

Observe that

$$L(\boldsymbol{j}z) = \lambda \cdot z, \quad \forall z \in H_1(\partial N),$$

where the dot denotes the (skew-symmetric) intersection pairing on $H_1(\partial N)$ defined using the orientation on ∂N *as boundary of* N. The subgroup $\boldsymbol{j}H_1(\partial N) \subset H_1(N)$ is mapped by L onto the subgroup $m_0\mathbb{Z} \subset \mathbb{Z}$. We obtain a short exact sequence

$$0 \to H_1^\tau(N) = \ker L \to H_1(N) \xrightarrow{L} \mathbb{Z} \to 0.$$

Any $\sigma_0 \in H_1(N)$ such that $\boldsymbol{Lk}(\sigma_0, \lambda) = 1$ produces a splitting of the above sequence

$$H_1(N) \cong H_1^\tau(N) \oplus \mathbb{Z}\langle\sigma_0\rangle.$$

Moreover, any element $c \in H_1(N)$ determines a morphism

$$c^\sharp = \boldsymbol{lk}_N(c, -) \in H_1\widehat{(N, \partial N)}.$$

The element c is completely determined by the quantities

$$c^\sharp \in H_1\widehat{(N, \partial N)}, \quad n(c) := \boldsymbol{Lk}_N(c, \lambda) \in \mathbb{Z}.$$

More precisely, we can write $c = [c] + c^\tau$

$$[c] = n(c)\sigma_0, \quad c^\tau := c - [c].$$

Define $c^* \in \operatorname{Hom}\big(H_2(N, \partial N), \mathbb{Z}\big) \oplus \operatorname{Hom}(H_1(N, \partial N), \mathbb{T})$

$$c^* = [c]^* + c^\sharp, \quad [c]^* := \boldsymbol{Lk}_N(c, -), \quad c^\sharp = \boldsymbol{lk}_N(c^\tau, -).$$

The correspondence $c \longleftrightarrow c^*$ is precisely the Poincaré duality.

Before we continue this line of thought we want to present a guiding example which will provide some intuition behind the above abstract constructions.

Example B.9. Suppose N is the complement of a knot $\mathcal{K}$ in a rational homology sphere M. Denote by r the order of $\mathcal{K}$ in $H_1(M)$. (If $H_1(M) = 0$ we set $r = 1$.) Since the linking form on M is nondegenerate there exists a knot $\mathcal{K}^\sharp \subset M$ such that

$$\boldsymbol{Lk}_M(\mathcal{K}, \mathcal{K}^\sharp) = \frac{1}{r}.$$

Then

$$H_1(N) \cong \{(\alpha, c) \in \mathbb{Q} \times H_1(M);\ \alpha = \boldsymbol{lk}_M(c, \mathcal{K}) \bmod \mathbb{Z}\}.$$

The pair $(1/r, \mathcal{K}^\sharp)$, which corresponds to the homology class in N carried by the knot $\mathcal{K}^\sharp$, can be taken as a generator of the free part of $H_1(N)$. We will denote this class simply by $\mathcal{K}^\sharp$. Observe that it corresponds to the choice $\sigma_0 \in H_1(N)$ explained in the preceding discussion. The torsion part of $H_1(N)$ is isomorphic to

$$\big\{(0, c) \in \mathbb{Q} \times H_1(M);\ \boldsymbol{lk}_M(\mathcal{K}, c) = 0\big\} \cong \mathcal{K}^\perp := \ker \boldsymbol{lk}_M(\mathcal{K}, -)\colon H_1(M) \to \mathbb{Z}_r.$$

We have seen in **2.**, page 218, that

$$H_1(N, \partial N) \cong H_1(M)/\langle \mathcal{K} \rangle.$$

The linking form on M induces a nondegenerate pairing

$$\ker \boldsymbol{lk}_M(\mathcal{K}, -) \times H_1(M)/\langle \mathcal{K} \rangle \to \mathbb{T}.$$

This is precisely the nondegenerate linking

$$\boldsymbol{lk}_N : H_1(N)^{\tau} \times H_1(N, \partial N) \to \mathbb{T}.$$

The curve $\lambda \in H_1(\partial N)$ which generates $\ker \boldsymbol{j}$ can be uniquely written as $m_0\lambda_0$ where λ_0 is primitive. If we set

$$\frac{\nu}{r} = \boldsymbol{lk}_M(\mathcal{K}, \mathcal{K})$$

then $m_0 = (\nu, r)$. Thus m_0 can be determined from λ and r using the equality

$$\lambda \cdot \mu = r,$$

where ∂N is oriented as boundary of N. We can describe

$$H_1(\partial N) \ni z = \alpha\mu \oplus \frac{n}{r}\lambda, \quad \alpha = n\, \boldsymbol{lk}_M(\mathcal{K}, \mathcal{K}).$$

The morphism $\boldsymbol{j}$ has the form

$$\alpha\mu + \frac{n}{r} \mapsto (\alpha, n\mathcal{K}) \in H_1(N).$$

Suppose we have chosen $\omega_0 \in H_1(N)$ such that $\lambda_0 \cdot \omega_0 = 1$. There is no unique choice but each such choice can be represented as

$$\omega_0 = a_0\mu + \frac{n_0}{r}\lambda, \quad a_0 - \frac{n_0\nu}{r} \in \mathbb{Z}.$$

The equation $\lambda_0 \cdot \omega_0 = 1$ implies

$$ra_0 = m_0$$

so that

$$\frac{m_0}{r} = \frac{n_0\nu}{r} \bmod \mathbb{Z}.$$

If we write

$$\nu = m_0\nu_0, \quad r = m_0 r_0, \quad (\nu_0, r_0) = 1$$

we deduce

$$\frac{1}{r_0} = \frac{n_0\nu_0}{r_0} \bmod \mathbb{Z}$$

so that

$$n_0\nu_0 = 1 \bmod r_0\mathbb{Z}, \quad a_0 = \frac{1}{r_0}$$

and

$$\boldsymbol{j}\omega_0 = \left(\frac{1}{m_0}\mu, n_0\mathcal{K}\right), \quad \boldsymbol{j}\lambda_0 = (0, r_0\mathcal{K}). \qquad \square$$

We can finally explain what do we need to know to compute how the linking form of a rational homology sphere changes by performing a Dehn surgery along a knot.

Suppose N is the complement of a knot $\mathcal{K}$ in a rational homology spheres N_1. $\mu \in H_1(\partial N)$ is a primitive curve such that

$$r := \lambda \cdot \mu > 0.$$

Fix $\kappa \in \partial N$ such that $\kappa \cdot \mu = 1$. We can assume that Dehn surgery is given by the identification

$$S^1 \times \partial D^2 \to \partial N,$$

$$S^1 \times \mathbf{pt} \mapsto \kappa, \quad \mathbf{pt} \times \partial D^2 \mapsto \mu.$$

If O denotes the center of D^2, we denote by $\mathcal{K}$ the image of $S^1 \times O$ in N_μ. At 1. we have shown that we have an extension

$$0 \to \mathbb{Z}\langle\mu\rangle \to H_1(N) \to H_1(N_\mu) \to 0$$

classified by the morphism

$$\boldsymbol{lk}_{N_\mu}(\mathcal{K}, -)\colon H_1(N_\mu) \to \mathbb{T}.$$

We have the following result.

Proposition B.10. *The linking form of N_μ is completely determined by the following data.*

I$_1$. $\mu, \kappa \in H_1(\partial N)$ *such that* $\kappa \cdot \mu = 1$.

I$_2$. *The Abelian groups $H_1(N)$, $H_1(N, \partial N)$ and the morphism $\boldsymbol{j}\colon H_1(\partial N) \to H_1(N)$. Fix a generator λ of* $\ker \boldsymbol{j}$ *and set $r = \lambda \cdot \mu$, $k = \lambda \cdot \kappa$. Fix $\Lambda \in H_2(N, \partial N)$ such that $\partial\Lambda = \lambda$ and denote by m_0 the positive integer such that $\lambda = m_0\lambda_0$ where $\lambda_0 \in H_1(\partial N)$ is a primitive class.*

I$_3$. *The Poincaré duality linking pairing*

$$\boldsymbol{lk}_N\colon H_1^\tau(N) \times H_1(N, \partial N) \to \mathbb{T}.$$

I$_4$. *A cycle $\sigma_0 \in H_1(N)$ such that $\sigma_0 \cdot \Lambda = 1$. Algebraically, this is equivalent to choosing a splitting*

$$H_1(N) = \textit{free part} \oplus \textit{torsion part}.$$

$\mathbf{I_5}$. *A positive integer r_0 and a cycle*

$$\nu = \alpha\mu + \frac{n}{r}\lambda \in H_1(\partial N)$$

such that

$$r_0\sigma_0 - \boldsymbol{j}\nu = 0 \in H_1(N).$$

Remark B.11. The preceding discussion and Example B.9 show that these data are completely determined by the *homological* properties of the pair $(N_1, \mathcal{K})$. In other words, the computation of the linking form of a rational homology sphere can be determined by performing only *homological* computations. This is certainly not the case for the Reidemeister torsion which is not a homotopy invariant. □

Proof. Using $\mathbf{I_1}$ and $\mathbf{I_2}$ we can now determine

$$H_1(N_\mu) := H_1(N)/\langle \boldsymbol{j}\mu\rangle$$

and thus the canonical extension

$$0 \to \mathbb{Z}\langle\mu\rangle \to H_1(N) \xrightarrow{\mathbf{proj}} H_1(N)/\langle \boldsymbol{j}\mu\rangle \cong H_1(N_\mu) \to 0.$$

The element $\mathcal{K} \in H_1(N_\mu)$ is the image of $\boldsymbol{j}\kappa$ in $H_1(N)/\langle \boldsymbol{j}\mu\rangle$. The above extension completely determines the morphism

$$\boldsymbol{lk}_{N_\mu}(\mathcal{K}, -)\colon H_1(N_\mu) \to \mathbb{T}.$$

Observe that $H_1^\tau(N)$ embeds in $H_1(N_\mu) = H_1(N)/\langle \boldsymbol{j}\mu\rangle$ as the kernel $\mathscr{K}_\mu$ of this morphism. We can now produce an isomorphism

$$H_1(N_\mu)/\langle\mathcal{K}\rangle \to H_1(N, \partial N) = H_1(N)/\mathrm{Range}\,\boldsymbol{j}$$

by going up-and-down along the diagram below (with exact diagonals)

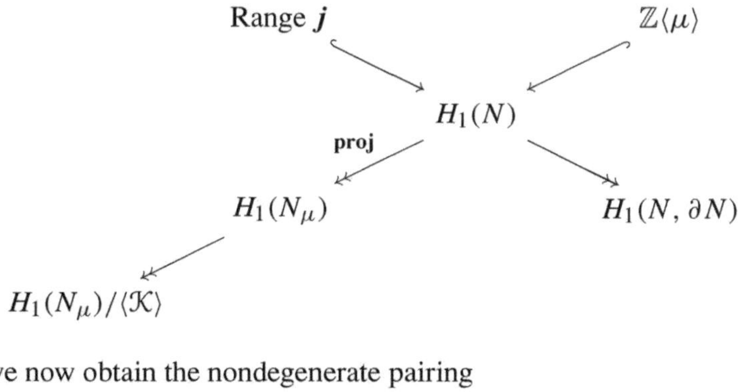

Using $\mathbf{I_3}$ we now obtain the nondegenerate pairing

$$\boldsymbol{lk}_{N_\mu}\colon \mathscr{K}_\mu \times H_1(N_\mu)/\langle\mathcal{K}\rangle \to \mathbb{T}.$$

In particular, this implies we can now compute all the pairings of the form $\boldsymbol{lk}_{N_\mu}(u, v)$ where $u, v \in H_1(N_\mu)$ and at least one of them is in $\mathcal{K}_\mu$.

To complete the determination of $\boldsymbol{lk}_{N_\mu}$ we will need to use the following elementary results.

Lemma B.12. *Suppose $\mathcal{K}^\sharp \in H_1(N_\mu)$ is such that*

$$\boldsymbol{lk}_{N_\mu}(\mathcal{K}, \mathcal{K}^\sharp) = 1/r \in \mathbb{T}.$$

Then for all $u, v \in H_1(N_\mu)$ we have

$$\boldsymbol{lk}_{N_\mu}(\mathcal{K}, u) = \frac{n(u)}{r}, \quad \boldsymbol{lk}_{N_\mu}(\mathcal{K}, v) = \frac{n(v)}{r}, \quad 0 \le n(u), n(v) < r$$

$$u_0 := u - n(u)\mathcal{K}^\sharp, \quad v_0 := v - n(v)\mathcal{K}^\sharp \in \mathcal{K}_\mu$$

and

$$\begin{aligned}\boldsymbol{lk}_{N_\mu}(u, v) &= \boldsymbol{lk}_{N_\mu}\big(u_0 + n(u)\mathcal{K}^\sharp, v_0 + n(v)\mathcal{K}^\sharp\big)\\ &= \boldsymbol{lk}_{N_\mu}(u_0, v_0) + \boldsymbol{lk}_{N_\mu}\big(n(v)u_0 + n(u)v_0, \mathcal{K}^\sharp\big) + n(u)n(v)\,\boldsymbol{lk}_{N_\mu}(\mathcal{K}^\sharp, \mathcal{K}^\sharp).\end{aligned}$$

The above result shows that in order to determine $\boldsymbol{lk}_{N_\mu}$ we need to find the self-linking number $\boldsymbol{lk}_{N_\mu}(\mathcal{K}^\sharp, \mathcal{K}^\sharp)$ for *some* solution $\mathcal{K}^\sharp$ of the equation

$$\boldsymbol{lk}_{N_\mu}(\mathcal{K}, \mathcal{K}^\sharp) = 1/r \in \mathbb{T}.$$

The cycle $\sigma_0 \in H_1(N)$ described in $\mathbf{I_4}$ descends to a solution $[\sigma_0] \in H_1(N)/\mathbb{Z}\langle\mu\rangle \cong H_1(N_\mu)$ of above equation. We can now represent the cycle ν of $\mathbf{I_5}$ as a linear combination

$$\nu = \alpha\mu + \frac{n}{r}\Lambda$$

so that

$$r r_0 \sigma_0 = r\boldsymbol{j}\nu = r\alpha\mu + n\lambda = n\lambda = n\partial\Lambda.$$

Thus

$$\boldsymbol{lk}_{N_\mu}(\sigma_0, \sigma_0) = \frac{n}{r r_0}(\sigma_0 \cap \Lambda) = \frac{n}{r r_0}. \qquad \square$$

The proof of the above proposition explicitly describes an algorithm for computing the change in the linking form under a Dehn surgery. As our next example will show, the concrete implementations of this algorithm can be computationally demanding.

Example B.13 (Surgery on $M := L(24, 23)$). Fix a generator $\boldsymbol{g}_0$ of $H_1(L(24, 1)) \cong \mathbb{Z}_{24}$ such that the linking form q has the description

$$q(x\boldsymbol{g}_0, y\boldsymbol{g}_0) = -\frac{xy}{24} \in \mathbb{T}.$$

Fix a knot $\mathcal{K}_0$ in this lens space representing $4\boldsymbol{g}_0$ in homology. Denote U_0 a small (open) tubular neighborhood of $\mathcal{K}_0$in M, and by μ_0 its meridian oriented such that

$$\lambda_0 \cdot \mu_0 = 6,$$

where the above intersection pairing uses the orientation on ∂U_0 *as boundary of* $N := M \setminus U_0$. Thus ord $(\mathcal{K}) = 6$ and

$$q(\mathcal{K}_0, \mathcal{K}_0) = -\frac{16}{24} = -\frac{4}{6} = -\frac{2}{3} \in \mathbb{T}.$$

The kernel of $\boldsymbol{j}\colon H_1(\partial N) \to H_1(N)$ is generated by a curve $\lambda \in H_1(\partial N)$. According to Remark B.5 we can choose a basis of $H_1(T)$ so that μ_0 has the coordinates $(0, 1)$ while λ has the coordinates $(6, 4)$. Then

$$H_1(N) \cong \Big\{(\alpha\mu_0, c\boldsymbol{g}_0) \in \mathbb{Q}\langle\mu_0\rangle \times \mathbb{Z}_{24}\langle\boldsymbol{g}_0\rangle;\ \alpha = -\frac{c}{6} \bmod \mathbb{Z}\Big\}.$$

Thus

$$H_1^{\tau}(N) = \ker \boldsymbol{lk}_M(\mathcal{K}_0, -) \cong 6\mathbb{Z}_{24} \cong \mathbb{Z}_4\langle \boldsymbol{u}_0 := 6\boldsymbol{g}_0\rangle.$$

Similarly

$$H_1(N, \partial N) \cong H_1(M)/\langle\mathcal{K}_0\rangle \cong \mathbb{Z}_4\langle\boldsymbol{v}_0\rangle,$$

where $\mathbf{v}_0$ denotes the generator defined as the image of $\boldsymbol{g}_0$ in $H_1(M)/\langle\mathcal{K}_0\rangle$. Then

$$\boldsymbol{lk}_N(\boldsymbol{u}_0, \boldsymbol{v}_0) = \boldsymbol{lk}_M(6\boldsymbol{g}_0, \boldsymbol{g}_0) = -\frac{1}{4}.$$

This equality produces $\mathbf{I}_3$. Observe that

$$\boldsymbol{lk}_M\big(-\boldsymbol{g}_0, \mathcal{K}_0\big) = \frac{1}{6}.$$

Thus $\sigma_0 := (1/6\mu_0, -\boldsymbol{g}_0) \in H_1(N)$ solves $\mathbf{I}_4$.

A nontrivial Dehn surgery on $\mathcal{K}_0$ is described by a primitive curve $\mu \in H_1(\partial N)$ such that $\mu \neq \mu_0$. Suppose $\mu = (1, 0) := -\frac{2}{3}\mu + \frac{1}{6}\lambda$, i.e. we perform a $(-2/3, 1)$-surgery. Then we can choose $\kappa := (0, 1) = \mu_0$. Observe that

$$\boldsymbol{j}\mu = \Big(-\frac{2}{3}\mu_0, \mathcal{K}_0 = 4\boldsymbol{g}_0\Big) \in H_1(N), \quad \boldsymbol{j}(\kappa) = (1, 0) \in H_1(N).$$

Hence $H_1(N_\mu)$ is defined by the extension

$$0 \to \Big\{\Big(-\frac{2n}{3}, 4n\boldsymbol{g}_0\Big);\ n \in \mathbb{Z}\Big\} \hookrightarrow \big\{(\alpha, c\boldsymbol{g}_0);\ \alpha + \frac{c}{6} \in \mathbb{Z}\big\} \to H_1(N_\mu) \to 0.$$

More explicitly, observe that we have the direct sum decomposition

$$H_1(N) = \mathbb{Z}\langle\sigma_0\rangle \oplus \mathbb{Z}_4\langle\boldsymbol{u}_0 := 6\boldsymbol{g}_0\rangle,$$

and we can write $\boldsymbol{j}\mu = -4\sigma_0$. We conclude that

$$H_1(N_\mu) \cong \mathbb{Z}_4\langle\sigma_0\rangle \oplus \mathbb{Z}_4\langle\boldsymbol{u}_0\rangle.$$

The extension

$$0 \to \mathbb{Z} \xrightarrow{(\cdot 4\sigma_0, 0)} \mathbb{Z} \oplus \mathbb{Z}_4 \to \mathbb{Z}_4\sigma_0 \oplus \mathbb{Z}_4\boldsymbol{u}_0 \to 0$$

is classified by the character

$$\chi : \mathbb{Z}_4\sigma_0 \times \mathbb{Z}_4 u_0 \to \mathbb{T}, \quad \chi(\sigma_0) = \frac{1}{4}, \quad \chi(\boldsymbol{u}_0) = 0.$$

The cycle $\boldsymbol{j}\kappa = 6\sigma_0 \oplus \boldsymbol{u}_0 \in H_1(N)$ projects to the cycle

$$\mathcal{K} = 2\sigma_0 \oplus \boldsymbol{u}_0 \in H_1(N_\mu),$$

and we have

$$\boldsymbol{lk}_{N_\mu}(\mathcal{K}, \sigma_0) = \chi(\sigma_0) = 1/4, \quad \boldsymbol{lk}_{N_\mu}(\mathcal{K}, \boldsymbol{u}_0) = \chi(\boldsymbol{u}_0) = 0.$$

We deduce that

$$\ker \boldsymbol{lk}_{N_\mu}(\mathcal{K}, -) \cong \mathbb{Z}_4\langle\boldsymbol{u}_0\rangle.$$

Now observe that $\sigma_0 \in H_1(N)$ descends to a generator of $H_1(N_\mu)/\langle\mathcal{K}\rangle \cong \mathbb{Z}_4$. It must therefore descend to a generator of

$$H_1(N, \partial N) \cong H_1(N)/\text{Range}\,\boldsymbol{j}.$$

We have identified $H_1(N, \partial N) \cong H_1(M)/\langle\mathcal{K}_0\rangle$ with the cyclic group of order 4 with generator $\mathbf{v}_0$ induced by $\boldsymbol{g}_0 \in H_1(M)$. Observe that $\boldsymbol{g}_0$ lifts to $-\sigma_0 \in H_1(N)$. Thus the isomorphism

$$\mathbb{Z}_4\langle\sigma_0\rangle \cong H_1(N_\mu)/\langle\mathcal{K}\rangle \cong H_1(N, \partial N) \cong \mathbb{Z}_4\langle\boldsymbol{v}_0\rangle$$

can be concretely described by the correspondence $\sigma_0 \mapsto -\boldsymbol{v}_0$. We conclude that the pairing

$$\boldsymbol{lk}_{N_\mu} : \ \ker \boldsymbol{lk}_{N_\mu}(\mathcal{K}, -) \times H_1(N_\mu)/\langle\mathcal{K}\rangle \to \mathbb{T}$$

has the form

$$\boldsymbol{lk}_{N_\mu}(x\boldsymbol{u}_0, y\sigma_0) = -\,\boldsymbol{lk}_N(x\boldsymbol{u}_0, y\boldsymbol{v}_0) = \frac{xy}{4}.$$

Hence the pairing

$$\boldsymbol{lk}_{N_\mu} : \ \ker \boldsymbol{lk}_{N_\mu}(\mathcal{K}, -) \times H_1(N_\mu) \to \mathbb{T}$$

is given by

$$\boldsymbol{lk}_{N_\mu}(\boldsymbol{u}_0, \sigma_0) = \frac{1}{4}, \quad \boldsymbol{lk}_{N_\mu}(\boldsymbol{u}_0, \boldsymbol{u}_0) = \boldsymbol{lk}_{N_\mu}(\boldsymbol{u}_0, \mathcal{K} - 2\sigma_0) = -\frac{1}{2}.$$

Since $4\sigma_0 = -\boldsymbol{j}\mu$ we deduce that $\boldsymbol{lk}_{N_\mu}(\sigma_0, \sigma_0) = 0$. Summarizing all of the above we deduce

$$\boldsymbol{lk}_{N_\mu}(a\boldsymbol{u}_0 + b\sigma_0, x\boldsymbol{u}_0 + y\sigma_0) = \frac{ay + bx}{4} + \frac{ax}{2}. \qquad \square$$

So far we have discussed only Dehn surgeries which produce rational homology spheres. We want to spend the rest of this section discussing the remaining case.

Suppose N is homeomorphic to the complement of a tubular neighborhood of a knot in a rational homology sphere M. Set $T := \partial N$. Orient T as boundary of N. Denote by $\lambda \in H_1(T)$ a longitude, i.e a generator[1] of the kernel of the inclusion induced map $H_1(T) \to H_1(N)$. Denote by m_0 the divisibility of λ, i.e. m_0 is the positive integer such that $\lambda = m_0\lambda_0$ where λ_0 is a primitive element of $H_1(T)$.

For any primitive class $c \in H_1(T)$ denote by N_c the closed three-manifold obtained from N by Dehn surgery with data c. If $\lambda \cdot c \neq 0$ the three-manifold N_c is a rational homology sphere. When $c = \lambda_0$ the three-manifold N_{λ_0} is a rational homology $S^1 \times S^2$. We denote it by N_0. We want to describe the homological invariants of N_0 in terms of the homological invariants of N. Denote by $\bar{U}$ the solid torus we attach to N to produce N_0. Note that

$$H_3(N_0) \cong H_3(N_0, N) = \mathbb{Z}, \quad 0 = H_1(\bar{U}, T) \cong H_1(N_0, N)$$

and

$$H_2(N) = 0, \quad H_2(N_0, N) \cong H_2(\bar{U}, T) \cong \mathbb{Z}.$$

As generator of $H_2(\bar{U}, T)$ we can take the disk D_μ spanning the meridian. The long exact sequence of the pair (N_0, N) now implies

$$0 \to H_2(N_0) \to H_2(\bar{U}, T)(\cong \mathbb{Z}) \to H_1(N) \to H_1(N_0) \to 0.$$

Since the generator D_μ of $H_2(\bar{U}, T)$ goes to the torsion class $\boldsymbol{j}(\lambda_0) \in H_1(N)$ we deduce that the image of $H_2(N_0)$ in $H_2(N_0, N)$ is generated by $m_0[D_\mu]$. Moreover, we have a short exact sequence

$$0 \to \langle \boldsymbol{j}\lambda_0 \rangle \to H_1(N) \to H_1(N_0) \to 0. \tag{B.5}$$

where $\langle \boldsymbol{j}\lambda_0 \rangle$ denotes the cyclic group of order m_0 generated by $\boldsymbol{j}\lambda_0$.

Consider now the long exact sequence of the pair $(N_0, \bar{U})$. Denote by $\Lambda \in H_2(N, \partial N)$ a relative 2-cycle bounding λ. Observe that $H_2(N, \partial N) \cong H_2(N_0, \bar{U})$ is generated by Λ. Since $H_2(\bar{U}) = 0$ we deduce

$$0 \to H_2(N_0) \to H_2(N, \partial N) \cong \mathbb{Z}\langle \Lambda \rangle \xrightarrow{\partial} H_1(\bar{U}) \to H_1(N_0) \to H_1(N, \partial N) \to 0.$$

The connecting morphism ∂ is trivial because $\partial\Lambda = \lambda = m_0\mu \to 0 \in H_1(\bar{U})$ so that we obtain an isomorphism

$$H_2(N_0) \cong H_2(N, \partial N) \cong \mathbb{Z}\langle \Lambda \rangle,$$

and a short exact sequence

$$0 \to H_1(\bar{U}) \to H_1(N_0) \to H_1(N, \partial N) \cong \operatorname{coker} \boldsymbol{j} \to 0. \tag{B.6}$$

[1] There are two generators, and a choice can be determined by fixing an orientation.

We denote by $\mathcal{K}_0$ the homology class in $H_1(N_0)$ carried by the core of $\bar{U}$ and by $i_{\mathcal{K}}$ the inclusion induced morphism $i_{\mathcal{K}}\colon H_1(\bar{U}) \to H_1(N_0)$. Hence

$$H_1(N, \partial N) \cong H_1(N_0)/\langle \mathcal{K}_0 \rangle.$$

The extension (B.6) defines a character of $H_1(N, \partial N)$ which, in view of the Poincaré duality on $(N, \partial N)$, can be identified with the linking by a torsion element in $H_1(N)$. Using (B.5) we deduce that this is given by the linking with λ_0. Note also that $\mathcal{K}$ is not a primitive class. It has divisibility m_0.

Dualizing the sequence (B.6) we deduce

$$1 \to \widehat{H_1(N, \partial N)} \to \widehat{H_1(N_0)} \xrightarrow{\hat{i_{\mathcal{K}}}} \widehat{H_1(\bar{U})} \to 1$$

Restricting the second map to the identity component of $\widehat{H_1(N_0)}$ we obtain a surjection

$$S^1 \cong \widehat{H_1(N_0)}_{\mathbf{id}} \overset{\hat{i_{\mathcal{K}}}}{\twoheadrightarrow} \widehat{H_1(\bar{U})}_{\mathbf{id}} \cong S^1.$$

Since the linking number of $\boldsymbol{j}(\mathcal{K}_0)$ and Λ (in N) is m_0 we deduce that the map $i_{\mathcal{K}}^{\sharp}$ is an m_0-cover. Denote by z the coordinate on $\widehat{H_1(\bar{U})}$, and by T the coordinate on the identity component of $\widehat{H_1(N_0)}$. The above map is described by $z = T^{m_0}$.

Dualizing the sequence (B.5) we obtain the short exact sequence

$$1 \to \widehat{H_1(N_0)} \to \widehat{H_1(N)} \to \mathbb{U}_{m_0} \to 1, \tag{B.7}$$

where $\mathbb{U}_{m_0}$ is the group of m_0-th roots of 1. Restricting to the identity components we deduce

$$\widehat{H_1(N_0)}_{\mathbf{id}} = \widehat{H_1(N)}_{\mathbf{id}}.$$

Denote by $\hat{\mathfrak{T}}_X$ the Fourier transform of the Reidemeister torsion of X. The surgery formulæ have the form

$$\hat{\mathfrak{T}}_{N_0} \cdot \hat{i_{\mathcal{K}}} \hat{\mathfrak{T}}_U = \hat{\mathfrak{T}}_N |_{\widehat{H_1(N_0)}}. \tag{B.8}$$

Observe now that since H is an Abelian group of positive rank the augmentation map

$$\mathfrak{aug}\colon \mathrm{Map}(H, \mathbb{C}) \to \mathrm{Map}(H/\operatorname{Tors}(H), \mathbb{C})$$

is precisely the integration along the fibers of $H \to H/\operatorname{Tors}(H)$. If $\hat{f}$ is the Fourier transform of $f \in \mathrm{Map}\,(H, \mathbb{C})$ then the Fourier transform of $\mathfrak{aug}(f)$ is the restriction of $\hat{f}$ to the identity component of $\hat{H}$.

If we now further restrict the surgery formula (B.8) to the identity component of $\widehat{H_1(N_0)}$ and then take the inverse Fourier transform we deduce

$$\mathfrak{T}_{N_0}^{\mathfrak{aug}}(T)(1 - T^{m_0})^{-1} \approx \mathfrak{T}_N^{\mathfrak{aug}}(T).$$

Recalling that

$$\mathcal{T}_N^{\mathrm{aug}}(T) \sim \Delta_N(T)(1-T)^{-1},$$

where $\Delta_N(T)$ denotes the Alexander polynomial of N, we conclude

$$\mathcal{T}_{N_0}^{\mathrm{aug}}(T) \sim \frac{\Delta_N(T)}{(1-T)(1-T^{m_0})}.$$

Bibliography

[1] M. F. Atiyah, R. Bott: A Lefschetz fixed point formula for elliptic complexes II. Applications, *Ann. of Math.* **88** (1968), 451–491.

[2] M. F. Atiyah, V. K. Patodi, I. M. Singer: Spectral asymmetry and Riemannian geometry I, *Math. Proc. Cambridge Philos. Soc.* **77** (1975), 43–69.

[3] S. Baldridge: Seiberg-Witten invariants of 4-manifolds with free circle actions, math.GT/9911051.

[4] J. M. Bismut, W. Zhang: An extension of a theorem by Cheeger and Müller, *Astérisque* **205**, 1992.

[5] G. E. Bredon: *Topology and Geometry*, Grad. Texts in Math. 139, Springer-Verlag, 1993.

[6] G. E. Bredon: *Sheaf Theory*, Grad. Texts in Math. 170, Springer-Verlag, 1997.

[7] G. W. Brumfiel, J. W. Morgan: Quadratic functions, the index modulo 8, and a $\mathbb{Z}/4$-Hirzebruch formula, *Topology* **12** (1973), 105–122.

[8] G. Burde, H. Zieschang: *Knots*, de Gruyter Stud. Math. 5, Walter de Gruyter, 2003.

[9] D. Burghelea, L.Friedlander, T. Kappeler: Witten deformation of the analytic torsion and the Reidemeister torsion, *Voronezh Winter Mathematical Schools*, Amer. Math. Soc. Transl. Ser. (2) 184, Amer. Math. Soc., Providence, RI, 1998, 23–39; also available at `http://www.math.ohio-state.edu/~burghele/`.

[10] D. Burghelea, L. Friedlander, T. Kappeler, P. MacDonald: Analytic and Reidemeister torsion for representations in finite type Hilbert modules, *Geom. Funct. Anal.* **6** (1996), 751–859.

[11] A. Carey, M. Marcolli, B. L. Wang: *Exact triangles in Seiberg-Witten-Floer homology. Part I: the geometric triangle*, math.DG/9907065, Comm. Contemp. Math., to appear.

[12] A. Cayley: On the theory of elimination, *Cambridge and Dublin Math. J.* **3** (1848), 116–120; *Collected Papers Vol.1*, Cambridge University Press, 1889, 370–374.

[13] T. A. Chapman: Topological invariance of Whitehead torsion, *Amer. J. Math.* **96** (1974), 488–497.

[14] J. Cheeger: Analytic torsion and the heat equation, *Ann. of Math.* **109** (1979), 259–322.

[15] W. Chen: Casson invariant and Seiberg-Witten gauge theory, *Turkish J. Math.* **21** (1997), 61–81.

[16] M. M. Cohen: *A course in simple homotopy theory*, Grad. Texts in Math. 10, Springer-Verlag, 1973.

[17] J. F. Davis, P. Kirk: *Lecture Notes in Algebraic Topology*, Grad. Stud. in Math., Amer. Math. Soc., 2001

[18] P. Deligne: Le déterminant de la cohomologie, in *Current Trends in Aritmetical Algebraic Geometry*, Contemp. Math. 67, Amer. Math. Soc, 1987, 93–177.

[19] G. DeRham, S. Maumary, M. A. Kervaire: *Torsion et Type Simple d'Homotopie*, Lecture Notes in Math. 48, Springer-Verlag, 1967.

[20] A. Dimca: *Singularities and the Topology of Hypersurfaces*, Universitext, Springer-Verlag, 1992.

[21] S. K. Donaldson: Topological field theories and formulæ of Casson Meng-Taubes, *Proceedings of the Kirbyfest*, Geom. Topol. Monogr. 2, 1999, 87–102; this volume is freely available at `http://www.maths.warwick.ac.uk/gt/gtmono.html`

[22] M. Farber: Combinatorial invariants computing the Ray-Singer analytic torsion, *Differential Geom. Appl.* **6** (1996), 351–366; dg-ga/9606014.

[23] M. Farber, V. Turaev: Poincaré-Reidemeister metric, Euler structures, and torsion, *J. Reine Angew. Math.* **520** (2000), 195–225; math.DG/9803137.

[24] R. H. Fox: Free differential calculus. I, *Ann. of Math.* **57** (1953), 547–560.

[25] R. H. Fox: Free differential calculus. II, *Ann. of Math.* **59** (1954), 196–210.

[26] R. H. Fox, G. Torres: Dual presentations of the group of a knot, *Ann. of Math.* **59** (1954), 211–218.

[27] R. H. Fox: A quick trip through knot theory, in *Topology of 3-Manifolds and Related Topics*, Proc. of the Univ. of Georgia Institute 1961, Prentice-Hall, 1962, 120–167.

[28] J. M. Franks: Knots, links and symbolic dynamic, *Ann. of Math.* **113** (1981), 529–552.

[29] W. Franz: Ueber die Torsion einer Ueberdeckung, *J. Reine Angew. Math.* **173** (1935).

[30] W. Franz: Abbildungsklassen und Fixpunktklassen dreidimensionaler Linsenräume, *J. Reine Angew. Math.* **185** (1943), 65–77.

[31] D. Freed: Reidemeister torsion, spectral sequences, and Brieskorn spheres, *J. Reine Angew. Math.* **429** (1992), 75–89.

[32] D. Fried: Periodic points and twisted coefficients, in *Geometric Dinamics*, Lecture Notes in Math. 1007, Springer-Verlag, 1983.

[33] D. Fried: Homological identities for closed orbits, *Invent. Math.* **71** (1983), 419–442.

[34] D. Fried: Lefschetz formulas for flows, in *The Lefschetz centennial conference, part II*, Contemp. Math. 58, III, Amer. Math. Soc. Providence, R.I. 1987, 19–69.

[35] D. Fried: Counting circles, in *Lect. Notes in Math.*, 1342, Springer-Verlag, 1988, 196–216.

[36] M. Furuta, B. Steer: Seifert fibred homology 3-spheres and the Yang-Mills equations on Riemann surfaces with marked points, *Adv. in Math.* **96** (1992), 38–102.

[37] I. M. Gelfand, G. E. Shilov: *Generalized Functions: Volume 1. Properties and Operations* Academic Press, 1964.

[38] I. M. Gelfand, M. M. Kapranov, A.V. Zelevinski: *Discriminants, Resultans and Multidimensional Determinants*, Mathematics: Theory and Applications, Birkhäuser, 1994.

[39] P. B. Gilkey: *Invariance Theory, the Heat Equation and the Atiyah-Singer Index Theorem*, 2nd Edition, CRC Press, Boca Raton, 1995.

[40] R. E. Gompf: $Spin^c$-structures and homotopy equivalences, *Geom. Topol.* **1** (1997), 41–50; `http://www.maths.warwick.ac.uk/gt/`

[41] R. E. Gompf, A. I. Stipsicz: *An Introduction to 4-Manifolds and Kirby Calculus*, Grad. Stud. Math. 20, Amer. Math. Soc., 1999.

[42] V. Guillemin, S. Sternberg: *Geometric Asymptotics*, Math. Surveys Monogr. 14, Amer. Math. Soc., 1990.

[43] S. Halperin, D. Toledo: Stiefel-Whitney homology classes, *Ann. of Math.* **96** (1972), 511–525.

[44] A. Hatcher: *Algebraic Topology*, Cambridge University Press, 2001.

[45] F. E. Hirzebruch: Hilbert modular surfaces, *Enseign. Math.* **19** (1973), 183–281.

[46] F. Hirzebruch, W. D. Neumann, S.S. Koh: *Differentiable Manifolds and Quadratic Forms*, Lect. Notes in Pure Appl. Math. 4, Marcel Dekker, 1971.

[47] M. Hutchings: Reidemeister torsion in generalized Morse theory, math.DG/9907066, *Forum Math.* **14** (2002), 209–245.

[48] M. Hutchings, Y.-J. Lee: Circle valued Morse theory, Reidemeister torsion and Seiberg-Witten invariants of 3-manifolds, *Topology* **38** (1999),861–888; math.DG/9612004.

[49] M. Hutchings, Y.-J. Lee: Circle valued Morse theory and Reidemeister torsion, *Geom. Topol.* **3** (1999), 369–396; `http://www.maths.warwick.ac.uk/gt/`

[50] M. Jankins, W. D. Neumann: *Lectures on Seifert Manifolds*, Brandeis Lecture Notes, 1983.

[51] R. Kirby, L. Taylor: Pin structures on low dimensional manifolds, in *Geometry of Low Dimensional Manifolds. 2*, London Math. Soc. Lecture Noters Ser. 151, Cambridge University Press,, Cambridge 1990, 177–242.

[52] P. Kirk, M. Lesch: The η-invariant, and spectral flow for Dirac-type operators on manifolds with boundary, math.DG/0012123.

[53] F. Knudsen, D. Mumford: Projectivity of moduli spaces of stable curves I. Preliminaries on Det and Div, *Math. Scand.* **39** (1976), 19–55.

[54] M. Kreck, S. Stolz: Non connected moduli spaces of positive sectional curvature metrics, *J. Amer. Math. Soc.* **6** (1993), 825–850.

[55] S. Lang: *Algebra*, Addison-Wesley, 3rd Edition, 1993.

[56] F. Latour: *Existence de 1-formes fermées non-singulières dans une classe de cohomologie de deRham, Inst. Hautes Études Sci. Publ. Math.* **80** (1994).

[57] S. Lefschetz: *Algebraic Topology*, Amer. Math. Soc. Colloq. Publ. 27, Amer. Math. Soc., 1942.

[58] C. Lescop: *Global Surgery Formula for the Casson-Walker Invariant*, Ann. of Math. Stud. 140, Princeton University Press, 1996.

[59] W. B. R. Lickorish: *An Introduction to Knot Theory*, Grad. Texts Math. 175, Springer-Verlag, 1997.

[60] Y. Lim: The equivalence of Seiberg-Witten and Casson invariants for homology 3-spheres, *Math. Res. Letters* **6** (1999), 631–644.

[61] Y. Lim: Seiberg-Witten invariants for 3-manifolds in the case $b_1 = 0$ or 1, *Pacific J. Math.* **195** (2000), 179–204.

[62] Y. Lim: Seiberg-Witten moduli spaces for 3-manifolds with cylindrical-end $T^2 \times \mathbb{R}^+$, *Comm. Contemp. Math.* **2** (2000), 461–509

[63] I. G. Macdonald: The Poincaré polynomial of a symmetric product, *Math. Proc. Cambridge. Phil. Soc.* **58** (1962), 563–568.

[64] S. MacLane: *Homology*, Band 114, Springer-Verlag 1963.

[65] M. Marcolli, B. L. Wang: Seiberg-Witten invariant and the Casson-Walker invariant for rational homology 3-spheres, math.DG/0101127, *Geom. Dedicata*, to appear.

[66] T. Mark: Torsion, TQFT, and Seiberg-Witten invariants of 3-manifolds, *Geom. Topol.* **6** (2002), 27–58; math.DG/9912147, `http://www.maths.warwick.ac.uk/gt/`

[67] C. R. F. Maunder: *Algebraic Topology*, Dover Publications, 1996.

[68] G. Meng, C. H. Taubes: $\underline{SW}$ = Milnor torsion, *Math. Res. Lett.* **3** (1996), 661–674.

[69] J. W. Milnor: Two complexes which are homeomorphic but combinatorially distinct, *Ann. of Math.* **74** (1961), 61–75.

[70] J. W. Milnor: A duality theorem for Reidemeister torsion, *Ann. of Math.* **76** (1962), 137–147.

[71] J. Milnor: Spin structures on manifolds, *Enseign. Math.* **9** (1963), 198–203.

[72] J. W. Milnor: Whitehead torsion, *Bull. Amer. Math. Soc.* **72** (1966), 358–428.

[73] J. W. Milnor: Infinite cyclic covers, in *Conference on the Topology of Manifolds*, J. G. Hocking, ed., Prindle, Weber&Schmidt Inc., 1968.

[74] W. Müller: Analytic torsion and R-torsion for Riemannian manifolds, *Adv. in Math.* **28** (1978), 233–305.

[75] A. Nemethi, L.I. Nicolaescu: Seiberg-Witten invariants and complex singularities, *Geom. Topol.* **6** (2002), 269–328; math.AG/0111298.

[76] A. Nemethi, L.I. Nicolaescu: Seiberg-Witten invariants and complex singularities II. Singularities with good $\mathbb{C}^*$-action, math.AG/0201120.

[77] W. Neumann: *Abelian covers of quasihomogeneous surface singularities*, Proc. Sympos. Pure Math. 40, Part 2, 233–244.

[78] W. D. Neumann: A calculus for plumbing applied to the topology of complex surface singularities and degenerating complex curves, *Trans. Amer. Math. Soc.* **268** (1981), 299–344.

[79] L. I. Nicolaescu: *Lectures on the Geometry of Manifolds*, World Scientific Publ. Co., 1996.

[80] L. I. Nicolaescu: Adiabatic limits of the Seiberg-Witten equations on Seifert manifolds, *Comm. Anal. Geom.* **6** (1998), 331–392; math.DG/9601107

[81] L. I. Nicolaescu: Seiberg-Witten invariants of lens spaces, Canad. J. Math. **53** (2001), 780–808; math.DG/9901071.

[82] L. I. Nicolaescu: *Notes on Seiberg-Witten Theory*, Grad. Stud. in Math. 28, Amer. Math. Soc., 2000.

[83] L. I. Nicolaescu: Seiberg-Witten invariants of rational homology spheres, math.DG/0103020.

[84] S. P. Novikov: Multivalued functions and functionals. An analogue of Morse theory, *Sov. Math. Dokl.* **24** (1981), 222–226.

[85] P. Ozsváth, Z. Szabó: *The theta divisor and the Casson-Walker invariant*, math.GT/0006194.

[86] A. V. Pajitnov: On the Novikov complex for rational Morse forms, *Ann. Fac. Sci. Toulouse Math.* **4** (1995), 297–338.

[87] A. V. Pajitnov: Simple homotopy type of the Novikov complex and Lefschetz ζ-function of the gradient flow, *Uspehi Mat. Nauk* **54** (1999), 117–170 (Russian); dg-ga/9706014 (English),

[88] H. Rademacher, E. Grosswald: *Dedekind Sums*, Carus Math. Monogr., Math. Assoc. Amer., 1972.

[89] D. B. Ray, I. M. Singer: R-torsion and the Laplacian on Riemannian manifolds, *Adv. in Math.* **7** (1971), 145–210.

[90] K. Reidemeister: Homotopieringe und Linseräume, *Hamburger Abhandlungen* **11** (1935).

[91] K. Reidemeister: Überdeckungen von Komplexen, *J. Reine. Angew. Math.* **173** (1935), 164–173.

[92] D. Rolfsen: *Knots and Links*, Math. Lect. Series, Publish and Perish, 2nd Edition, 1990.

[93] J. J. Rotman: *Notes on Homological Algebra*, Van Nostrand Reinhold Co., 1970.

[94] M. Rueff: Beiträge zur Untersuchung de Abbildungen von Mannifaltigkeiten, *Compositio Math.* **6** (1938), 161–202.

[95] D. A. Salamon: Seiberg-Witten invaraints of mapping tori, symplectic fixed points and Lefschetz numbers, in *Proceedings of the Sixth Gökova Geometry-Topology Conference*, Turkish J. Math. **23** (1999), 117–143; `http://www.math.ethz.ch/~salamon/`

[96] N. Saveliev: *Lectures on the Topology of* 3*-Manifolds. An Introduction to the Casson Invariant*, de Gruyter Textbook, Walter de Gruyter, 1999.

[97] P. Scott: The geometries of 3-manifolds, *Bull. London Math. Soc.* **15** (1983), 401–487.

[98] H. Seifert: On the homology invariants of knots, *Quart. J. Math. Oxford* **1** (1950), 23–32.

[99] H. Seifert, W. Threlfall: *A Textbook of Topology*, Academic Press, 1980.

[100] J. P. Serre: *A Course in Arithmetic*, Grad. Texts in Math. 7, Springer-Verlag, 1973.

[101] L. Siebenmann: Topological manifolds, in *Proceedings Internat. Congr. Math.* 1970, Gauthier-Villars, Paris, 1971.

[102] N. E. Steenrod: Homology with local coefficients, *Ann. of Math.* **44** (1943), 610–627.

[103] N. E. Steenrod: *The Topology of Fibre Bundles*, Princeton University Press, 1951.

[104] C. H. Taubes: SW $\Rightarrow$ Gr: from the Seiberg-Witten equations to pseudo-holomorphic curves, *J. Amer. Math. Soc.* **9** (1996), 845–918.

[105] C. H. Taubes: Counting pseudo-holomorphic submanifolds in dimension 4, *J. Differential Geom.* **44** (1996), 819–893.

[106] C. H. Taubes: Gr $\Rightarrow$ SW: from pseudo-holomorphic curves to Seiberg-Witten solutions, *J. Differential Geometry* **51** (1999), 203–334.

[107] C. H. Taubes: Seiberg-Witten invariants and pseudoholomorphic subvarieties for self-dual harmonic 2-forms, *Geom. Topol.* **3** (1999), 167–210; `http://www.maths.warwick.ac.uk/gt/`

[108] C. H.Taubes: The Seiberg-Witen-invariants and 4-manifolds with essential tori, *Geom. Topol.* **5** (2001), 441–519; `http://www.maths.warwick.ac.uk/gt/`

[109] G. Torres: On the Alexander polynomial, *Ann. of Math.* **57** (1953), 57–89.

[110] V. G. Turaev: The Alexander polynomial of a three-dimensional manifold, *Math. Sb.* **97** (1975), 341–359. (in Russian)

[111] V. G. Turaev: Reidemeister torsion and the Alexander polynomial, *Math. Sb.* **101** (1976), 252–270.

[112] V. G. Turaev: Reidemeister torsion in knot theory, *Uspehi Mat. Nauk* **41** (1986), 97–147 (Russian); English transl. *Russian Math. Surveys* **41** (1986), 119–182.)

[113] V. G. Turaev: Euler structures, nonsingular vector fields and torsions of Reidemeister type, *Izv. Akad. Nauk* **53** (1989), 607–643; *Math. USSR Izv.* **34** (1990), 627–662.

[114] V. G. Turaev: Torsion invariants of $Spin^c$-structures on 3-manifolds, *Math. Res. Lett.* **4** (1997), 679–695.

[115] V. G. Turaev: A combinatorial formulation for the Seiberg–Witten invariants of 3-manifolds, *Math. Res. Lett.* **5** (1998), 583–598.

[116] V. G. Turaev: Surgery formula for torsions and Seiberg-Witten invariants of 3-manifolds, math.GT/0101108.

[117] V. G. Turaev: *Introduction to Combinatorial Torsions*, Lectures in Mathematics, ETH Zürich, Birkäuser, 2001.

[118] K. Walker: *An Extension of Casson's Invariant*, Ann. of Math. Stud. 126, Princeton University Press, 1992.

[119] C. T. C. Wall: Quadratic forms on finite groups, and related topics, *Topology* **2** (1964), 281–298.

[120] J. H. C. Whitehead: *On incidence matrices, nuclei and homotopy types*, Ann. of Math. ,**42** (1941), 1197–1237.

[121] G. H. Whitehead: *Elements of Homotopy Theory*, Grad. Texts in Math. 61, Springer-Verlag, 1978.

[122] E. T. Whittaker, G. N. Watson: *A course of Modern Analysis*, Cambridge University Press, 1996.

Symbols

Index

Bei Fragen zur Produktsicherheit wenden Sie sich bitte an:
If you have any questions regarding product safety,
please contact:

Walter de Gruyter GmbH
Genthiner Straße 13
10785 Berlin
productsafety@degruyterbrill.com